Service Provision

WILEY SERIES IN COMMUNICATIONS NETWORKING & DISTRIBUTED SYSTEMS

The 'Wiley Series in Communications Networking & Distributed Systems' is a series of expert-level, technically detailed books covering cutting-edge research and brand new developments in networking, middleware and software technologies for communications and distributed systems. The books will provide timely, accurate and reliable information about the state-of-the-art to researchers and development engineers in the Telecommunications and Computing sectors.

Other titles in the series:

Wright: *Voice over Packet Networks* 0-471-49516-6 (February 2001)
Jepsen: *Java for Telecommunications* 0-471-49826-2 (July 2001)
Sutton: *Secure Communications* 0-471-49904-8 (December 2001)
Stajano: *Security for Ubiquitous Computing* 0-470-84493-0 (February 2002)
Martin-Flatin: *Web-Based Management of IP Networks and Systems* 0-471-48702-3 (September 2002)
Berman *et al.*: *Grid Computing. Making the Global Infrastructure a Reality* 0-470-85319-0 (March 2003)

Service Provision

Technologies for Next Generation Communications

Edited by

Kenneth J. Turner
University of Stirling, UK

Evan H. Magill
University of Stirling, UK

David J. Marples
Telcordia Technologies, UK

John Wiley & Sons, Ltd

Telephone (+44) 1243 779777

Email (for orders and customer service enquiries): cs-books@wiley.co.uk
Visit our Home Page on www.wileyeurope.com or www.wiley.com

Other Wiley Editorial Offices

John Wiley & Sons Inc., 111 River Street, Hoboken, NJ 07030, USA

Jossey-Bass, 989 Market Street, San Francisco, CA 94103-1741, USA

Wiley-VCH Verlag GmbH, Boschstr. 12, D-69469 Weinheim, Germany

John Wiley & Sons Australia Ltd, 33 Park Road, Milton, Queensland 4064, Australia

John Wiley & Sons (Asia) Pte Ltd, 2 Clementi Loop #02-01, Jin Xing Distripark, Singapore 129809

John Wiley & Sons Canada Ltd, 22 Worcester Road, Etobicoke, Ontario, Canada M9W 1L1

British Library Cataloguing in Publication Data

A catalogue record for this book is available from the British Library

ISBN 0-470-85066-3

Typeset in 10/12pt Times by Integra Software Services Pvt. Ltd, India
Printed and bound in Great Britain by Antony Rowe Ltd, Chippenham, Wiltshire
This book is printed on acid-free paper responsibly manufactured from sustainable forestry
in which at least two trees are planted for each one used for paper production.

Contents

List of Contributors

The contributors to the book, their affiliations at the time of writing, and their current email addresses are as follows:

Farooq Anjum
Telcordia Technologies, USA
fanjum@telcordia.com

John-Luc Bakker
Telcordia Technologies, USA
jbakker@telcordia.com

Gordon S. Blair
Lancaster University, UK
gordon@comp.lancs.ac.uk

Marcus Brunner
NEC Europe, Germany
brunner@ccrle.nec.de

Graham M. Clark
Marconi Communications, UK
gcsfts@fish.co.uk

Munir Cochinwala
Telcordia Technologies, USA
munir@research.telcordia.com

Stephen Corley
British Telecommunications, UK
steve.corley@bt.com

Geoff Coulson
Lancaster University, UK
geoff@comp.lancs.ac.uk

Wayne Cutler
Marconi Communications, UK
wayne.cutler@marconi.com

James M. Irvine
University of Strathclyde, UK
j.m.irvine@strath.ac.uk

Pierre C. Johnson
Consultant, Canada
pierrejohnson@mac.com

Chris Lott
Telcordia Technologies, USA
lott@research.telcordia.com

Evan H. Magill
University of Stirling, UK
ehm@cs.stir.ac.uk

David J. Marples
Telcordia Technologies, USA
dmarples@iee.org

Peter Martin
Marconi Communications, UK
petemartin@ntlworld.com

Alistair McBain
Marconi Communications, UK
alistair.mcbain@marconi.com

Erich S. Morisse
Consultant, USA
esmorisse@yahoo.com

Robert Pinheiro
Telcordia Technologies, USA
bob@bobpinheiro.com

James T. Smith
Consultant, USA
james@jtsmith2.us

Hyong Sop Shim
Telcordia Technologies, USA
hyongsop@research.telcordia.com

Simon Tsang
Telcordia Technologies, USA
stsang@research.telcordia.com

Kenneth J. Turner
University of Stirling, UK
kjt@cs.stir.ac.uk

John R. Wullert
Telcordia Technologies, USA
wullert@research.telcordia.com

Preface

Subject of this Book

Communications is a very broad subject. At one time, a distinction might have been made between computer communications (data networking) and telecommunications (public voice networking). However these two domains are rapidly converging, with the same network technologies being used to support data, voice, video, and other media.

Communications services provide useful facilities to end-users. Services therefore take a user-oriented rather than network-oriented view. Services are the financial basis of communications since they are directly responsible for operator revenues. The major developments in communications technology during the past decade have been driven by the services that can be sold to customers; technology in itself is not of interest to end-users. Operators also differentiate themselves in a highly competitive market by the services that they sell. A sound understanding of services is thus vital for anyone working in the communications domain.

The focus of the book is on technical issues. It deals with the technologies that support the development of services, as well as the networking aspects needed to support services. Commercial issues are important, such as pricing and selling communications services, but are not the subject of this book.

Aim of this Book

The aim of this book is therefore to present the broad sweep of developments in communications services. Because of the breadth of the subject, it can be difficult even for practitioners to understand what is happening outside their own specialist niches. Rapid technological changes also make it difficult to keep abreast. To fully understand current developments would require participation in many standards organizations, would need monitoring of many projects, and would necessitate many hours of reading emerging standards (many of them hard to come by except for committee members). This book aims to alleviate much of the difficulty by bringing together in one place the combined expertise of the contributors.

The book is designed to help anyone with a technical interest in communications services. This includes communications engineers, strategists, consultants, managers, educators and students. Although the book is written for the practicing engineer, it would also be suitable for self-study or as part of a graduate course on communications. It is assumed that the reader has a computing background and a general knowledge of communications.

Book Structure

The book has been divided up into self-contained topics. Each chapter can therefore be read in isolation. Although there is a logical progression in the book, chapters can be read in any order. The book can therefore be used for familiarization with the field, reading the chapters in the order written; the book can also be used for reference, reading individual chapters for more depth on particular topics.

However certain topics such as call processing, quality of service, service architecture, and feature interaction tend to be recurring themes. They are touched on where relevant, but can be studied in more depth in the specialist chapters on these topics.

Part I (Network Support for Services) considers how networks support communications services. *Chapter 1 (Introduction and context)* sets the scene for the whole book. It overviews the nature of communications services and the approaches taken to developing them. *Chapter 2 (Multimedia technology in a telecommunications setting)* explains how communications services have grown rapidly to include multimedia as a central provision. The issues surrounding multimedia are explored, presenting the key standards that support multimedia services. *Chapter 3 (Call processing)* explains the major aspects of call processing. A variety of call models is introduced, along with switch-based services. Call processing is explained for (advanced) intelligent networks and softswitches. *Chapter 4 (Advanced Intelligent Networks)* explains the origins of intelligent networks and the motivation for their development. The architecture and major functions are presented for the (Advanced) Intelligent Network. A variety of Intelligent Network services is introduced, including internationally standardized services and service features. *Chapter 5 (Basic Internet technology in support of communication services)* introduces the main areas in which Internet technology impacts service provision. Approaches to Internet Quality of Service are discussed. A topical solution for Internet Telephony is explored. Directory-enabled Networks are described as a means of achieving better management interoperability. Home Networks and Active Networks are discussed as examples of emerging Internet-based networks. *Chapter 6 (Wireless technology)* reviews a range of wireless solutions supporting a variety of different geographical areas. The evolution of cellular systems is traced through the first, second and third generations, with particular emphasis on the emerging Third Generation mobile systems. Other wireless systems addressed include broadcast networks and local wireless networks.

Part II (Building and analyzing services) considers the architecture, creation, development, and analysis of services. *Chapter 7 (Service Management and Quality of Service)* focuses on the management of services, using Service Level Agreements and Quality of Service as the basis. Mechanisms are discussed in depth for managing and monitoring Quality of Service. *Chapter 8 (Securing communication systems)* deals with the important question of security as it affects services. The basic principles of cryptography are explained, along with the kinds of threat to which communications systems are subject. Mechanisms are introduced for authentication, non-repudiation, and access control. Digital cash is used as a case study to illustrate how security issues arise and are dealt with. *Chapter 9 (Service creation)* examines service creation as pioneered in the (Advanced) Intelligent Network. The Telcordia SPACE system is used as a concrete example of a Service Creation Environment. Service creation is also considered for Internet services such as Internet Telephony and the Web. The important topics of service integration and service introduction are discussed. *Chapter 10 (Service architectures)* starts by reviewing two early efforts relevant to service architecture standardization: Open Systems Interconnection and the Distributed Computing Environment. The evolution of distributed services is then discussed in the context of subsequent standards: Open Distributed Processing, the Telecommunications Information Networking Architecture, and the Common Object Request Broker Architecture. The chapter also highlights the increasing importance of the middleware paradigm for service provision. *Chapter 11 (Service capability APIs)* focuses closely on specific programming interfaces for creating services: the Telecommunications Information Networking Architecture, Java APIs for The Integrated Network, and various Parlay APIs. *Chapter 12 (Formal methods for services)* explains the nature of a formal method, and how it can be used to help service design. As examples, it is explained how services were formalized for Open Systems Interconnection and Open Distributed Processing. General classes of formal method are discussed for modeling communications services. *Chapter 13 (Feature interaction: old hat or deadly new menace?)* explores feature interaction in general, and in the context of telephony. Reactions to the problem are presented from the perspective of researchers, operators, and vendors. The changing nature of feature interaction is discussed in the context of how services are evolving.

Part III (The future of services) looks at the evolution of communications services and their supporting technologies. *Chapter 14 (Advances in services)* examines factors such as convergence and context that are driving new kinds of service. Presence-based services, messaging services, and service discovery are considered, all augmented by wireless connectivity. Home-based services are also discussed in depth. *Chapter 15 (Evolving service technology)* explores new techniques being used to create communications services: software agents, constraint satisfaction, Artificial Neural Networks, and Genetic Programming. *Chapter 16 (Prospects)* rounds off the book by evaluating the approaches developed for communications services. It discusses trends in the kinds of service that will be deployed and how they will be created. A mid- to long-term projection is presented of how services will evolve.

Appendix 1 Abbreviations collects in one place the major acronyms used throughout the book.

Appendix 2 Glossary briefly defines some key terms used in the book.

Appendix 3 Websites covers the major online sources for organizations discussed in the book, such as standardization bodies and trade associations.

Finally, the *Bibliography* gives details of all articles referenced throughout the book, and provides further reading for more detailed study.

Reviewers

Most chapters were reviewed by contributors to the book. However, a number of other individuals reviewed drafts of some chapters. The editors warmly thank the following: Daniel Amyot (University of Ottawa, Canada), Derek Atkins, Lynne Blair (Lancaster University, UK), Chris Brightman (Telcordia Technologies, USA), Elaine Bushnik (Consultant), Mario Kolberg (University of Stirling, UK), Stan Moyer (Telcordia Technologies, USA), Stephan Reiff-Marganiec (University of Stirling, UK), Mark Ryan (University of Birmingham, UK), Carron Shankland (University of Stirling, UK), Richard Sinnott (National E-Science Centre, UK), Charlie Woloszynski (Telcordia Technologies, USA).

Acknowledgments

The editors are grateful to all the contributors for rising to the challenge of the book, and to the reviewers for their careful reading of chapters. The editors also thank Birgit Gruber and Sally Mortimore of John Wiley for their faith in the book, and for their support of the publishing process.

Pierre Johnson would like to give credit to Andrew Patrick (National Research Council of Canada) and Paul Coverdale (Nortel Networks) for their work on QoS targets based on human factors, and for helping bring many of these concepts together in the QoS target model. Peter Dodd also deserves mention for helping to nurture this work.

Gordon Blair and Geoff Coulson acknowledge the support offered by the FORCES consortium (EPSRC Grant GR/M00275) in the development of the ideas presented in Chapter 10.

Peter Martin thanks Marconi Communications for supporting the work on Genetic Programming for service creation. John-Luc Bakker and Farooq Anjum are grateful to Ravi Jain (NTT DoCoMo) for his contributions to Chapter 11.

Evan Magill is grateful to Muffy Calder (University of Glasgow), Mario Kolberg (University of Stirling), Dave Marples (Telcordia Technologies), and Stephan Reiff-Marganiec (University of Stirling) for their strong and direct influence on the ideas in Chapter 13.

Part I

Network Support for Services

This part of the book considers how networks support communications services. The chapters deal with the impact on networking of multimedia, call processing, (Advanced) Intelligent Networks, the Internet, and wireless technology.

Chapter 1 (Introduction and context) sets the scene for the whole book. It begins by introducing the nature of communications services. The field of communications services is explored using the structure of the book as a roadmap. Network support for services is discussed, showing how aspects such as traditional telephony, Intelligent Networks, wireless telephony, the Internet and multimedia have developed. Techniques for building and analyzing services are reviewed. Several service architectures have been standardized, along with Application Programming Interfaces. Service creation and associated analytic methods are presented. New developments in services and service technologies give a glimpse into the future of communications services.

Chapter 2 (Multimedia technology in a telecommunications setting) explains how communications services have grown rapidly from simple telegraphy to a wide range of present-day services. Multimedia underlies may of these new services. The nature of multimedia is introduced, along with the market drivers for multimedia services. Issues surrounding multimedia include charging mechanisms and Quality of Service support. Key standards for multimedia services are reviewed, mainly those developed by the International Telecommunications Union, the Internet Engineering Task Force, and the International Organization for Standardization. The main elements of multimedia services are introduced. Quality of Service issues are then discussed. Typical multimedia services are presented as examples. Interworking is covered between multimedia services and traditional voice services. Multimedia aspects of terminal devices and user interfaces are explored. The chapter concludes with a brief review of future developments in multimedia services.

Chapter 3 (Call processing) opens with an historical review of call processing in telephone switching. Key aspects of call processing are then explained: interfaces, number translation, route selection, user preferences, resource control, and call data recording. A variety of call models is introduced, ranging from a simple half call model, through multi-segment models, to the more complex models required in mobile communication. Switch-based services are discussed for the Public Switched Telephone Network and for the Integrated Services Digital Network. Problems due to feature interaction are mentioned briefly. Call processing is investigated for (Advanced) Intelligent Networks. By way of comparison, the approach taken by softswitches is also explained in detail. The chapter closes with an evaluation of call processing and how it is likely to develop in future years.

Chapter 4 (Advanced Intelligent Networks) explains the origins of Intelligent Networks and the motivation for their development. Closely related but differing approaches are discussed for the Advanced Intelligent Network designed by Bellcore (now Telcordia Technologies) and the Intelligent Network standards from the International Telecommunications Union. The architecture and major functions are presented for the (Advanced) Intelligent Network. The key factor is the separation of switching from service control. The structured architecture and specialized functions ensure a modular approach. For example, it is explained how four separate planes deal with different levels of abstraction: the service plane, the global functional plane, the distributed functional plane, and the physical plane. The major elements of the architecture are discussed: the Service Switching Point and its associated call models, the Service Control Point, and the Intelligent Peripheral. A variety of Intelligent Network services is introduced, including internationally standardized services and service features. The chapter is rounded off by an evaluation of Intelligent Networks and their future evolution.

Chapter 5 (Basic Internet technology in support of communication services) introduces the main areas in which Internet techniques impact service provision: the Internet itself, Internet telephony, Directory-Enabled Networks, Home Networks, and Active Networks. Approaches to Internet Quality of Service are discussed, particularly Integrated Services, Differentiated Services, and Multi-Protocol Label Switching. Issues surrounding Internet telephony are explored, with the Session Initiation Protocol and its supporting protocols serving as a topical example. Directory-Enabled Networks are a means of achieving better interoperability at the level of management data. Internet technology in the home is discussed through the example of the Open Services Gateway initiative. The approach of Active Networking is considered as a significant advance in how networks can be made more programmable and more flexible. The chapter ends with a summary of how various Internet technologies are already playing a vital role in service provision.

Chapter 6 (Wireless technology) introduces the main aspects of wireless networks. It reviews a range of wireless solutions supporting a variety of different areas: personal, local, suburban, rural, and global. Cellular systems are a major technology, and their evolution is traced through the first, second, and third generations. As the evolving standard, the goals and challenges of Third Generation mobile technology are discussed in detail. Important considerations in cellular networks include handover, location management, network architecture, location technologies, satellite systems, and market trends. Private Mobile Radio and Broadcast networks are also discussed to round out the picture. Local wireless networks are explored: Bluetooth, cordless technologies such as the Digital Enhanced Cordless Telecommunications, Wireless Local Loop, and Wireless Local Area Networks. The chapter finishes with an evaluation of wireless technologies and their future prospects.

1

Introduction and Context

Kenneth J. Turner[1], Evan H. Magill[1] and David J. Marples[2]

[1]University of Stirling, UK, [2]Telcordia Technologies, USA

1.1 Communications Services

The term 'service' is a very broad one. An everyday sense of the word is assistance for someone else, e.g. a booking service or a delivery service. The same general meaning carries over into technical domains such as communications. As will be seen throughout the book, services appear in many forms. The following examples illustrate the richness and diversity of communications services:

- audio services such as telephony, audio-conferencing, mobile telephony, recorded weather forecasts, voice-controlled flight enquiries, online music;
- image services such as facsimile, video-conferencing, video-on-demand, streaming video;
- distributed data services such as file sharing, meeting schedulers, electronic whiteboards, online gaming, electronic newspapers, storage area networks;
- Web-based services such as travel booking, mobile Web access, e-business, e-commerce, e-learning, e-medicine.

This is a growing list. The big advantage of communications services is that they can be delivered to users wherever they are, whenever they want.

Communications services provide facilities to end-users. They therefore focus on the end-user view rather than the network view. A service is thus an abstraction of the underlying network, including its protocols and resources. As an example, consider the Plain Old Telephone Service. The subscriber simply expects to dial other subscribers and speak to them. This simple service is easy to understand and to describe. However, it requires a very complex infrastructure in the telephone network to make it work. This is completely invisible to the end-user. In fact, the hallmark of an effective service is that its users need know nothing about how it is realized.

Communications services often have a distinct commercial aspect: a service is something that is packaged and sold to customers. A telephone subscriber, for example, might pay for a call answering service or a call forwarding service. What a marketing person might call

Service Provision – Technologies for Next Generation Communications. Edited by Kenneth J. Turner, Evan H. Magill and David J. Marples
 ISBN: 0-470-85066-3

a service could correspond to a number of individual technical services. For example, someone who pays for a charge-card service might also benefit from bulk call discounts, itemized billing and short-code dialing – all individual services from a technical perspective.

Services provide the primary income for network operators, so they lie at the financial heart of the communications sector. Operators and suppliers invest heavily in support for accurate billing of service use. Services and tariffs are also vital for operators to differentiate themselves in what is an increasingly competitive market.

The following sections provide an overview of communications services, using the chapters of this book as a routemap.

1.2 Network Support for Services

A study of the timeline for communications systems (see Chapter 2) will show that communications services have grown extremely rapidly in the past 150 years. Data services (telegraphy) were followed by voice services (telephony), audio services (radio), image services (facsimile), video services (TV), text services (telex), mobile services (paging, telephony), Internet services (email, file transfer, remote access, telephony), and Web-based services (just about anything prefixed with 'e-'). Many services were initially supported by specialized networks. However, the convergence of computing and communications has meant that new services can be supported over existing networks, and that current services can be provided over shared networks. IP-based networks, using standards from the Internet Protocol family, are emerging as a common infrastructure for many communications services. Even broadcast wireless services such as radio and TV may succumb to this trend.

Telephony has been the driver for many service developments. The advent of Stored Program Control telephone exchanges (see Chapter 3) made it practicable to do much more than merely establish an end-to-end path. More flexible call handling was introduced in the 1980s. However, the intermingling of basic switching (establishing an end-to-end path) and additional services (such as free numbers) made it difficult to develop and maintain new services. With the standardization of Intelligent Networks in the 1990s (see Chapter 4), a clean separation was obtained between switching and services. Signaling emerged as an important issue in its own right, allowing advanced services to be built using more sophisticated control mechanisms.

In parallel, the Internet has been developing since the 1970s. Unlike telephone networks, which are primarily voice-oriented, the Internet began with data as its main focus. However, the increasing digitization of nearly all signals has allowed voice to be carried with almost the same ease (see Chapter 5). Indeed the Internet has shown itself appropriate for carrying many kinds of media. It is significant that many traditional telephony developers are now building networks using Internet technologies as a common infrastructure for voice, image, and data.

However, mobile communication (see Chapter 6) has provided a new impetus for telephony developments. The early analogue networks for mobile telephony have also progressed to digital form. Third Generation networks will exhibit a significant element of Internet technology in their design.

1.3 Building and Analyzing Services

As communications services have become more widespread and more vital, Quality of Service (see Chapter 7) has become a major issue – particularly for services such as audio and video that must respect real-time constraints despite the volume of data. For commercial use, Service Level Agreements are now commonplace as a means of defining the standard of service that the customer should expect. This in turn requires mechanisms for monitoring and controlling Quality of Service.

Information pervades all modern organizations, and has significant commercial value. Protecting it through physical security is relatively straightforward. However, widespread distributed access through communications services poses many new kinds of threat (see Chapter 8). Sophisticated protocols have been developed for securely distributing information, authenticating this information and its users, and providing accurate records of how information is manipulated and used. Even the ordinary user needs such mechanisms to allow safe online payment and to protect personal information.

Communications services are typically very complex to design and to maintain. Techniques and tools for service creation were first developed for telephony services (see Chapter 9). Service Creation Environments have tended to be vendor-specific, making it difficult to re-use the design of a service with another kind of switch (end office or exchange). Although Intelligent Networks have helped to improve portability, telephony services remain rather specialized and switch-specific. There is hope, however, that Internet-based services will be more portable.

Several standardization efforts have focused on architectures to support services (see Chapter 10). The standards for Open Systems Interconnection were developed to support interoperability in a heterogeneous networking environment. Open Systems Interconnection also defined a Basic Reference Model that gave prominence to communications services. A further architecture was standardized for Open Distributed Processing, dealing with distributed systems and not just networking. The service concept is also important in Open Distributed Processing, but takes a more general form.

Client-server approaches have emerged as a common basis for providing communications services. The Distributed Computing Environment is an example of middleware as an infrastructure supporting distributed communications services. The Common Object Request Broker Architecture was developed for distributed interworking, including legacy systems that need to interoperate with new systems.

Pragmatic, industry-led developments have also included Application Programming Interfaces for communications services (see Chapter 11). The Telecommunications Information Networking Architecture was explicitly designed around communications services as a central element. Microsoft's Telephony Application Programming Interface exposed communications interfaces so that more complex services could be built. Several similar interfaces were designed by other manufacturers. The general approach has been exploited in Computer–Telephony Integration and in call centers. The widespread use of Java led to the definition of Java APIs for The Integrated Network.

The Parlay Group pioneered the concept of opening up networks to third party service provision. Traditionally, services have been provided *within* the core of networks. This has several advantages, including service provision by a single authority that can exercise tight control over quality and reliability. Increasingly, services are being provided at the *edge* of networks. These services may be offered by third parties. However, it is becoming feasible for end-users to define their own services. This is close to the Internet philosophy, where the core of the network is deliberately kept simple in order to provide scalable, high-volume data transfer. In an Internet view, services are mainly provided in the end systems.

Communications services are often complex, and need to interwork across heterogeneous systems. It is therefore important that the services be defined precisely and, if possible, be checked for accuracy. Formal methods (see Chapter 12) have been used for this kind of task. Several Formal Description Techniques have been standardized in the context of communications systems. Communications services have attracted the use of a number of formal methods. Some of this work has been architectural in nature, most has focused on service specification, and some has investigated rigorous analysis of services.

A particular form of service analysis deals with service compatibility – the so-called feature interaction problem (see Chapter 13). A modern switch will typically contain code for hundreds of features (the building blocks of services). Unfortunately, it is very difficult to design features that do not interfere with each other. As the number of features grows, the risk of unexpected interference rises. Both formal and informal methods have been developed to detect and resolve such interference. Although feature interaction has been studied extensively in traditional telephony, the issue arises in other domains and is likely to be problematic in newer kinds of communications service.

1.4 The Future of Services

Users are increasingly contactable wherever they are: in the office, at home, on the move, at leisure. They receive communications via fixed-line phone, cellular phone, voicemail, pager, facsimile, email, PC, and PDA. A major growth area in recent years has been instant messaging: the user is notified of 'buddies' who are online and available to chat. Presence-based services, availability control, and wireless connectivity (see Chapter 14) are becoming the major drivers in anytime, anywhere communication. Services for the home are also becoming a reality. Home Network technologies and the Open Services Gateway initiative are allowing end-users and third parties to provide new services in the home.

New approaches to service provision are also gaining ground (see Chapter 15). Software agents are already common as a means of supporting users in local and distributed information retrieval and processing. Constraint satisfaction is a general-purpose technique, but has also been found useful in dealing with conflict among services. Artificial Neural Networks have been applied to many problems of a pattern recognition character, including applications in communications. Genetic Programming is an optimization technique that has also been applied to communications services.

Despite the high-tech industry downturn of the early 2000s, it is clear that communications services will continue to grow dramatically (see Chapter 16). A number of trends are already evident. Services will become a lot more of everything: more widespread, more complex, more personal, more programmable, more pervasive, more interlinked, more dispersed. It is hoped that the readers of this book will be encouraged to contribute towards the development of future communications services.

2

Multimedia Technology in a Telecommunications Setting

Alistair McBain

Marconi Communications Ltd, UK

Since the opening of the first telegraphy networks in the middle of the 19th century many additional telecommunications services have been developed, with new ones continuing to be added. The resulting explosion of telecommunications services over time is illustrated in Figure 2.1, which maps some of the key services onto a timeline for telecommunications. It can be observed that not only are the numbers of services increasing, they are also becoming more complex with additional audio, video, and data content. This trend looks set to continue into the future, with advances in technology being harnessed to ensure that greater functionality is delivered with greater user friendliness.

Historically, each new service has been supported by a single medium ('monomedia') that is often carried on a dedicated network infrastructure, leading to what is sometimes referred to as a 'stove pipe' network architecture. This is clearly not a sustainable approach as ever more new services emerge in the future, especially since the nature of many of these new services is impossible to predict.

The solution is to move away from the 'dedicated network per service' approach towards so-called converged networks that are capable of supporting a range of services on a common network infrastructure. In turn, this approach eases the introduction of services that are amalgams of distinct services, such as voice and video, that were previously unrelated and supported on separate networks. The word 'multimedia' has emerged to describe such new complex services.

This chapter explores the impact that multimedia is having on telecommunications services. It starts by considering definitions for multimedia. Then it examines the market drivers that are creating the demand for these new services. A short overview of the appropriate standards is given, leading to a review of the components that make up multimedia services, using some common services as examples. The need to interwork with traditional services, notably voice, is emphasized. The chapter concludes with some thoughts on multimedia terminal equipment and future trends. It should be remembered that multimedia services and technologies are still evolving, so this chapter can present only a snapshot of current activities.

Service Provision – Technologies for Next Generation Communications. Edited by Kenneth J. Turner, Evan H. Magill and David J. Marples
 ISBN: 0-470-85066-3

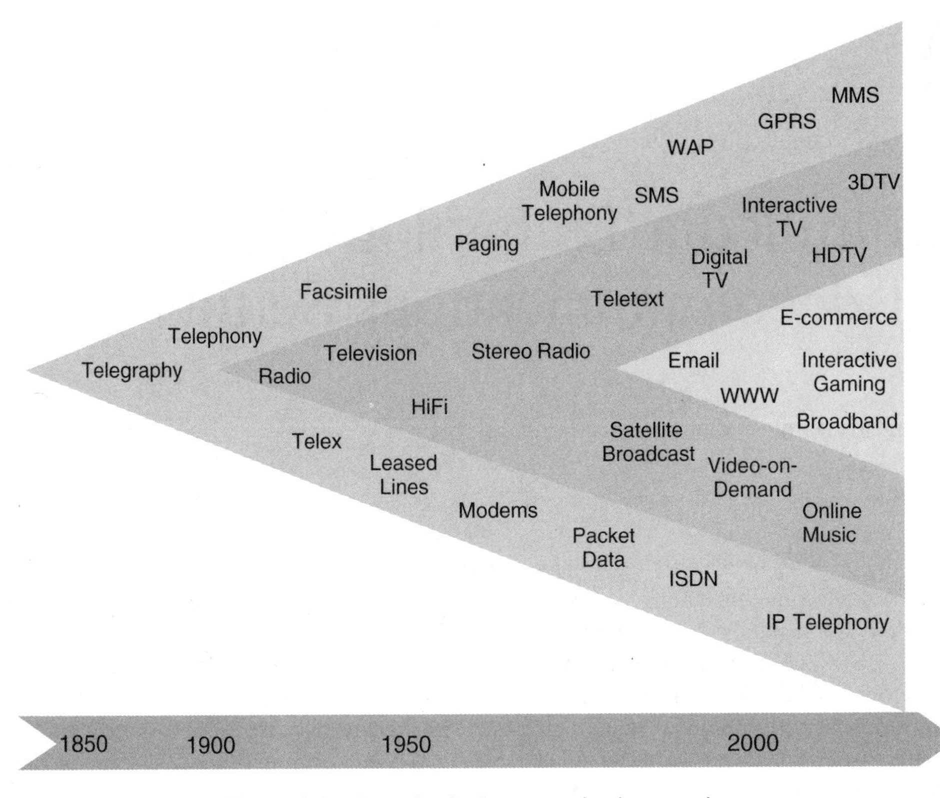

Figure 2.1 Growth of telecommunications services.

2.1 Definition of Multimedia

'Multimedia' has become a very popular concept over recent years, but the term is used very loosely. A generally accepted definition of multimedia is the combination of two or more different media types, one of these usually being audio. The popular image of a multimedia call is video telephony involving the use of voice and video. Voice and data, for example a virtual whiteboard, is another common combination.

In passing, it can be observed that there are many alternative definitions for 'multimedia' given in the literature, for example (Whyte 1995):

> *'Multimedia communications, through a range of technologies, including communication over a distance, aspires to provide a rich and immediate environment of image, graphics, sound, text and interaction, which assists in decision-making or environmental involvement'.*

A more formal definition is provided by the Telecommunications Sector of the International Telecommunications Union (ITU-T) in Recommendation I.113. This defines a multimedia service as '*a service in which the interchanged information consists of more than one type, such as text, graphics, sound, image and video*' (ITU 1997a).

ITU-T Recommendation F.700 (ITU 2000a) considers multimedia services from both end-user and network provider viewpoints:

'From the end user's point of view, a multimedia telecommunication service is the combination of telecommunication capabilities required to support a particular multimedia application. Such a service is usually considered to be independent of the network(s) providing these capabilities.

From the Network Provider's point of view, a multimedia telecommunication service is a combination or set of combinations of two or more media components (e.g. audio, video, graphics, etc.) within a particular network environment in such a way as to produce a new telecommunication service. This telecommunication service is considered to be fully dependent on the specific capabilities of the networks utilized.

Additionally, the user wishes to have the same type of presentation and mode of operation for all services needed to support similar applications; and the Service or Network provider desires to have common protocols and interfaces to accommodate interoperability for a variety of technical implementations.'

Multimedia services can be considered at a number of levels. Pure broadcast services such as television are examples of the simplest, level one, category of multimedia. This chapter will be confined to considering what are known as the second and third levels of multimedia. Second level multimedia relates to services which exhibit a degree of interaction between the user and the network; these are also referred to as interactive multimedia. The third level of multimedia is hypermedia where the user is able to navigate through a structure of linked elements.

The concept of a multimedia service dates back to around 1980 when it was regarded as an evolutionary step from the ISDN (Integrated Services Digital Network) concept that was being developed at this time. Much of the early work on multimedia services was undertaken in Japan, see, for example, Kitahara 1982. A major driver was the limitation of conventional text-based communication media in handling the Japanese character set. This problem had already led to the widespread adoption of the fax service in Japan during the 1970s.

2.2 Market Drivers for Multimedia

2.2.1 User expectations

The way that people regard communications services has undergone a radical change in the past decade. Cell (mobile) phones, personal numbering and freephone services are just a few of the innovations that have challenged traditional telephony. Meanwhile, the Internet has totally transformed data communications, bringing email and browsing services to the mass market. For video, low-rate coding and low-cost high-capacity storage have opened the way for a host of personal video entertainment services. The widespread use of Personal Computers has served to bind many of these innovations.

At home, ISDN, ADSL (Asynchronous Digital Subscriber Loop) and cable modem technologies are now delivering bandwidths far in excess of that required for simple voice calls. Cable and satellite television systems are also becoming much more closely coupled with the traditional voice network.

Users are becoming more mobile. The phenomenal growth in SMS (Short Messaging Service) text messages sent by cell phones is a precursor to the greater bandwidth capacities that will be available for non-voice services in future generations of mobile networks. Increasingly, users are expecting the same degree of connectivity whether they are in the office, at home, or even on the golf course.

Working habits are also changing. Telecommuting (home working) is becoming common, although perhaps not yet to the degree that some have predicted; see Allen and Wolkowitz 1987 for a discussion of the issues. One of the reasons for this slow take-up is the absence of a suitable telecommunications infrastructure. While ISDN goes some way to resolving this, the ability to handle video and the fast transfer of large data files remains an essential requirement.

These, and other issues, are increasing user expectations so that, in many cases, a basic voice service is now no longer considered adequate. Multimedia PC usage has conditioned users to expect multimedia features on the public network. It is anticipated that these multimedia services will be delivered over a common, or converged, network.

While multimedia is well suited for business and educational purposes it is the application to the leisure market that promises to support the mass-market roll out of multimedia. Besides supporting entertainment services there is scope for a whole range of multimedia gaming services. Such games are designed to be highly addictive, allowing users to interact with other players who can be located anywhere in the world. High bandwidth links support a virtual reality gaming environment, and players can communicate directly by speech as well as through their game characters.

2.2.2 Need for new revenue streams

With increasing competition between service providers, pricing, even for sophisticated telecommunications services like mobile ones, is approaching commodity levels. As a result, service providers are seeking new ways to differentiate their service offerings. One such way is by the introduction of new multimedia services.

2.2.3 Effect of the Internet

There can be no doubt that the most significant event in the recent history of telecommunications has been the emergence of the Internet as a vehicle for communication and accessing shared information. The growth of the Internet has been well documented (e.g. at http://www.netsizer.com/). As a result, there can now be no doubt that IP (Internet Protocol) has become well established as the *de facto* standard within the data communications world. In particular, IP delivered across Ethernet is now a common local infrastructure. The influence of IP continues to spread, for example we are now witnessing the emergence of voice services carried over IP networks. Furthermore, the PCs used to access the Internet are fast evolving into multimedia terminal devices.

2.2.4 Content

Fundamental to the success of a multimedia enabled network is the availability of suitable content. Such content should make good use of the multimedia capabilities of the network, as well as exciting the user and stimulating new uses of the network.

Speaking at a 1993 conference, an executive from the computer gaming company Sega summarized this view:

> *'The games platform is only an enabler. Apart from price, three further things are required: content, content, content'.*

Multimedia content can be hosted on servers, for example video-on-demand applications, or it can be interactive between users as for online gaming.

2.2.5 *Increasing mobility*

The widespread adoption of mobile telephony has conditioned users to expect telecommunications services without the need for a fixed access point. The introduction of enhanced second generation and third generation mobile networks is enabling services such as mobile videophone and MMS (Multimedia Messaging Service).

MMS is, as its name suggests, the ability to send messages comprising a combination of text, sounds, images, and video to MMS-capable handsets. MMS applications, such as photo sharing, instant postcards, and cartoon messaging, are largely targeted at non-business users. This is the high-volume market that was previously exploited during the 1990s with mobile telephony and SMS. To this end, a new generation of handsets has emerged. These feature video support, including a built-in digital camera and high-resolution color display.

In reality, there are likely to be two classes of mobile multimedia service: those that are used 'on the move', and those that are accessed 'on the pause'. On the move services (when the user is actually in motion) are likely to be limited to lower bandwidth applications such as telephony and SMS. A mobile user is unlikely to be able to digest higher bandwidth services (e.g. video), and in any case it is difficult to provide high capacity radio access to a truly mobile user. On the other hand, mobile users are often stationary for periods of time (paused), perhaps in an airport departure lounge or at a roadside service area. This is when the 'mobile' user is likely to make use of high-bandwidth services such as email and file transfer. The provision of suitable access bandwidth infrastructure is much easier to provide at these dedicated locations.

2.2.6 *Converged networks*

The success of the Internet as a vehicle for new service delivery has led to a widely held view that future telecommunications services will be hosted on 'converged networks' that will carry voice, IP data and all other services on a common IP infrastructure (McBain 2001). This approach can be readily extended to support multimedia services.

This leads to a vision of the future where all services, including telephony, are IP-based and accessed over a standard Ethernet interface. In many ways, this will be the final realization of the ISDN dream of a single user interface that will access all services carried over a common network.

There are, however, a number of technical issues that must be addressed in order to turn this vision into reality. For example, it will be important to ensure that the necessary Quality of Service requirements are met within an IP-based multiservice environment. The development of MPLS (Multi-Protocol Label Switching) provides one way to successfully address these concerns by adding traffic engineering to the IP routing protocols.

2.2.7 *Charging for multimedia services*

The successful introduction of multimedia services within a public network environment will largely depend upon the choice of charging regime that is deployed. The complexity of multimedia means that there is potentially a wide range of parameters that could form the basis for charging, including:

- time of day;
- call duration;
- mean bandwidth;
- peak bandwidth;
- distance;
- data error rate;
- data transfer delay and variation.

Clearly, this multitude of charging options provides potential for multimedia service providers to offer service differentiation. Traditional end-user charging for communications services has been based on many of these elements, particularly bandwidth and distance. Charging for multimedia services, on the other hand, is likely to involve a much higher degree of per-service charging that reflects the perceived value of the service that is being provided. For example, the cost of watching a film via VoD (Video-on-Demand) should be comparable to the cost of renting the film on DVD or videotape.

An alternative approach to billing is to accept that a flat rate subscription removes the need for the (not inconsiderable) overheads that are associated with transaction tracking and measurement. For this reason, some of the music download and VoD suppliers are currently implementing such a model called SVoD (Subscription Video-on-Demand).

2.2.8 Public network performance

It is very important that the approach adopted for a future IP-based multimedia network is 'carrier class' (very high quality) and can support public network services. While there are many multimedia solutions available today for enterprise applications, they generally do not scale to provide adequate throughput, nor do they provide the availability required to form the basis of a revenue-earning public network. This implies that a public multimedia network must conform to certain requirements, the main ones being highlighted below:

IP-based with QoS: As stated above, while the new network will be IP-based, this must be IP with QoS (Quality of Service) guarantees. This important requirement is considered further below.

Support for a new range of applications: There are already many different services offered by today's networks, and this number will continue to grow. Many of these new services are likely to be multimedia in nature. It is obviously impossible to predict the characteristics of these new services, which are sometimes referred to as AYUS (As Yet Unknown Services), nonetheless it is important to try to anticipate their requirements when designing the new network.

High availability: In order to carry lifeline telephony and other critical services the network will need to achieve at least 99.999 % (five nines) availability. This requires that there are no single points of failure within the network. In order to achieve this, the equipment deployed should support features such as hot-swappable hardware and online software upgrades.

Scalable: The design of the network must be economic in both small and large configurations as well as supporting in-service growth. High traffic levels should be handled without degrading service.

Based on global standards: It is obviously desirable to have a standards-based approach to public networks. While recognizing that the ITU, through its Recommendations, has defined the traditional telecommunications infrastructure, it is important to also acknowledge the role that other bodies such as the IETF (Internet Engineering Task Force) are now taking in shaping the Internet and the applications that run over IP networks.

Some of the key standards for multimedia are discussed in Section 2.3.

2.2.9 Ensuring Quality of Service (QoS)

It goes without saying that the provision of adequate QoS support is essential to transport multimedia traffic over a common protocol. Unfortunately, traditional IP has a poor reputation in this regard. For historical reasons it was designed to support only best-efforts traffic, and the performance of the public Internet serves to reinforce this view. It is not possible to guarantee the delivery of real-time services such as voice over a conventional IP network in the presence of other data traffic. Additions to the basic IP, such as IntServ (Integrated Services) and DiffServ (Differentiated Services), have gone some way towards meeting this need. Another approach to the provision of QoS within an IP core network has been the more recent development of MPLS.

2.3 Standards for Multimedia Services

Stable and implementable standards are a prerequisite for the successful adoption of multimedia services. Many standards bodies are active within the world of communications. As a consequence there are many standards, particularly in an area such as multimedia that embraces communications, computer, and broadcast technologies. The two principal standards-making bodies involved in defining multimedia services are the ITU and the IETF, but there are many other relevant standards activities within other bodies such as IEC (International Electrotechnical Commission) and ISO (International Organization for Standardization). This section considers the significant standards that define multimedia services.

2.3.1 International Telecommunications Union (ITU)

The ITU is affiliated to the United Nations, and is a formal body with representation at a country level. The relevant part of the ITU for multimedia matters is the ITU Telecommunications Standardization Sector (ITU-T) that is concerned with the many aspects of telephony and data communications. ITU-T publish agreed standards in the form of Recommendations. The main recommendations that relate to multimedia are introduced below.

F.700: Framework Recommendation for audiovisual/multimedia services (ITU 2000a)

This is the key ITU-T recommendation that relates to multimedia. It considers multimedia services to be built from six potential media components:

- audio;
- video;

- still pictures;
- graphics;
- text;
- data.

These media components are discussed further in Section 2.4. Other recommendations in the F.700 series define several multimedia services.

H.320: Infrastructure of audiovisual services – Systems and terminal equipment for audiovisual services (ITU 1999)

H.320 defines videoconferencing over ISDN connections. Since the use of dialup connections is much more flexible compared to leased lines, the introduction of H.320 has led to the widespread adoption of videoconferencing.

H.323: Packet-based multimedia communications systems (ITU 2000b)

H.323 was first published in 1996. It defines a multimedia conferencing protocol that includes voice, video, and data conferencing. It is intended for use over packet-switched networks such as LANs and the Internet that do not provide a guaranteed Quality of Service. H.323 is broad and comprehensive in its scope, while being flexible and practical in its applicability. The standard is based on the IETF specification for RTP (Real-time Transport Protocol, (IETF 1996)), with additional protocols to allow call signaling, data, and audiovisual communications.

H.323 has proven to be an extremely scalable solution that meets the needs of both service providers and enterprises, with H.323 products ranging from software and hardware implementations of the protocol stacks to wireless phones and video conferencing hardware.

There are a number of annexes to H.323 that provide detail on using the protocol for applications such as fax and mobility.

2.3.2 Internet Engineering Task Force (IETF)

In contrast to the ITU, the IETF is a totally open body where anyone may participate. As its name suggests, the IETF has been instrumental in defining and evolving the Internet through a series of documents that are published as RFCs (Requests for Comment). Studies within the IETF are progressed within Working Groups of experts, the main Working Groups concerned with multimedia topics being as follows.

RFC 1889: A Transport Protocol for Real-Time Applications (RTP, (IETF 1996))

RTP (Real-time Transport Protocol) was the result of the work of the Audio/Video Transport Working Group of the IETF. This group was formed to develop a protocol for real-time transmission of audio and video over UDP (User Datagram Protocol) and IP multicast. RFC 1889 also has an associated profile for audio/video conferences and payload format. RTP provides end-to-end network transport functions suitable for applications transmitting real-time data such as audio, video, or simulation data over multicast or unicast network services.

RFC 2326: Real-Time Streaming Protocol (RTSP, (IETF 1998))

Streaming media refers to a sequence of moving images and/or audio that is sent in compressed form over the Internet and played to the user on arrival. This means that a user does not have to wait to download a large file before seeing the video or hearing the sound. Rather, the media is sent in a continuous stream and is played as it is received. The user needs a suitable player, which is a program that decompresses and sends video data to the display and audio data to speakers. The player may be integrated into the user's Web browser or may be downloaded as a plug-in.

RTSP was developed by the Multiparty MUltimedia SessIon Control (MMUSIC) Working Group. MMUSIC is chartered to develop Internet standards to support Internet teleconferencing sessions. RTSP is an application-level protocol for control over the delivery of data with real-time properties. RTSP provides an extensible framework to enable controlled, on-demand delivery of real-time data such as audio and video. Sources of data can include both live data feeds and stored clips. The protocol is intended to control multiple data delivery sessions, to provide a means for choosing delivery channels such as UDP, multicast UDP and TCP (Transmission Control Protocol), and to provide a means for choosing delivery mechanisms built on RTP.

RFC 3261: Session Initiation Protocol (SIP, (IETF 2002d))

This is another significant standard that has emerged from the MMUSIC Working Group. It was developed within the IETF as an alternative approach to the ITU-T H.323 protocol. SIP is a text-based protocol, similar to HTTP (HyperText Transfer Protocol) and SMTP (Simple Mail Transfer Protocol), and is deployed in order to initiate interactive communication sessions between users. Examples of these sessions are voice, video, chat, interactive games, and virtual reality.

SIP is a signaling protocol for initiating, managing, and terminating real-time interactive sessions between Internet users. Like H.323, SIP defines mechanisms for call routing, call signaling, capabilities exchange, media control, and supplementary services. However, SIP is generally regarded as being simpler and more extensible than H.323, though in the medium term SIP is being developed to interoperate with the longer-standing H.323 protocol suite. The position of SIP was considerably strengthened when Microsoft announced support for SIP in the Windows XP operating system. SIP has also been adopted by the 3GPP (Third Generation Partnership Project) to support multimedia services in the next generation of mobile networks.

2.3.3 International Organization for Standardization (ISO)

ISO is responsible for a wide range of standardization activities. Several of its many committees have played a pivotal role in defining multimedia services. These include MHEG (Multimedia and Hypermedia Information Coding Expert Group), JPEG (Joint Photographic Experts Group) and MPEG (Moving Picture Experts Group). The activities of these groups are summarized in the paragraphs below.

MHEG

MHEG is charged with producing a standardized coding scheme for multimedia data structures that can be supported by different applications, services and platforms. This is specified in ISO/IEC 13522 (ISO/IEC 1997a). MHEG can be regarded as an intermediate transfer format for multimedia objects between different internal coding schemes. The MHEG model is independent of any set of presentation, processing, storage, or distribution techniques. It uses ASN.1 (Abstract Syntax Notation 1) to allow the specification of complex nested data structures that may be encoded to a bit stream according to standard rules. Any computer that supports an ASN.1 decoder may then interpret these bit streams.

JPEG

The JPEG committee is tasked with producing standards for continuous-tone image coding. The best known standard that has been produced by JPEG is ISO/IEC 10918 (ISO/IEC 1994a), which is a family of standards covering still image compression. These standards are also published by the ITU-T as Recommendation T.81 (ITU 1992a). This is a complex set of specifications. In reality most JPEG implementations deploy only a basic subset of the standard, transferring coded images using a file format known as JFIF (JPEG File Interchange Format). JPEG coding provides a compression ratio of around 20:1 without appreciable quality loss.

JPEG2000

JPEG2000 is the next generation of still image coding standard, defined in ISO/IEC 15444 (ISO/IEC 2000a). It was designed to support more recent applications such as digital photography. The algorithms support a smooth transition from lossy to lossless coding as the available bandwidth changes. This is achieved by the use of multi-scale wavelets instead of more conventional block-based DCT (Discrete Cosine Transform) coding. Furthermore, it is possible to selectively define the quality of individual parts of the image. This has applications in fields such as medical imaging.

MPEG-1

MPEG was established in order to produce agreed standards that allow for the compression, decompression, processing and coded representation of moving pictures, audio and their combination. To date MPEG has agreed four standards.

The first task of MPEG was to standardize a video coding algorithm that could support video coded at 1.5 Mbit/s or less and be stored digitally, i.e. on a computer disk. The standard was designed to be generic, and is intended be applied to other applications. It comprises a range of tools that can be configured to meet any particular application. MPEG-1 is based on a hybrid DCT/DPCM (Discrete Cosine Transform/Differential Pulse Code Modulation) coding scheme with motion compensation similar to the H.261 coding standard (ITU 1993a). Further refinements in prediction and subsequent processing have been added in order to support random access as found in digital storage media. MP3 (MPEG-1 layer 3) is an audio-only coding standard based on MPEG-1 that is widely used to distribute music content. The MPEG-1 standard was first published as ISO/IEC 11172 (ISO/IEC 1992).

MPEG-2

The second MPEG initiative was to build on the success of MPEG-1 by developing a coding technique that would allow broadcast quality (CCIR 601) TV pictures to be carried at 10 Mbit/s or less. This study began in 1990, and was extended in 1992 to include HDTV (High-Definition TV), which was to have been the subject of a subsequent MPEG-3 program. The resulting standard is ISO/IEC 13818 (ISO/IEC 1994b).

In common with MPEG-1 the video-coding scheme used in MPEG-2 is generic. It is similar to that of MPEG-1, but has additional features. It offers reduced complexity for implementations that do not require the full video input formats supported by the standard. So-called Profiles, describing functionalities, and Levels, describing resolutions, are included within MPEG-2 to provide a range of conformance levels. Usable compression ratios of around 50:1 can be obtained with MPEG-2 coding.

MPEG-4

Work on the MPEG-4 standard started in 1994, building on MPEG-1 and MPEG-2, while taking advantage of further improvements in coding algorithms. The main point of interest of MPEG-4, however, is that it is designed to support multimedia services, particularly for mobile applications. MPEG-4 incorporates a number of key features.

Support for both fixed and mobile applications: In keeping with the increasing trend towards greater mobility within a network, MPEG-4 is designed to support mobile and wireless applications as well as the more traditional wire-line ones.

Compression efficiency: Improved coding efficiency, in particular at very low bit rates below 64 kbit/s, continues to be an important feature of the MPEG standards. Bit rates for the MPEG-4 video standard are between 5–64 kbit/s for mobile or PSTN (Public Switched Telephone Network) video applications and up to 2 Mbit/s for TV and film applications.

Universal accessibility and robustness in error prone environments: Multimedia audio–visual data needs to be transmitted and accessed in heterogeneous network environments, possibly under the severe error conditions that may be associated with mobile use. Although the MPEG-4 standard is independent of the network (physical layer), the algorithms and tools for coding audio–visual data need to be designed with awareness of individual network peculiarities.

High interactive functionality: Future multimedia applications will call for extended interactive functionalities to assist user needs. In particular, flexible, highly interactive access to and the manipulation of audio–visual data will be of prime importance. It is envisioned that, in addition to conventional playback of audio and video sequences, users will need to access the content of audio–visual data in order to present, manipulate and store the data in a highly flexible way.

Coding of natural and synthetic data: Next-generation graphics processors will enable multimedia terminals to present both audio and pixel-based video data, together with synthetic audio/video in a highly flexible way. MPEG-4 will assist the efficient and flexible coding and representation of both natural (pixel-based) and synthetic data.

The MPEG-4 standard was published as ISO/IEC 14496 (ISO/IEC 2001a).

MPEG-7

MPEG is also currently involved in some related activities. MPEG-7 is concerned with the definition of a Multimedia Content Description Interface. This is an application-independent content representation for information search. It aims to create a standard for describing the multimedia content data that will support some degree of interpretation of the information's meaning. This can be passed onto, or accessed by, a device or computer code. A further refinement is MPEG-21 which aims to describe an open framework for the integration of all resources and components of a delivery chain necessary to generate, use, manipulate, manage, and deliver multimedia content across a wide range of networks and devices. MPEG-21 will define any new standards that are required.

2.3.4 The Open Mobile Alliance (OMA)

The Open Mobile Alliance was formed in June 2002 by nearly 200 companies involved in mobile communications. It was created by merging the efforts of the Open Mobile Architecture initiative and the WAP (Wireless Access Protocol) Forum.

OMA is the main body that is driving the development of multimedia applications for use within a mobile environment. The OMA produced the first standards for MMS (Open Mobile Alliance 2001). MMS builds on earlier mobile protocols such as the WAP by adding support for standard image formats such as GIF (Graphics Interchange Format) and JPEG, and video formats such as MPEG-4. MMS also supports audio formats, for example MP3 and MIDI (Musical Instrument/Integration Digital Interface).

2.3.5 Proprietary standards

In addition to the open coding standards described above, there are also several proprietary standards that have achieved widespread usage for streaming audio and video applications.

DivX (www.divx.com)

DivXNetworks have implemented an audio/video codec (coder–decoder) that is based on MPEG-4.

Windows Media Player (www.microsoft.com/windows/windowsmedia)

The Microsoft Windows Media Player has proven popular, not least because of its Microsoft pedigree. Version 9 is claimed to have a significant bandwidth advantage over MPEG-4. It is also the first streaming technology to be incorporated into professional broadcast equipment.

QuickTime (www.apple.com/quicktime)

Developed by Apple, QuickTime was first introduced in the early 1990s to provide multimedia support on the Macintosh computers, and was later introduced onto the PC platform. QuickTime version 6 also supports MPEG-4 encoding.

RealNetworks (www.realnetworks.com)

RealNetworks was founded in 1995 with a mission to supply codecs that are optimized for the support of streaming audio and video. Their current player is RealOne. Several coding algorithms are supported, including MPEG-4.

2.4 Multimedia Services and their Constituent Media Components

2.4.1 Media components

Section 2.3.1 listed the six media components that have been identified by the ITU-T Recommendation F.700 (ITU 2000a). The following paragraphs consider the key characteristics of these components in more detail.

In principle, a multimedia service can then be considered to comprise some combination of these media components. Media components can be added and dropped during the life of the call. As an example a call can be initiated as a voice only connection; a video component may then be introduced during the call. It is also quite possible that a call may start with one component (e.g. voice) and finish with a different one (e.g. file transfer).

In the future it is conceivable that further media components will be added if suitable sensors and drivers can be developed. Examples could include olfactory and tactile components.

Audio

The audio service media component covers the entire aural spectrum from about 20 Hz up to around 20 kHz. However, the most significant part of this spectrum is the band between 300 Hz and 3 kHz that is required to carry the human voice. A voice connection represents the fundamental telecommunications service. It is difficult to imagine a future where the vast majority of calls between people will not continue to be pure voice, or at least with a substantial voice component. A voice service can thus be characterized by the need to transfer human speech with a bandwidth of around 3 kHz, and with no more than a short (less than around 100 ms) delay. It is this latter requirement that qualifies voice to be considered as a real-time service. Traditional voice coding is performed at 64 kbit/s, but acceptable voice quality can be achieved with 8 kbit/s or less.

ITU-T F.700 considers five audio quality levels, ranging from 'allowing the presence of the speaker to be detected' up to 'hi-fi or CD quality'.

Video

There are a number of different video services including conferencing, surveillance and entertainment. These may be either broadcast or viewed on-demand. The services have a range of requirements in terms of bandwidth and service quality. ITU-T F.700 again considers five quality levels from 'allow movement to be detected' up to 'high-definition television'.

An entertainment video service, for example, requires high bandwidth but is not too critical of delay. On the other hand, a video conference service needs a relatively low bandwidth but has the same real-time delay requirements as voice. In all cases video connections require a low error rate connection. Any errors are instantly noticeable and, since video is a real-time service, errors cannot be corrected by retransmission.

High-quality video requires perhaps 10 Mbit/s of bandwidth. However, as is the case with voice, improvements in coding technology are reducing the required bandwidth for video connections. The standard video coding algorithms that are deployed within multimedia services are those developed by MPEG. The MPEG-2 coding standard, for example, can achieve acceptable performance using around 2.5 Mbit/s, while the more recent MPEG-4 standard can achieve the same performance using less than 1 Mbit/s.

Still pictures

In general, still picture services are not real-time and are of a significantly higher image resolution than video media components.

The most widespread still picture service is facsimile (fax). Fax is a stand-alone service and cannot be readily integrated with other service components into a multimedia service. More recently, with the adoption of digital scanning and picture storage techniques, many other still picture formats have emerged which are more compatible with a multimedia approach. The problem with the electronic storage of a still picture is the very high capacity required for a high-resolution image. However, since there is much redundant information it is possible to greatly reduce, or compress, the image without markedly affecting the picture quality. The key compression standards that are deployed are the JPEG family, described earlier in Section 2.3.3.. The bit rate required for a JPEG encoded picture depends on the picture size and resolution, but typical file sizes range from a few kbytes up to a few Mbytes.

Graphics

Unlike still pictures which are images of the real world, graphics are 'drawn,' either by human or computer. A graphics media component is usually coded as a set of vectors, each of which defines a line in terms of its attributes such as its start and end points within the graphics frame.

Like the still picture media component, a graphics service media component is normally not a real-time service.

Text

The original telecommunications services were textual, encoding plain text using a simple code such as Morse code that allowed the text to be transmitted over the low bandwidth transmission links that were then available. Later text-based services included telegrams and telex. Today, these have all but been replaced by electronic mail (email) that, as an IP application, integrates well into multimedia services. However, the recent dramatic growth in SMS has served to demonstrate that simple text services still fill a useful role for instant messaging and 'chat' applications.

Textual services provide a bandwidth-efficient means of communication that works best with the limited character sets that are associated with European languages. It is not so easy to use text services with the much more complex character sets used by some Asian languages. This has encouraged the development of services such as fax.

Data

Perhaps the simplest definition of a data media component is that it encompasses anything that is excluded from the definitions of the other components. Examples of data services include:

- web browsing;
- file transfer;
- file sharing;
- telemetry.

Bandwidths required for data services range from a few bits per day for a telemetry application to applications such as file transfer that can consume all of the bandwidth that is available on a particular connection.

2.4.2 Attributes of a multimedia connection

Quality of Service

QoS (Quality of Service) deals with a collection of parameters that characterize a connection. Each media component of a multimedia call will require different QoS support that the network will be required to accommodate.

From an end-user point of view the principal factors which affect the perceived QoS include:

Delay: The delay across a network is obviously a critical issue for real-time services such as voice. It is true that transporting voice over any packet network will introduce a packetization delay into the connection. However, there is a degree of phobia here that dates back to the times when echo cancellation was expensive to implement. When it is considered that the delays within digital mobile telephone networks require the use of echo control on every connection, it can be appreciated that this is no longer either a significant cost or a source of concern. In reality studies show that, with echo cancellation, a round trip delay of the order of 300 ms can be tolerated for a speech connection. This is supported by the ITU, who recommend a maximum one-way delay of 150 ms as being acceptable for most user applications (ITU 2000c).

Delay variation or jitter: Delay variation is concerned with how much the delay can vary over a time period. This is an issue not only for voice and other real-time services but also for 'near real-time' services such as video-on-demand, where the absolute delay is not important but playback buffers could empty if the delay variation exceeds certain limits.

Media component synchronization: The relative delay of each media component should be limited and reasonably constant in order to provide synchronization between the components. This is of particular importance for services that include audio and video components in order to provide 'lip synchronization' between a sound and the movement that is seen to be causing that sound. It is recommended that this delay be less than 100 ms (ITU 2000d).

Error rate (packet loss): Packet loss affects many services, and it is of particular concern for video services. The coding techniques deployed mean that individual packet loss effects may last for many video frames over several seconds. For example, a packet loss rate as low as 3 % may translate into a video frame error rate of 30%.

Quantization distortion: When an analogue signal (e.g. voice or video) is converted into a digital form, then quantization distortion is introduced due to the digital sample not being exactly equal to the analogue value which lies at the center of a sample (quantization) interval. The level of distortion will vary according to sample accuracy and the interval between samples. Furthermore, if the signal undergoes successive analogue to digital conversions then the additive quantization distortions may reach an unacceptable level.

Subjective measurements: Perhaps the most important QoS criteria are those that determine how an end-user perceives the service. This is not easy to quantify since there are many uncertain factors, such as the user's expectations and the references used for comparison. There are several recognized methods for determining the subjective quality of a video channel. Some rely on measuring the number of JND (Just Noticeable Difference) units between the test system and a reference source. DSCQS (Double Stimulus Continuous-Quality Scale) (ITU 2002) is a method which expresses subjective ratings on a five-point scale, 1–5 representing (respectively) bad, poor, fair, good, and excellent correspondence between the test and reference videos.

Multimedia services cover the spectrum in terms of bandwidth requirement. In general, data services are not real-time and so can tolerate error protection by conventional retransmission methods.

Figure 2.2 maps some services onto a matrix by considering delay requirements and tolerance to errors. It can be noted that no services are identified as requiring a low delay while being able to tolerate a high error loss.

Bandwidth: The bandwidth required by a multimedia service will depend on both the components to be supported and the QoS level that is provided. In general, video components will require the highest bandwidths, while text may need only a very low bit rate. Each constituent part of a multimedia connection must be able to support the required bandwidth. Often the capabilities of the access segment will limit the end-to-end bandwidth that is available for a connection.

Bandwidth on demand: Multimedia calls should be able to add and drop service components during the duration of the call. Even within a single media component it should be possible to modify the bandwidth deployed. For example, the image quality (and hence the bandwidth

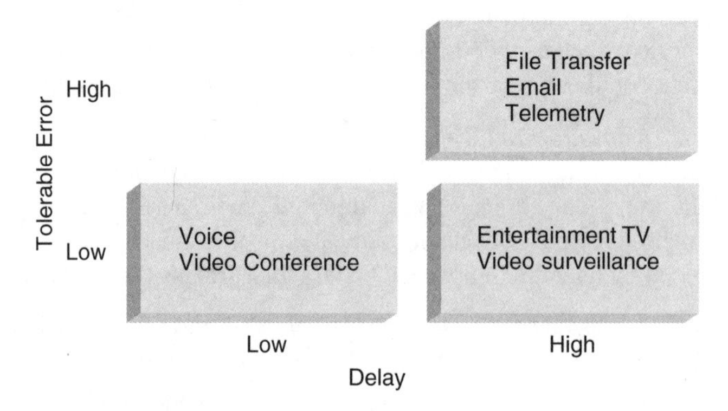

Figure 2.2 Error loss versus delay for some services.

required) might change during the course of a call. It should not be a requirement to plan this at the start of the call. The initial bandwidth may be for only the voice connection, but the bit rate should be adjustable during the call, provided that the network resources are available.

2.4.3 Examples of multimedia services

While there are a large number of potential multimedia applications, many have been implemented only in trials. Smith and Jones (1995) provide an overview of the design methodology for multimedia services. Vaughan (1996) provides a good overview of the tools and equipment that are available to develop and use multimedia services.

This section briefly considers some typical multimedia applications in terms of their main attributes. There are many potential multimedia services. With the exception of 'click to dial' the examples here are ones that are in common use. They are mainly applications that operate over IP, although video conferencing tends to use other technologies such as TDM (Time Division Multiplexing) or ATM (Asynchronous Transfer Mode).

Video multicast

Video multicast is a method for distributing broadcast television channels built on IP multi-casting. A user who wishes to watch a particular channel is added to the appropriate IP multicast group, provided that this is allowed by the user's service profile.

Using MPEG-2 coding, it is possible to achieve broadcast television quality over as little as 1 Mbit/s; however this depends on a number of factors. There is always a trade-off between delay and performance. Another concern is the frequency of scene changes. Sports events, for example, which involve almost continuous camera panning require higher bandwidths for satisfactory performance. In general, bandwidths in the region of 5 Mbit/s can be expected to provide adequate performance under most circumstances, but a bit rate of up to 10 Mbit/s or more might be needed in order to support high-definition TV quality.

Video conferencing

Video conferencing is an example of a multimedia service that has been deployed since the mid-1980s. The service evolved to address a real need, that is to provide an effective substitute to face-to-face meetings, thus eliminating the costs of travel. To be effective it is necessary for even a basic system to support perhaps six participants at each end of the link. In order to be usable, i.e. to be able to identify the facial expressions of the participants, this implies the use of a real-time video component with a resolution of the order of broadcast television. Multi-party video conference calls are also possible, either deploying a split screen to show all the distant parties simultaneously, or using voice switching to select the appropriate video stream to display.

There has been a steady reduction in the bandwidth required to support video conferencing as real-time coding technologies have improved. Early systems used at least 2 Mbit/s to support a rather poor quality monochrome video component. Today, dial-up systems operate at ISDN rates of 64 kbit/s or multiple 64 kbit/s channels.

Microsoft NetMeeting

Microsoft NetMeeting (http://www.microsoft.com/windows/NetMeeting/) is an application that has been bundled with Microsoft Windows that can support voice, video, and whiteboarding, along with program sharing. NetMeeting functionality is now incorporated into the Microsoft Messenger application that forms part of Microsoft Windows XP. While the voice and video performance that can be achieved may be less than ideal, NetMeeting at least shows the potential for multimedia calls. NetMeeting is a multimedia application that aims to integrate all of the multimedia capabilities of the PC using H.323 signaling (ITU 1998a).

The bandwidth that is required to support a NetMeeting session will depend on which multimedia components are in use for a call. Since NetMeeting was designed to share an office LAN with other traffic, it is not intended to be bandwidth intensive. An average bit rate of around 100 kbit/s is typical for a NetMeeting session. This does mean, however, that the voice and video performance are constrained not just by the low bandwidth available but also by the lack of specialized coding/decoding hardware within the average PC.

NetMeeting can include all possible service components: audio, video, still pictures, graphics, text, and data.

Click to dial

This service refers to the integration of IP telephony with Internet Web-oriented applications. An example would be a user browsing the site of a retailer, wishing to speak to a customer service agent. By clicking a link on the Web page a voice call is established to the retailer's call center, and the user is connected to an agent who can provide the necessary assistance.

Click to dial is essentially an audio service that is launched from a graphics application. A refinement would be to incorporate a video component.

MBone (Virtual Multicast Backbone on the Internet)

MBone is an IP multicast technology that supports the distribution of real-time multimedia information. It allows an event to be broadcast over the Internet in a multimedia format; indeed the MBone originated from an effort to multicast audio and video from an IETF meeting that took place in 1992. It has since largely remained as a research tool.

MBone operates as a virtual network within the Internet, using routers that are able to provide multicast support. It allows isochronous (evenly timed) media components such as real-time audio and video to be transferred over the Internet to multiple destinations. Other multimedia applications such as whiteboarding are also supported.

Online gaming

An online game is one that is played while connected to the Internet. Many of these games support multiplayer features that allow a user to play against other people either networked locally or online.

A common requirement is for a game to allow the player to become immersed within a virtual, fantasy world. While the details vary from game to game, the basic requirement is for video quality graphics together with sound. Players are also often able to communicate with each other using audio and sometimes video links. Since play must take place in real time, the requirements on the supporting multimedia network can be quite onerous.

Online gaming applications have exhibited high growth, particularly in some Asia Pacific countries where this application has been instrumental in driving the roll out of ADSL access technology.

The role of audio

While only a sample of the array of possible multimedia services has been considered, it is interesting to note that an audio media component is common to all of these services. Indeed, there are few multimedia services that do not involve an audio component. It is therefore important that interworking is in place to allow the multimedia audio component to connect to traditional voice networks, as discussed in the following section.

2.5 Interworking between Multimedia and Traditional Voice

The requirement to support interworking is always a major issue when implementing a new telecommunications service. In particular, the need to interwork with the PSTN and to support 'legacy' telephony is a key requirement. From a commercial point of view it is important to remember that over 90 % of service provider revenues still arise from carrying voice traffic. Even in the new multimedia world, voice will remain the largest revenue earning service.

Voice traffic is assumed to include 'lifeline telephony', so any multimedia network that is deployed to carry voice must provide suitable availability.

Interworking is achieved using devices that are known as gateways. Gateways undertake a number of functions:

- *Terminate the traditional voice network*: Typically, connections from the legacy network will consist of E1 or T1 links with individual voice connections being carried as Pulse Code Modulated streams within 64 kbit/s timeslot channels. Call control is implemented over ITU-T Signaling System No. 7 (ITU 1993b).
- *Transcode the voice channels*: It is not a requirement to change the speech coding from the G.711 64 kbit/s standard that is deployed within the legacy network, however, advantage is usually taken to transcode into a more bandwidth-efficient technique. There is, of course, the trade-off that has to take place between the bandwidth saving and the coding delay. The main coding algorithms that are currently deployed are as follows.
  — G.726 – speech codec for 5.3 kbit/s and 6.4 kbit/s (ITU 1990);
  — G.728 – speech codec for 16 kbit/s (ITU 1992b);
  — G.729 – speech codec for 8 kbit/s or 11.8 kbit/s (ITU 1996a).
- *Provide echo cancellation*: The delays that are associated with packetizing voice data require echo cancellation to be introduced into the speech path for the speech to remain intelligible. In practice, transcoding and echo cancellation functions are usually achieved using the same DSP devices.
- *Handle fax and other modem connections*: Fax and modem calls masquerade as voice calls so that they can carry data across a PSTN. It is necessary that the gateway can identify these calls and handle them accordingly in order not to lose the data that is being carried within the modem connection. Identification is made either from the associated signaling if

the call originated within an ISDN environment, or by simple detection of the modem tones within the speech path. Once identified, it is important not to compress the data or to apply echo cancellation. The modem link is terminated within the gateway, and the raw data is carried transparently across the multimedia network. These are other tasks that are performed by the DSPs (Digital Signal Processors) within the gateway.

- *Terminate the multimedia network*: Examples of interfaces that could be deployed in a multimedia network include Ethernet (at 100 Mbit/s or Gigabit rates), POS (Packet Over SONET) and ATM.
- *Management and control*: In general, gateways are 'dumb' devices. The associated call control intelligence resides in a separate Media Gateway Controller function. The architecture for the media gateway protocol is specified by the IETF within RFC 2805 (IETF 2000a). The control interface between the gateway and the gateway controller was first defined by the IETF as RFC 3015 (IETF 2000b). This RFC was incorporated by the ITU as Recommendation H.248 (ITU 2000e).

2.6 Terminal Equipment and User Interfaces

There have been many attempts to introduce multimedia terminals over the years, however none can claim to have been a success in terms of mass acceptance. Many of the early multimedia terminals were ISDN based (e.g. Davies and McBain 1988) and were often proprietary. They typically comprised a video-phone, perhaps with an integrated ASCII text keyboard. However, this situation is set to change for two reasons. The first is the emergence of usable standards for multimedia components that have been outlined above. These standards provide some reassurance that equipment from different vendors and service providers will interoperate and will not suffer from premature obsolescence.

The second significant change is the almost universal acceptance of the PC within both home and office. Although, unlike the Apple Macintosh, the internal architecture of the PC was not optimized for the support of real-time multimedia services, it is likely that future multimedia terminal equipment will evolve from the PC due to its ubiquity. The deployment of Microsoft Windows coupled with Intel-based hardware has become a *de facto* user terminal standard. It is the power of this platform (in its current form) to support video applications that is now driving forward multimedia applications. To this end, most PCs are now equipped with high-performance video and sound cards along with high-definition displays and speakers. Many are also fitted with microphones and video cameras.

The adoption of the PC as a multimedia terminal was assisted by the establishment of the Multimedia PC Working Group which defined the minimum levels of hardware and software considered necessary to run multimedia applications. This is contained in the MPC (Multimedia PC) standard that specifies a target machine for which developers can design multimedia applications. These standards are upgraded as required in order to take account of hardware improvements.

There are other possibilities for a consumer multimedia terminal. Developments in digital television coupled with increasing interactivity suggest an alternative starting point could be the combination of a television receiver coupled with an intelligent STB (Set-Top Box). Indeed, the boundary between a domestic television and a multimedia enabled PC is becoming ever more blurred. The term 'interactive TV' is used to cover the range of new services that are now being made possible as a result of this convergence. For example, there could be links from TV screens to Web-based forms for online shopping, the ability to vote on a TV topic, and an on-screen instant messenger allowing viewers to chat online with friends watching the same TV program.

With the increasing digitization of content, from music (e.g. CD, MP3) to video (e.g. MPEG-2, DVB, DVD and DivX), we are witnessing emerging capabilities to receive, store, and access multimedia content of many sorts from a single device in the home: possibly a PC, possibly an STB, possibly a custom 'media center'.

Reception (delivery): At one time, radio required a radio receiver, video broadcasts needed a TV, and Web pages made use of a PC. These days, such boundaries have gone. Radio transmissions are also available on digital TV and as continuous streams over the Internet, video can also be streamed to PCs, and access to Web pages is possible via the TV.

Storage: The latest STBs now contain hard drives that can be used to store video. Interfaces are available to use the hard drives in PCs for the same functionality. Many people have also downloaded their CD collections to their PC hard drive, and perhaps DVD movies will be next. Some products are now available that effectively act as a home entertainment hub (as opposed to STBs and/or PCs evolving to provide this functionality).

Access: A key issue is that whilst the majority of terminal devices remain fixed, the end-user is often mobile within his home. Manufacturers such as Microsoft are keen to overcome such barriers with the use of wireless technologies such as 802.11. The idea is to allow access to the content from any TV in the home, whilst the intelligence that provides the content handling lies elsewhere with the ubiquitous PC (even if is confined to the back bedroom or study).

There are also some exciting developments concerning handheld terminal devices. In addition to third generation mobile applications such as MMS that have been discussed previously, PDAs (Personal Digital Assistants) are evolving to include elements of multimedia support. These devices are now moving away from keyboards towards more natural methods for inputting commands and textual data, such as using handwriting and voice recognition. An additional benefit of this change is that the terminals themselves can reduce in size since they are no longer required to incorporate bulky keyboards.

Another factor that is driving the new generation of multimedia terminals is the development of transducer technologies that are physically small and feature a low power consumption. In particular, CCD (Charge-Coupled Device) cameras and color LCDs (Liquid Crystal Displays) have enabled many portable terminals to support video applications.

Finally, the concept of 'usability' is becoming an increasingly important issue as telecommunications terminal equipment becomes ever more complex. In short, terminals should be intuitive to use and be built around standard user interfaces, with help options available to provide assistance if required.

2.7 The Future

Multimedia services are successful largely because they allow a more user-friendly interaction over a communications network. This user friendliness is largely made possible by the availability of low-cost bandwidth and processing power. It is clear that the growth of multimedia services is now starting to have a significant impact on both business and leisure activities, and that this trend is set to continue.

While predicting the future is always dangerous, it is worth bearing in mind a couple of points. The first is that the number of services that are supported by telecommunications networks continues to grow. This trend shows no sign of changing. The second point is that new services often take longer to become established than at first thought. There are several examples in the recent past of services that have exhibited a long gestation period, such as video-on-demand and video telephony. Whilst technology constraints are a major inhibitor, part of this problem also relates to achieving a critical mass of users against the background of an industry that can be conservative and suspicious of new approaches.

However, there are exceptions. If a service meets a real user need and is tariffed appropriately, then its growth can be meteoric. The recent growth of the Internet and of mobile telephony amply demonstrates this effect.

3

Call Processing

Graham M. Clark and Wayne Cutler
Marconi Communications Ltd, UK

3.1 The Beginnings of Call Processing

The art or science of call processing dates back to the earliest history of telecommunications networks. A Kansas City telephone operator is alleged to have diverted calls away from the undertaking services of Almon B. Strowger, causing him to embark on the development of automatic switching. This operator was a most intimate practitioner of call processing. Only with the operator's intervention could the desires of the caller be brought together with the network resources and terminal equipment to establish end-to-end communication paths. Whilst the technologies of the network have changed dramatically over the years, there is still in every public switching network a function that enables the end-user's wishes to be realized.

Strowger's desire was to make sure that calls destined for him actually reached his telephone rather than that of his competitor. The technology that he invented was based upon electrical pulses generated from the telephone, causing relays in the switch to select the destination line. This actually combined the logical decision-making function that is now recognized as call processing with the physical operation of switching the path between the calling and called parties. What was previously done by the operator's interpretation (or misinterpretation) of the caller's instructions, followed by the patching of wires across a manual distribution board, was now done as a combined electromechanical function. Provided that the equipment was installed and wired correctly, there was no possibility of error arising between the logical and physical operations. Strowger filed his original patent for an electromechanical switch in 1889, with the first such telephone exchange entering service in 1892.

Strowger's technology endured for many years, enhanced at various times to enable local, national, and international call handling. However, electromechanical selectors are bulky, noisy, and consume much power. They work well for the establishment of simple connections, but do not readily allow call routing to be influenced by any criteria other than the dialed digits. The addition of extra circuitry at key points between the selectors enables some special services to be provided, but these were not that widespread and often relied on

Service Provision – Technologies for Next Generation Communications. Edited by Kenneth J. Turner, Evan H. Magill and David J. Marples
 ISBN: 0-470-85066-3

the electrical ingenuity of the local network maintenance staff. Over the years, a number of alternative technologies progressively appeared that reduced the size and power consumption of equipment. However, for many years telephone exchanges simply automatically selected a path through a switch to enable an end-to-end path. By the late 1960s, electronic-based systems appeared which used stored computer programs to control the underlying switch. These were the first SPC (Stored Program Control) exchanges. The 1970s saw an evolution in SPC exchanges, taking advantage of the growing power and reliability of processors to do much more than merely control an underlying switch. The resulting digital exchanges provided capabilities for supplementary services to enhance the basic connection of calls. So it became common to offer users a range of additional call processing services that could, for example, record, redirect, and filter their calls. The first such digital exchanges emerged towards the end of the 1970s.

Throughout this time the basic system requirement was still to establish a two-way path for voice communications within the PSTN (Public Switched Telephone Network). This changed in the last two decades of the 20th century, when there was significant demand to carry data services over the PSTN. The resulting ISDN (Integrated Services Digital Network) placed new demands on call processing, which had to be aware of the different types of traffic being carried by the network and the different treatment to be applied to each. A similar time-frame saw the deployment of mobile networks, initially providing regional and national coverage but quickly developing to support international roaming. Again it was the call processing component of the systems that was responsible for providing the enhanced functionality to enable user and terminal mobility. Now the emerging demand is for higher bandwidths, multiple connections and mixed media streams, using fast and efficient cell and packet transport technologies. The user expectations of quality, reliability, and service have increased in line with the developments in technology. While simple fixed voice communication is still a major revenue stream, it no longer defines the basic communication paradigm. Call processing has been adapted and enhanced to address an increasing diversity of communications services, and its role continues to be no less important, despite advances in underlying network technologies and user terminal equipment.

3.2 Key Attributes of Call Processing Systems

The partitioning of functions within a modern central office (exchange) varies between different manufacturers, and is often related to the architecture of the central processor, the line transmission and signaling hardware, and the switch fabric. However, there is generally a core set of functions that is identified as the call processing component of the system. This may be defined as *that set of functions which is responsible for determining how end-user demands should be satisfied using the available network resources*. Prior to the introduction of digital technology, the user demands were conveyed solely via the dialed digits. Subsequent to the introduction of digital technology, it was possible for the user's demands to be further represented via data held on the exchange (e.g. divert all incoming calls to a pre-programmed destination number). Under these circumstances, call processing can be said to be responsible for *overall call management based on the originating user's wishes (as represented by dialed digits and/or user data) within the constraints of the terminating user's wishes (as represented by user data) and the available network resources*. If there are conflicts between the wishes of the originating and terminating subscriber, then call processing must manage the conflict accordingly. The management of such conflict is part of *feature interaction* (see Section 3.4.3 and Chapter 13).

In most cases call processing is realized entirely in software. This permits maximum flexibility for in-service enhancement and reconfiguration, which is essential in a mission-critical real-time application. Increasingly, the software has to support some form of portability between platforms in order to operate in multi-vendor, multi-technology environments. However, functionally, call processing has to maintain an abstract view of the hardware resources available in order to control the set-up and tear-down of calls or sessions, and to manage the allocation of connection-related resources. The level of that abstraction is one characteristic that varies between systems, as will be seen in the later descriptions of call models. The attributes described below are those that may be expected in any system's definition of call processing.

3.2.1 User and network interfaces

To achieve its purpose of controlling the establishment of calls, call processing must interface to users of the network in order to take in call requests from originating parties, alert terminating parties to new call arrivals, complete the connection of call paths, and tear down completed calls. Where calls transit between offices, equivalent interfaces are required across the network. Events and information received from these interfaces provide the basic prompts for call processing to initiate and progress call handling. The actual transfer of signaling information to and from the transport hardware is not usually considered to be part of the call processing domain. However, call processing may need to be aware of the signaling system being used on a particular user line or network link to format its output to match the signaling protocol in use, or to restrict its features to those that the signaling system supports.

Until the introduction of ISDN, the signaling capabilities of user terminals were limited to the application of analogue line conditions like voltage breaks and tone transmissions. However, given that the service offered by the networks was voice telephony, these signals were sufficient to carry all the user-defined call variables, i.e. the called party number, to the local call processing function. Even when supplementary services were provided to voice telephony (i.e. additional capabilities beyond simply providing a two-party basic call connection), these simple analogue signals were capable of exchanging sufficient information between the end-user and the central office. In this case, the system would typically use a tone or announcement to prompt appropriate digit input from the user. This form of signaling is termed in-band since it uses the same frequency range and transmission path as the one that carries voice during a telephone call. Stimulus signaling is a mechanism by which a limited set of user signals is interpreted by call processing according to the state of the call and the user profile information held by the system. Thus a flash-hook (recall) signal may be used to initiate a three-way call or to pick up a waiting call, depending on the context of its use. Similarly, a particular digit signal may have different meanings in different circumstances, such as selecting a party during a three-way call.

Because ISDN is designed to provide a wide range of voice and data services, ISDN user signaling needs to carry far more information about the type of call to be established as well as the requirements in terms of minimum bandwidth guarantees and interoperability. Therefore an intelligent digital signaling interface is required between the ISDN user terminal and the central office. In 1984, specifications for DSS1 (Digital Subscriber Signaling System

Number 1, ITU 1984a) were published by ITU-T (International Telecommunications Union – Telecommunications Sector). This signaling system provides protocols for circuit and frame-based services, supplementary services, and management. The signaling for supplementary services is explicit to the service being requested or controlled. This is known as functional signaling. It has benefits in the standardization of services and terminal devices, but is seen by many to restrict the innovation of new services. Because DSS1 has the capability to carry far more information than analogue user signaling, call processing for ISDN becomes more complex, with many more factors affecting the way that calls are handled. ISDN calls typically require two or three times the system processing resources of simple voice calls.

Like the user interface, signaling within the network has PSTN (Public Switched Telephone Network) and ISDN variants. The first digital central offices often had to work to analogue network interfaces, which suffered similar restrictions as analogue user interfaces in terms of the amount of information they could carry. In 1980, CCITT (now ITU-T) published in its 'Yellow Book' series the specification for Signaling System Number 7, a common-channel digital signaling system. SS7, as it is known, provides a layered hierarchy of protocols that through addition and enhancement have become the foundation for all circuit-mode communication signaling ever since. Call processing particularly concerns layer 4 of the protocol stack. For the PSTN this is represented by TUP (Telephony User Part, ITU 1980), and for the ISDN by ISUP (Integrated Services User Part, ITU 1988). The basic ISUP call message flow is similar to TUP, but is able to convey a larger amount of information between the subscribers during the establishment of the call. In practice TUP appears as a large number of national signaling variants, reflecting the network scenarios and service objectives of different operators. In the UK, the resulting national user part has been refined to the point of being able to carry virtually all of the information carried by ISUP. ISUP also typically appears as a large number of national signaling variants, although the differences from the core standard tend to be less than is the case for TUP.

Because there are different signaling systems present on both the user and network interfaces of most networks, one important aspect of signal handling within call processing is the interworking of signaling types. In the simplest cases this involves transferring parameter information from an encoded field in one message to an equivalent field in another message. However, it is more often the case that some information carried by one signaling system is not handled by another signaling system. In this situation the incompatibility can have an impact on the basic service and supplementary services offered on a particular call. Examples of this include the provision of default rather than requested bearer services, and the non-availability of number presentation services. Some central office systems reduce the complexity of call processing by limiting the number of interworking scenarios supported. A general trend in the past decade has been to withdraw support for analogue network signaling systems as SS7 becomes widespread.

3.2.2 Number translation and route selection

At the heart of call processing is the translation of the requested called party number to a destination. In public telephone networks, the numbering scheme is based on the E.164 standard (ITU 1984b) that defines a regulated number structure which consists of CC (Country Code), NDC (National Destination Code), and SN (Subscriber Number). This structure enables global routability within the public telephone network(s). On a given node, the corresponding destination may be a user line identity if the terminating line is on the same office, or an inter-office route identity if the called party is hosted on another central office. The translation is dependent on the numbering plans of the network as a whole and on any local numbering plans that apply. The latter are less of an issue today because there has been a progressive rationalization of numbering in most countries; previously, variations like short codes to reach adjacent numbering areas were quite prevalent. These variations could be handled by maintaining separate translation tables or by preprocessing the digit strings to a standard format to allow a common translation structure to be used for all calls. Preprocessing was also a necessity when interworking between electromechanical and digital systems, because the number formats sent by the older systems were often modified by their own translation mechanisms. For example, Strowger switches absorb the dialed pulses for each digit that they translate, and these may need to be reconstituted by the recipient call processing.

The way in which number translation is implemented depends on the type of data structure used to represent the digit strings. The most common are tree structures that mimic the structure of Strowger selectors and are efficient in terms of computer memory for well-populated number ranges. Hash tables have advantages in processing and storage for moderately populated ranges. Direct string look-ups can be useful where overlapping digit strings or sparse populations are involved. In some systems a combination of these structures is used to balance efficiencies within different number ranges and relative sparseness. One of the traditional measures of performance in switching systems is post-dialing delay, i.e. the elapsed time between the end of dialing and the ringing of the destination line. The design of the number translation data and the way it is accessed during the progress of a call can have a significant impact on this delay, and on the number of calls that can be processed under peak load.

The results of digit translation determine how the call should be routed or handled. There are three main categories of translation result commonly encountered in networks, though the occurrence of each type varies in proportion between central offices that have directly connected subscribers (Class 5 nodes or Local Exchanges) and central offices that do not have directly connected subscribers (Class 4 nodes or Trunk Exchanges).

The first category identifies a user or group of users to whom the call is directed. The user groups include PBXs (Private Branch Exchanges) and Centrex (Central Exchange Services). In handling individual PSTN users there is generally a one-to-one relationship between the user and a physical line, whereas in basic rate ISDN there is a requirement to select one of the user's lines to deliver the call. PBXs have a number of physical lines serving a community of users. The central office switch identifies the PBX and selects an available line, but is not usually aware of the routing of the call once it reaches the PBX equipment. In contrast, a Centrex user is a member of a logical group but has a dedicated line. If the basic number translation function only identifies the group, there will be a secondary translation based on the Centrex extension number to determine a specific user and line within the group.

The second category of translation result identifies remote destinations that have to be reached via other switches. There will be one or more routes that can carry the traffic toward the required destination. It should be noted that the subsequent node may or may not be the destination office, i.e. in turn it may have to forward the call on one of its routes to the final destination. Indeed, modern digital telephone networks are typically arranged in a hierarchical structure. As an example, at the time of writing, the UK network comprises about 700 Class 5 nodes, each of which is connected to two different Class 4 nodes. There are in total about 100 Class 4 nodes, which are meshed together. Therefore, a typical call between two users on different Class 5 nodes is routed via two Class 4 nodes. The translation information provides criteria for selecting a route appropriate to a particular call. Typically, the criteria are based on priority ordering, statistical distribution, or service preference. For redundancy reasons, there will typically be more than one route choice towards a destination. If the call attempt on the initially chosen route is unsuccessful, due to network congestion being encountered either at this node or at a subsequent node, then an alternative route may be selected for a subsequent attempt. This procedure of using a list of possible routes to avoid network congestion is known as re-routing. It may be performed a number of times on a single call until either the call is successful or the possible list of alternative routes is exhausted.

The final result category is for those calls that require some specialized treatment, such as re-direction announcements, information services, or IN (Intelligent Network) services. The type of treatment required and the resources necessary to provide it are generally identified as part of the translation process.

3.2.3 User subscription data

The introduction of digital technology enabled a user's preferences to be encapsulated via a block of data held on the exchange. This data is interrogated by call processing and is used in conjunction with the dialed digits to influence call handling. This data holds the subscribed service set applicable to the individual user (e.g. diversion, call waiting). Prior to the introduction of the Intelligent Network (see Section 3.5), the service set was realized on the exchange itself (see Section 3.4). Subsequent to the introduction of the IN, the service set could be a mixture of switch-based and IN services. However, in practice the former still tend to heavily outnumber the latter due largely to the fact that the switch-based services pre-date the IN and had already been deployed.

3.2.4 Resource and switch control

Although call processing is essentially a software function, it has the responsibility for allocating hardware resources to calls. Depending on the precise system architecture, this can include channels on links from peripheral switches (concentrators) to central switches, channels on inter-office trunks, digit receivers, echo controllers, tones, announcements and multi-party conference bridges. Digit translation and the processing of features determine that certain types of resource are required in a call. For instance, the routing tables for a particular digit string may indicate connection via a particular inter-office route, to a local subscriber line, or to an internal recorded announcement. Also, supplementary services may require the connection of in-band digit receivers.

Call processing maintains a data representation of the call-related resources, which includes any conditions for the use of these resources and selection criteria for their allocation. This data has to be kept synchronized with the status of the actual resources so that the latter do not become overallocated or locked up because of software malfunction. The form of the resource data in call processing may be a simple counter of the number of available resources, detailed resource knowledge based on a logical representation of the resource, or explicit knowledge of particular physical entities. The most complex resource allocation relates to circuit selection on inter-office routes. While this typically takes a logical view of the resources, it needs details about the availability of circuits, including their maintenance and call states, any traffic reservation that may be applied to them, and the selection criteria to be used to avoid or resolve dual seizure (i.e. a condition when a call attempt is simultaneously initiated at both ends of a single circuit). From the allocation of resources and determination of the connections required, the establishment of paths through the switching fabric is requested. At the point when a call is answered, which is normally the point when charging starts, the end-to-end connection must have been established. However, during the set-up of a call there is often a backward path enabled to allow in-band tones and announcements to be connected to the caller. IN services may require a forward path for collection of in-band digits. All of this path manipulation is under the control of call processing.

One function that falls between routing and resource control is the control of bearer interworking. The need for this arises when the incoming and outgoing circuits are being carried over different transmission technologies, e.g. analogue, PCM (Pulse Code Modulation), ATM (Asynchronous Transfer Mode), or packet switching. Interworking may simply involve the insertion of appropriate equipment to provide appropriate mapping/transcoding between the different technologies, or it may require some negotiation of bearer capabilities in order to accommodate the call on a low or restricted bandwidth bearer. Call processing is aware of which interworking scenarios are permitted and what may be negotiated, and also controls any transcoding equipment like any other connectivity resource.

3.2.5 Generation of call recording data

In electromechanical systems, charging for calls was generally based on timed pulses generated by the switching hardware to increment simple counter devices, or meters, associated with each subscriber line. Each pulse represented a chargeable unit, and all units were charged at the same rate. Variations in charging were accomplished by allowing variation in the rate of pulse generation, depending on the time of day and the broad categorization of call destination. Therefore the cost of a unit was constant, but the value of a unit in terms of the call duration time it represented varied according to the call parameters. This method was not sufficiently flexible to permit many of the charging features that are commonplace today.

For operators to offer the vast selection of tariffs, discounts, and feature-related charges that are often the basis of their competitive positioning, much more detailed knowledge is required of each individual call. Although the call details are subject to combinations of online and offline processing in order to determine actual charges, the raw call data originates from call processing. This includes the origination and destination numbers, the duration of the call, the basic charge rate that applies, and any special resources used. Additional information is required if supplementary services or IN services are used within a call. To meet these requirements, call processing has to gather data for the whole duration of the call, and ensure that it is stored and passed on without loss or error. The accuracy of timing parameters must meet the requirements of the local regulator, and must never result in a user being overcharged for the service received.

Similar per-call data from call processing is used to derive statistics on the loading of the system, the volume of traffic handled, and the traffic behavior and demand characteristics. Some detailed statistics require additional data from call processing. Systems tend to generate such detailed statistics only periodically. Call processing typically makes the data available either when explicitly requested or else based on its own internal data (e.g. output relevant data every *n*th call).

3.3 Switch Architectures and Call Models

As call processing software was developed for digital systems during the 1970s, it became clear that the software complexity involved in replicating and adding to the functionality of electromechanical systems could not be handled without employing some form of decomposition. Decomposition first involved the analysis of the problem space, and secondly the design of run-time software entities. Both were influenced by the overall system architecture adopted by each manufacturer, and by the processor configurations employed. Until the mid-1990s the majority of telecommunications manufacturers designed their own processing platforms, operating systems, and even programming languages to achieve their objectives for resilience and reliability. These set the framework within which their call processing realization would exist, grow, and be progressively refined.

Another factor in the decomposition is the way in which the concept of a telephone call is visualized as a collection of linked entities. This is known as the call model. The detail of the call model varies between manufacturers, because it is always influenced by their processing environment and hence their system realization. Also, call models have not remained static, but have undergone stages of migration and refinement to meet the evolving requirements of telecommunications. However, most telecommunications practitioners recognize the existence of three main classes of call model: the half call model, the three segment model, and the multi-segment model.

The objective of a call model and its associated data and software entities is to break down the call into a set of independent 'objects' that can be linked to establish a communications session. Each object encapsulates a range of functions that correspond to a subset of the total functionality required for handling a given call, whilst the links between them correspond to a logical interface. The variations among call models arise because of different approaches to the definition of these objects. A number of object-oriented analysis techniques would allow only real-world physical entities as candidates for modeled objects. However, experience has shown that the complexity of call processing can benefit, in some cases, from the definition of purely software functions as modeled objects. To be effective, a model must easily allow the introduction of new objects. These may modify the relationships between existing objects, though in general new objects should not force changes on existing objects.

3.3.1 Half call model

The half call model (shown in Figure 3.1) takes the view that, ignoring the functions of the central office itself, there are two real-world entities involved in a call. These are the calling

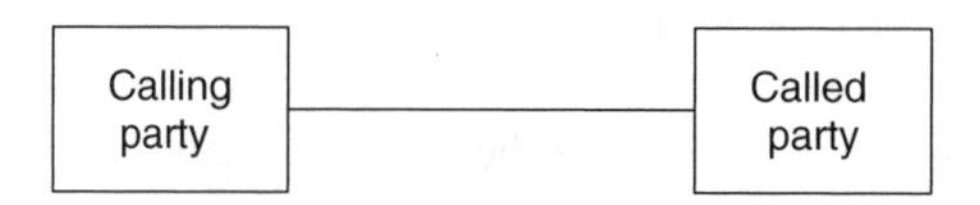

Figure 3.1 Half call model.

party and associated line, and the called party and associated line. All functions are therefore considered to be associated with the calling or the called party.

This type of model derives much of its origins from the application of multi-processor architectures to telephone systems, and the consequent desire to partition software into smaller run-time entities. Single processor systems typically had a single call processing entity. While this model was the first step in modeling the problem space, it did not address the addition of new objects well. A particular limitation was its inability to handle three-party calls. In addition, the model is not symmetrical. The calling party object has to handle not only the originating signaling and features but also the digit translation, routing, and the switch control. The called party object is concerned with only the called party features and signaling. Consequently, even as a software entity model this does not necessarily result in a balanced load between multiple processors.

3.3.2 Three segment model

In terms of representing the roles within a call, the three segment model (shown in Figure 3.2) looks more like the pre-Strowger configuration as it has the calling and called party (user A and user B), plus an intermediary association segment whose responsibilities are much like the manual operator. The user segments continue to handle the features of the user and line, including signaling and any associated features. However, the association segment links them for the duration of a call by providing the digit translation, routing, and switch control.

In drawing up a model for a particular call, the user segments embody all the call characteristics and functions of the users involved in the call. They are specific instances of the user object for A and B. The association segment, on the other hand, is a common resource drawn from a generic pool of association objects and does not embody any user-specific functions. This model therefore provides a clearer decomposition than the half call model. Although it does not specifically address the introduction of new objects, it simplifies the addition of functions to existing objects.

3.3.3 Multi-segment model

Even the three segment model has shortcomings when dealing with multi-party calls and with a proliferation of user features. Furthermore, ISDN required more flexibility in the relationship between users and lines, and IN required the insertion of a service switching function in the call. Therefore, a number of switching system manufacturers looked for further stages of decomposition to address these issues. The resultant multi-segment model is shown in Figure 3.3. The additional segments in the model reflect a mixture of physical objects and logical functional entities.

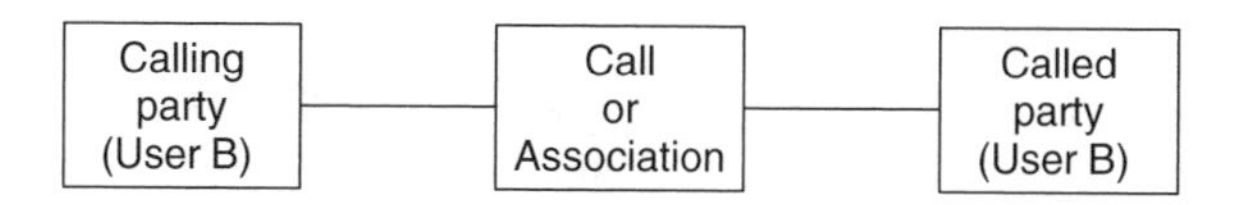

Figure 3.2 Three segment model.

Figure 3.3 Multi-segment model.

The separation of line-related functions from user-related functions removes any rigid associations between a user and the physical line serving them. The user segment is responsible for applying the user preferences based on the user data. At this stage the model can still be said effectively to reflect physical entities. It can be extended further on the same basis by recognizing that there are attributes and functions associated with groups of users that may be relevant in some call configurations. This can be represented by the addition of a separate optional segment (the User Group segment) in the model rather than by incorporating them within the user or association segments. This further separation means that the model can readily handle a user as a direct-dial line or as a member of a multi-user group. The latter may have its own selection and re-direction features encapsulated in the User Group segment. Figure 3.4 models a call first to User B, and secondly via a logical user group of which User B is a member. In both cases, the user segment representing User B would apply the user's preferences based on the user data. In the latter case, the group segment would be responsible for applying any group preferences as specified in the corresponding group-related data.

The final step in the refinement of the multi-segment model occurs when the segments identified from analysis of physical and conceptual entities are re-defined as owning only the attributes and functions that apply to a basic call with no supplementary services. Then, any additional functions, such as supplementary services, appear as additional segments in the model. Three such examples are shown:

- the terminating half of a call with both a call forwarding and a call waiting service (Figure 3.5);
- a call half with a three-party service (Figure 3.6);
- an Intelligent Network call with a Service Switching Function (Figure 3.7).

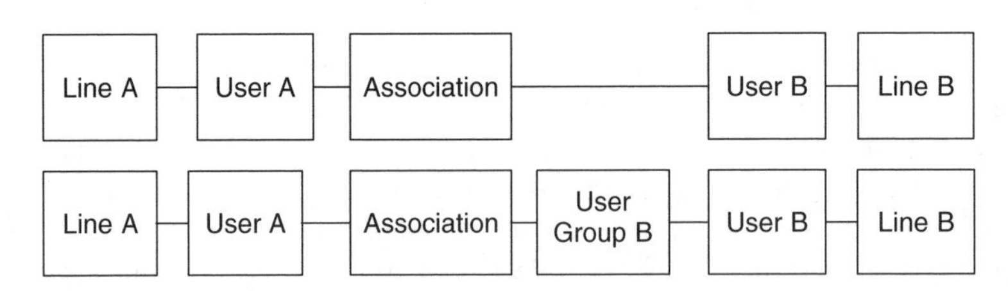

Figure 3.4 Inclusion of user group segment.

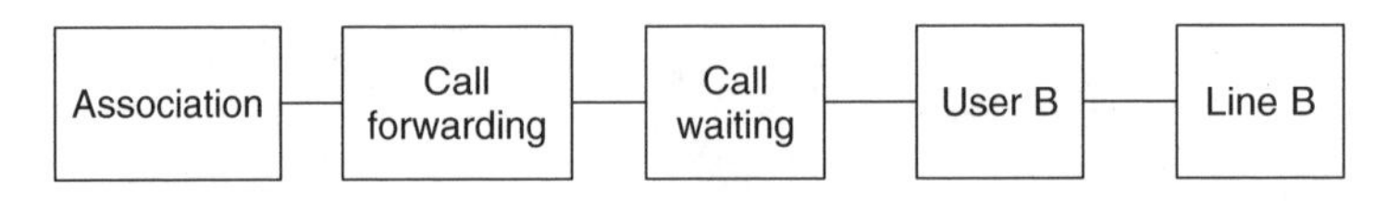

Figure 3.5 Call waiting segment in terminating call half.

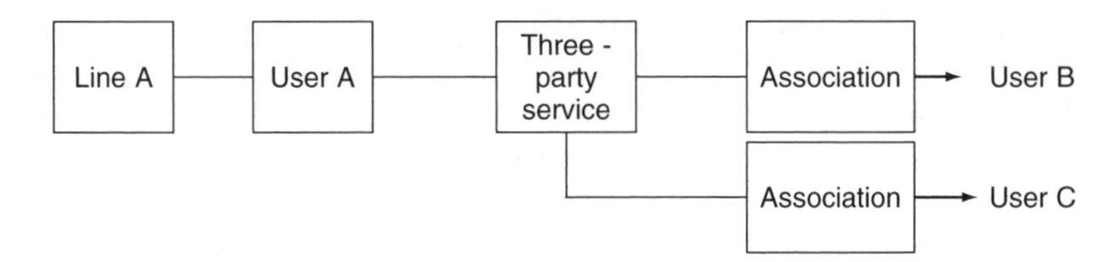

Figure 3.6 Three-party segment.

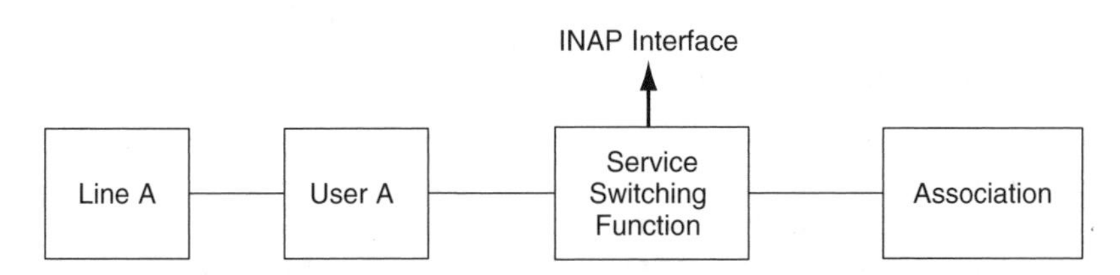

Figure 3.7 Access to Intelligent Networks.

If User B subscribes to Call Forwarding and/or Call Waiting, and the call conditions are such that these features apply, the model is configured with segments representing the features. These segments are considered to embody all the logic and data specific to the features.

The three-party segment is used to establish three-way conference calls within the model (as shown in Figure 3.6). User A is effectively involved in two calls, to Users B and C, each requiring a path and connection control. Hence the model includes two Association segments, with the three-party segment providing the branching between the two. The control of the conference bridge hardware to achieve a multi-party connection is also the responsibility of the three-party segment.

An IN service is effectively a feature whose operational data and logic are remote from the central office. Nevertheless, the multi-segment model can incorporate IN by representing the SSF (Service Switching Function) as a segment like any other feature. The positioning of the SSF in Figure 3.7 is appropriate for an originating user or dialed digit based IN service. A terminating IN service would be most likely represented by an SSF segment in the terminating call half.

3.3.4 Mobile switches

The introduction of mobile telephony required a new type of central office, namely the MSC (Mobile Switching Center). An MSC has physical links to other digital central offices. These enable it to make and receive calls from/to the fixed network. An MSC has a number of associated BSCs (Base Station Controllers) which in turn manage a number of BSs (Base Stations). The BS has a radio interface and is responsible for detecting the presence of users (or specifically their handsets) within its area/cell and for informing the BSC/MSC accordingly. In this way, the mobile network is able to maintain a location database for the mobile users. In addition, each mobile user is assigned an E.164 number that corresponds to a home MSC. This number is called the MSISDN (Mobile Station ISDN Number). When a user is detected in an area looked after by another MSC, there is a mechanism by which the Visited MSC informs the Home MSC of this fact. The Visited MSC assigns a temporary number called the MSRN (Mobile Station Roaming Number) to the visiting user and informs the Home MSC of this temporary number. The Home MSC maintains an HLR (Home Location Register) that contains the MSRN. The MSRN is a routing number that is used only internally to the network and is not visible to the end-users. In turn, the Home MSC passes subscription parameters to the Visited MSC so that the correct features can be applied to that user. Calls terminating at the mobile user are routed via the MSISDN to a Gateway MSC (which accesses the HLR to obtain the MSRN) and subsequently onto the Visited MSC. In practice, the Gateway MSC is the Home MSC. Calls originating from the user are handled by the Visited MSC and are not routed via the Home MSC. All features are applied at the Visited MSC.

Mobile switching systems had significant new requirements and implications for call models. Prior to the introduction of mobile switching systems, it was accepted that a user was served by a particular exchange line and hence by a particular central office. As mentioned previously, ISDN changed the user-to-line relationship slightly by providing two bearer channels per user (basic rate) or 30 channels per user (primary rate). Mobile systems had to go a significant step further, allowing a user to be (in theory) served by any line on any MSC in any network. Additionally, the user is able to move between lines and MSCs during a call. To support this flexibility the call processing architecture must be able to create and delete dynamically an instance of a user as the user moves between MSCs. In addition, it must also be able to associate the user instance with any line hosted on the switch, and re-assign that association during a call. The most widespread standard for the requirements of mobile telephony is GSM (Global System for Mobile Communications). GSM supports the creation of user instances in new locations via a VLR (Visitor Location Register). However, there are few examples of this being implemented as a separate physical entity. Instead, it is integrated into the MSC database. The multi-segment call model has been shown to support mobile requirements well as it allows the addition of new segments for handover, paging, and other mobile-specific functions. The following examples show segment configurations for GSM handover both within and between an MSC.

Intra-MSC handover (Figure 3.8) involves one user moving from a BS served by one BSC to another one controlled by a different BSC. As far as the user is concerned the same call continues despite the change of radio resources. Therefore the same user segment persists throughout the call, and the user should be unaware of the handover action. The call can be modeled by the introduction of a handover segment. This has the responsibility for managing all of the handover-specific actions and, where necessary, for duplicating signaling from the user segment to both of the BSC interface segments.

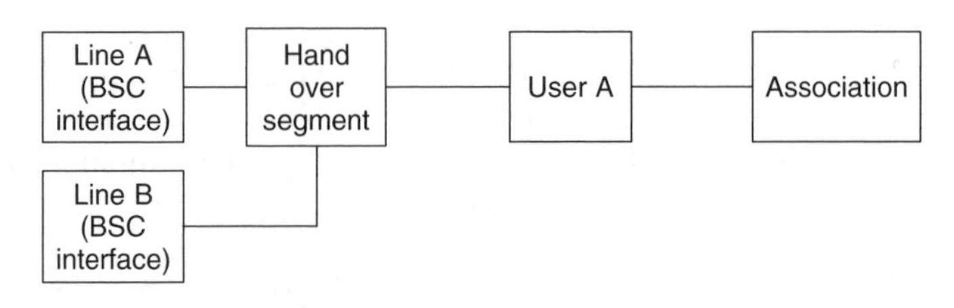

Figure 3.8 Intra-MSC handover.

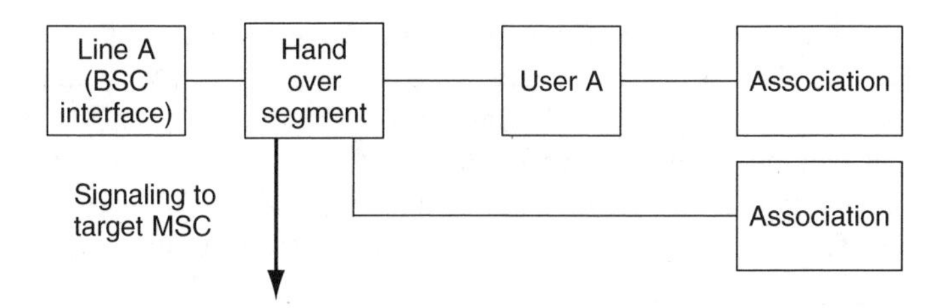

Figure 3.9 Inter-MSC handover.

Inter-MSC handover is more complex, as it involves direct signaling to the target MSC to prime the radio resources, and also the establishment of a terrestrial channel from the source MSC to the target MSC. However, by encapsulating these functions within the handover segment, no other part of the model is affected by these requirements. The additional association segment simply sees a request for the set-up of a call to a destination that is a line on the target MSC, and routes the call appropriately. The handover signaling interface is directed to an equivalent handover function in the target MSC. This is illustrated in Figure 3.9.

3.3.5 Implementation of call models

The previously illustrated models have been developed to aid the analysis of call processing systems. The translation of any of them into run-time entities varies from system to system. In general, irrespective of the applied call model, the identified objects are typically mapped to run-time software processes with the links between them corresponding to a proprietary message interface. The extent to which any particular manufacturer has refined its call model through the various stages depends on both technical and economic factors. The continuity of service required from telecommunications systems means that change nearly always has to be achieved through evolution of hardware and software, which may render some architecturally desirable changes infeasible. However, a number of systems have successfully implemented the multi-segment model, each in a slightly different way.

Marconi's System X call control is probably the closest to the analytical model, as it implements each of the model segments as a separate run-time process and links them in a chain structure via a common interface. The chain for a particular call is built up as the call progresses, and can be dynamically modified during the call as new call events occur. As signaling messages are received at either end of the chain, they are processed by the recipient segment before being passed along the chain. Each segment in the chain then has the option of passing the message along unmodified, modifying the message, absorbing the message, or replacing the message with one or more different ones. Messages that pass to the end of the chain generally result in the generation of user or network signaling. The common interface, which is not specific to any external signaling protocol, enables the introduction of new segment processes without impact on existing ones. The ordering of segments in the chain dictates the segment precedence for receipt of messages, and hence contributes to the resolution of feature interaction (Schessel 1992, Jackson and Zave 1998).

Nortel's Generic Services Framework (Utas 1995) also adopts a modular framework to implement a multi-segment model, but assembles the segments in a two-dimensional structure.

In one plane it links the basic components of the model, i.e. the line, user, and association segments. From each of these 'parent' segments are linked orthogonal chains of dependent feature segments. The processing of a signaling message by any one of the parent segments is not completed until any necessary processing by all of its dependent segments is completed. Only then will the parent pass on the message to the next segment in the parent chain. Again the ordering of the feature segments is designed to support the resolution of interactions.

3.4 Switch-Based Services

The functions required within call processing to establish, maintain, and disconnect calls are described in Section 3.2. These support what is often referred to as POTS (Plain Old Telephone Service). The introduction of computerized digital central offices during the 1970s and 1980s paved the way for a wide range of new call-related features to be offered to users. These features were intended to enhance the experience of using the network, and to enable users to have some choice in the way that the network handled their calls. The majority of these features are realized as extensions to call processing that break into the normal flow of POTS processing at appropriate points to modify the progress of the call in some way. Some features apply only a temporary modification to the flow of call data and then revert to the normal progression of events and actions. Other features cause a completely divergent sequence to be followed. One of the main reasons for the development of more complex call models was the need to have a framework where new features could be added in a controlled fashion, with minimal impact on the pre-existing call processing function. Even in systems that have not formally adopted a multi-segment model, the modular encapsulation of features within a simpler model has been crucial in enabling the ongoing addition of new features.

Each service has been developed for a particular manufacturer's switching system, constrained by the architecture, software languages, and processing environment of the system. Thus, the same feature developed by several manufacturers for the same network operator may display minor variations in its operation between the different versions. This is in addition to the fact that the operator has to purchase the feature separately from each manufacturer in order to roll it out to the whole network.

3.4.1 PSTN services

Despite the limitations of the user interface, PSTN call control is able to support a wide range of supplementary services under user control. The degree to which these are exploited by networks depends on the strategy adopted by the network operators. Some who wished to use the services as a lever for the widespread generation of additional calls were keen to roll them out in a form that could be accessed by the majority of users with basic multi-tone handsets. Other operators saw supplementary services as a lever to migrate users to ISDN, and therefore did not go to great lengths to make them available to PSTN users.

Even without the signaling capabilities of ISDN, call processing systems are able to support an extensive range of services, but have to do so in a way that makes all dialogue between the user and the service an exchange of in-band information. From the user side this is not recognized as anything different from the input of digits for call origination. However, within the system the digits are handled by different software (and sometimes hardware) components. To convey information to the user, the system has to use either tones or recorded announcements. Tones and announcements with very general wording, e.g. 'the service you have requested is unavailable', are easily re-used for different services but are not sufficient to guide users through the unique attributes of each new service. Therefore, each deployed service generally requires the provision of specific new announcements. Among these are some that have to be flexible in order to announce times, dates, and number strings. Call processing usually regards tones and announcements as system resources under its control, so for each service it must be aware of the applicable announcements and the procedures to drive them.

3.4.2 ISDN services

Although there were various national initiatives to provide digital user interfaces and associated user signaling protocols, it was not until the development of ISDN (Integrated Services Digital Network) under the various regional and international standards bodies that there was any common approach. The standardization of ISDN protocols included the definition of a set of services for which explicit messages and information elements were included on the user interface. In theory, this meant that every conformant network within a particular geographical region could support a common set of ISDN services. The principle of a defined service set with associated functional signaling has been used in mobile network standards in order to allow network interworking and inter-network roaming. The chief criticism of such a rigid service definition is that it does not readily allow operators to differentiate their service offerings. Both ISDN and mobile networks have looked to IN technology to assist this situation (see Section 3.5.4), because that enforces only a common set of capabilities upon which differentiated services can be built.

ISDN services make use of the ability to exchange information between the user and the network in signaling during the set-up, supervision, and tear-down of a call. They also rely on the ability of user terminal equipment to display information, and the ability of the terminal equipment to provide processing of signaling in order to present a simple but functional interface to the user. There are some differences between the ISDN supplementary services defined by ETSI (European Telecommunications Standards Institute) and ANSI (American National Standards Institute), but examples from the ETSI set serve as a useful indicator of the type of services included by both organizations. Some services relate to the presentation of calling and connected numbers to the user, and the restriction of that service to protect confidentiality and security. Knowing the identity of the caller enables a user not only to choose between accepting and rejecting a call, but also explicitly to request that the call be forwarded to another destination (i.e. a call deflection or diversion service). Call forwarding services are also available in the ISDN, and differ from their PSTN equivalents due to the ability to carry additional call history information (e.g. diverting party number(s)) thanks to the relative richness of ISDN signaling.

3.4.3 Feature interaction

The increasing number of switch-based services focused the need to control the way in which features in a call interact with one another, and how any potential conflict between the wishes of the calling party and the wishes of the called party are to be managed. This is the role of feature interaction. The overall objective of feature interaction is to combine features within a call in such a way that the resultant behavior is predictable, presents a valid service-enhancing experience to the user, and does not risk the integrity of the system. Within all of the architectures described in this chapter, there are techniques for introducing new features into call processing and for managing the interactions between them. The half call and three segment models generally require new features to be programmed into one of the existing software entities. This is done so that the execution order of the modified logic reflects the desired relationship between the new features and the pre-existing ones. The multi-segment model permits new features to be added via new objects. Therefore, the feature interaction solution adopted by any switch manufacturer is strongly influenced by the existing call model. Chapter 13 discusses the topic of feature interaction in greater detail.

3.5 Call Processing for Intelligent Networks

Intelligent Networks are described in detail in Chapter 4, so this section highlights some of the issues that have a particular impact on call processing. That said, the whole objective of the IN is to minimize the impact of new features on the call processing functions of switching systems. This is done by presenting a set of generic capabilities that can be utilized by any new service. However, before that objective can be satisfied, these generic enabling functions have to be supported by call processing. In some cases the success or failure of an IN solution has been dictated by the ability of a system to provide these functions. Because in most cases IN is an add-on to existing systems, it is generally accepted that the standards represent a degree of compromise between what is theoretically possible and what the various vendors' call processing entities could readily support. Consequently the standards are open to some interpretation, and that does not assist the objective of widespread interoperability.

3.5.1 Triggering

IN standards like AIN (Advanced Intelligent Network, Telcordia 1995), CS 1 (Capability Set 1, ITU 1993i) and CS 2 (Capability Set 2, ITU 1997c) identify a number of points in a call as DPs (Detection Points). These are generally associated with externally visible call events, e.g. call initiation, call answer and call release, but also include some internal events like the completion of digit translation. When call processing encounters detection points it may be required temporarily to suspend its operation in order to invoke a new IN service or to progress a service that is already active in the call. For the triggering of a new service, call processing has to associate that service with a detection point and to provide mechanisms to check that the call criteria demanded by the service have been met. Only then can a service switching entity be instantiated for the call, and the initial invocation of the service be made at the service control point.

3.5.2 Service switching

The service switching function is responsible for translation of IN operations into call processing events and vice versa, and for transferring call information to and from the IN service. The implementation of this function, and the type and number of services that it can support, is again dependent on the architecture adopted. In some architectures, the service switching can be visualized as a layer between the call processing and the service control, whereas other architectures incorporate service switching as an element of call processing.

3.5.3 *The IN and feature interaction*

The introduction of Intelligent Networks brought additional feature interaction requirements for call processing. In any real-world system, there is clearly a need to control the interaction of both switch-based and IN features. Since IN features were defined in terms of formal DPs, the implication is that all features ought to be handled in terms of IN DPs in order to have a common mechanism by which their triggering and interaction may be controlled. The extent to which IN-like triggering is applied to switch-based features varies between manufacturers. However, there are implementations (e.g. Marconi's System X) that do use formal DPs and triggers as a common control mechanism for the superset of switch-based and IN features. It is clear that the introduction of any IN service with interaction implications for existing features must be integrated into the feature interaction logic of the supporting switching system before the new service can be deployed. This fact has proved a significant barrier to the original IN vision of fast feature roll out. However, there is no doubt that the IN has conferred considerable benefit in terms of simple number translation services. These have no interaction implications and can utilize a centralized number translation database on the SCP (Service Control Point).

3.5.4 *The IN and mobility*

As stated previously, the features of a mobile subscriber are applied by the visited MSC. This has the drawback that the available feature set is dependent upon the visited MSC supporting all of the subscribed features at the home MSC. This can result in a mobile user effectively losing feature capability during roaming. This is clearly undesirable. One attempt to solve this issue was based on the use of IN to provide the user features. This led to CAMEL (Customized Applications for Mobile Network Enhanced Logic, ETSI 1998), which was based on IN CS 1 plus appropriate extensions.

3.6 Softswitches

Call processing has reached a high level of maturity in public and private switched telephony network environments. These networks have provided the most ubiquitous long-distance communications medium for many decades. The networks use circuit switching and limit connections to 64 kbps (or in some cases, multiples of 64 kbps). However, the latter part of the 20th century saw an explosion in data networks, principally caused by the growth of the Internet. There also emerged a number of technologies such as ATM (Asynchronous Transmission Mode) and IP (Internet Protocol) to satisfy the demands of these data networks. Both ATM and IP provide connections of significantly higher bandwidth, and introduce the switching of information via cells/packets instead of via dedicated circuit-switched paths. The growth of data traffic has meant that, in developed countries, the total volume of data traffic has exceeded that of voice traffic. Current trends are that both voice and data volumes are increasing but that the rate of increase for data is much greater. This means that, as time progresses, the volume of data traffic will increasingly dwarf that of voice traffic. This fact leads to two different motivations for carrying voice over a packet network rather than via a dedicated circuit network:

- In the enterprise or business environment, IP-based networks are ubiquitous. There is a motivation to use the IP network to also carry voice rather than having to maintain a separate voice network (e.g. via a PBX).
- From a network operator point of view, as the volume of data increasingly dwarfs that of voice traffic, it makes sense to have a single data network and carry voice as a minority usage (in terms of the overall percentage bandwidth) and high premium service (in terms of revenue earned) on that network.

Despite these changes in the underlying transport technology, there is still a significant ongoing role for many of the functions of call processing in the establishment of communication associations between endpoints. Call processing has to be migrated from its legacy of narrowband bearers and circuit switches in order to function in these new environments. The result of this migration is encapsulated in the softswitch.

3.6.1 Softswitch definition

The ISC (International Softswitch Consortium) offers the following simple definition of a softswitch (International Softswitch Consortium 2002a):

> *a softswitch is a software-based entity that provides call control functionality.*

Although this is a very brief statement, it does capture the two key attributes of the softswitch: it is a software entity with no explicit hardware dependency, and its function is the control of calls. The implication of the former is that COTS (Commercial Off the Shelf) hardware ought to be used rather than in-house processing platforms, as was typically the norm for digital telephone exchanges. The latter is open to refinement depending on the nature of the calls to be handled in a particular scenario.

The ISC has produced a reference architecture (International Softswitch Consortium 2002b) to define the constituent functions of a VoIP (Voice Over Internet Protocol) network, and to apply a consistent nomenclature to them. The document lists those functions that may be considered part of the Media Gateway Controller (elsewhere known as a call agent, call controller, call server, or softswitch). This list is summarized below:

- *Media Gateway Control Function*: controls Trunking Gateways, Access Gateways, and Media Servers;
- *Call Agent Function*: originates, terminates, and processes signaling messages from peers, endpoints, and other networks;
- *Routing Function*: provides routing decisions for intra- and inter-network calls;
- *Call Accounting Function*: produces details of each call for billing and statistical purposes;
- *Interworking Function*: provides signaling interworking between different endpoint protocols;
- *SIP (Session Initiation Protocol, IETF 1999a) Proxy Server Function*: provides routing and address translation for SIP endpoints.

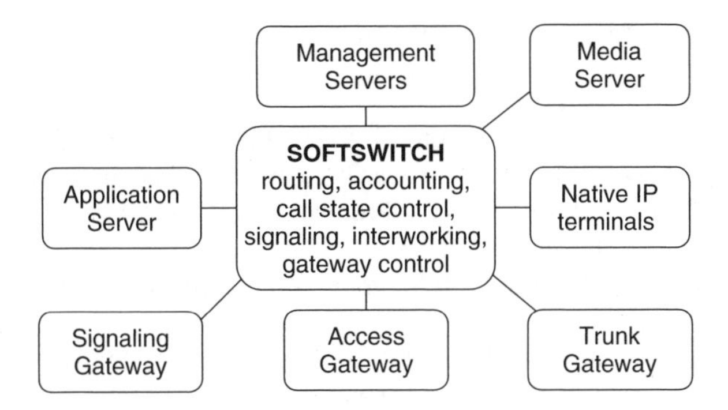

Figure 3.10 Architectural context for softswitch.

The inclusion of these functions does not imply any implementation structure or functional distribution, nor does it make them all obligatory within a softswitch. The requirements of different networks will determine the functions that need to be present. However, anything within a VoIP network that is part of call control, rather than media/signaling transport, applications or management, falls within the scope of the softswitch.

Figure 3.10 provides an architectural context for a softswitch. Essentially, it may be regarded as a 'decomposed telephone exchange' and includes a number of typical constituent parts that are interconnected via an IP network. The functionality of these elements is as follows:

- *Signaling Gateway*: terminates the physical media for the PSTN signaling (typically SS7) and transports the signaling to the softswitch over IP.
- *Trunk Gateway*: terminates the physical media for the NNI (Network-Network Interface) PSTN bearer and transcodes the media into the appropriate packet/cell format for transporting across the packet network. The PSTN bearer is typically a 2.048 Mbits/s digital trunk. The trunk gateway is controlled by the softswitch.
- *Access Gateway*: similar to a trunk gateway except that it terminates a legacy PSTN/ISDN UNI (User-Network Interface). In practice, there are many types of access gateway, reflecting the many different types of legacy UNI.
- *Media Server*: responsible for providing resources such as tones, announcements and conference bridges in a softswitch network. It may be controlled by the softswitch or by an application server.
- *Application Server*: responsible for providing high-level applications within the softswitch network. It encompasses both legacy IN SCPs as well as IP-based servers that (typically) use SIP.
- *Management Server*: responsible for providing the management of the elements within the softswitch network. Typically, a number of such servers would exist since a single server is unlikely to manage all of the elements.

One of the key goals of this decomposed architecture is to be able to 'mix and match' best-in-class products from different manufacturers. In order for this to be achieved, it is necessary to define standard protocols between the different elements. Examples of such protocols are:

- MGCP (Media Gateway Control Protocol, IETF 1999b) and H.248 (ITU 2000e) to enable gateway control between the MGC (Media Gateway Controller) and access/trunk gateways;
- SNMP (Simple Network Management Protocol, IETF 2002a) between the management server and managed element;
- INAP (IN Application Part) and SIP for application server access (between the softswitch and application server);
- SIGTRAN (Signal Transport, IETF 2000c, IETF 2001a, IETF 2002b) suite of protocols to enable signaling transport between the signaling gateway and the MGC, and between the access gateway and the MGC;
- SIP (IETF 1999a) and H.323 (ITU 1996b) as UNI protocols (between the softswitch and the native IP terminals);
- SIP as an NNI protocol between the softswitch and a SIP-based network;
- SIP-T (SIP for Telephony, IETF 2001b, IETF 2002c) and BICC (Bearer-Independent Call Control, ITU 2001) as inter-softswitch protocols.

3.6.2 Capabilities

The ISC reference architecture encompasses all the functionality required from a VoIP-based public network, equivalent to existing circuit-switched public networks. However, the realization of softswitches has been driven by the Internet telephony and enterprise IP telephony community on the one hand, and by public network operators on the other. Because the expectations of these communities are somewhat different, the capabilities of available softswitches and the extent to which they meet the above definitions vary depending on their backgrounds.

In the context of Internet and enterprise IP telephony, softswitches are fairly lightweight entities that rely on the underlying router network to provide Quality of Service differentiation for time-critical traffic. Service availability is generally achieved by providing multiple instances of the softswitch within the network so that outage of a single instance should not cause any noticeable service disruption. Working in a bounded and explicitly defined environment like a corporate network, a softswitch is likely to require support and interworking for only a very limited set of signaling protocols, and may suffice as a pure SIP or H.323 entity. To integrate flexibly into a public network, support for a range of IP, ISDN, and PSTN protocols becomes necessary. The required capacity of a softswitch in an enterprise environment may not need to be more than a few hundred connected lines and a few tens of simultaneous calls, and the set-up times can be less demanding than in a public network.

Public telephone networks are licensed bodies that have to meet regulatory requirements on the Quality of Service they deliver, the accuracy of charging, and the availability of service (particularly for calls to emergency services). The safeguards to meet these requirements have been engineered into Class 5 circuit switches over many years. Softswitches targeted at public networks also have to guarantee deterministic behavior in all circumstances to meet the same stringent requirements. They tend to be capable of high line capacity and throughput, and have in-built procedures for handling component failure. In addition they perform resource management for the bearer streams by controlling the media flows at the edges of their area of responsibility. Commentators from an IP-centric background might argue that many of these functions can be handled by the inherent capabilities of the router network. Transactions over IP require bandwidth only when there is information to be sent, and there is no reservation of end-to-end bandwidth. Traffic is handled essentially on a first-come, first-served basis, and networks are provisioned to ensure that for the majority of the time the expected traffic will not suffer unacceptable delay. Protocols and procedures do exist at the router level for the prioritization of traffic, so while there is a broad mix of traffic, that which is delay-tolerant gives way to that which is not. Practitioners from a telecommunications background tend to argue that in a network that is principally carrying high quality traffic, e.g. real-time voice and video, the softswitches need to be involved in the enforcement of Quality of Service. A softswitch can use its view of network resources to avoid the overbooking of network capacity, and working with its peers can offer end-to-end guarantees across different parts of a network.

A desirable feature of softswitches, particularly in public networks, is that they should enable the fast development and deployment of new services. This concept is not a new one; indeed it was one of the key drivers for IN in the 1980s. However, the advent of the softswitch is seen, particularly by operators, as an opportunity to overcome some of the drawbacks of IN in respect of interoperability, and to enhance their own ability to create services without the need for vendor intervention and without significant run-time overheads. The first objective is typified by the activities of the Parlay Group to define application interfaces that are richer in scope and capability than the IN, dealing with multimedia services as well as voice services. The second objective requires that call control itself is available for direct modification by the operator. While this has been likened to the freedom of an individual to load software onto a personal computer, it seems to carry significant risk for operators, vendors, and ultimately the network users.

As an essentially software entity, a softswitch could in theory be portable between different computing platforms. Portability of software is attractive because it allows owners to use their preferred platform provider and to take advantage of new developments in computer technology without having to purchase new software with each upgrade. Historically, most Class 5 switches employed highly customized processing platforms in order to guarantee performance, resilience, and reliability, and were therefore not portable. IN systems had greater scope for portability as they had less peripheral equipment to interface to the processor, but in general vendors tailored their products to one or maybe two platforms because of the variations in the processing environments. Therefore, while it is a desirable objective, softswitch portability is unlikely to become widespread unless a ubiquitous processing environment exists across all target platforms. At the time of writing, most computer hardware vendors are pitching their products at the telecommunications market on the basis of differentiated features rather than homogeneity.

3.6.3 Developments

The initial softswitch implementations fell into two broad categories, namely small VoIP systems in the enterprise environment and Class 4 replacement nodes that used VoATM (Voice over ATM), in the network operator space. The former were concerned with using existing data networks to carry voice, including (where possible) providing an alternative to the PSTN network to avoid long-distance call charges. The latter were specifically concerned with providing legacy voice services but using ATM as a transport technology, with the voice being carried over AAL1 CBR (ATM Adaptation Layer Protocol 1, Constant Bit Rate). Subsequently, network operators have also drifted towards VoIP as their choice of transport technology due to the ubiquitous nature of IP and the fact that it readily supports multimedia services (of which voice is one such service). Whilst the collapse of the technology sector has slowed down the roll out of softswitch technology, it is now increasingly likely that the next generation network will be based on softswitch technology, and will support multimedia services over IP. Multimedia services are discussed fully in Chapter 2.

In terms of protocol standardization, the ITU led the way with the H.323 standard (ITU 1996b), with the initial version being published in 1996. This includes a call control protocol H.225 (ITU 1996e) and a separate bearer control protocol H.245 (ITU 1996f). It also supports setting up multimedia sessions. The earliest VoIP systems were based on the H.323 standard. Subsequently, SIP has emerged as the preferred protocol for multimedia session control.

The IETF (Internet Engineering Task Force) also defined MGCP (Media Gateway Control Protocol, IETF 1999b) to enable control of IP gateways. MGCP was used to control the first generation of IP voice gateways. In some cases, MGCP was refined by other bodies for real network usage, e.g. the PacketCable consortium (PacketCable 1999). Subsequently, the H.248 standard (ITU 2000e) was jointly developed by the IETF and ITU as the preferred standard for gateway control. In addition, the SIGTRAN Working Group in the IETF defined a suite of protocols to enable reliable transport of legacy signaling within an IP network.

The desire for network operators to support their legacy service set over a packet transport technology has led to two competing standards in VoIP networks, namely SIP-T (SIP for Telephony) and BICC (Bearer-Independent Call Control). SIP-T provides a mechanism for encapsulating legacy signaling within SIP: typically, but not exclusively, ISUP. BICC was based on ISUP plus extensions to support an ATM/IP -based bearer. BICC CS 1 (ITU 2000i) concentrated on VoATM. BICC CS 2 (ITU 2000j) added support of VoIP. Currently, work on BICC CS 3 is in progress.

In addition, the last few years have seen the emergence of a number of consortia concerned with the realization of a deployable NGN (Next Generation Network). Examples include the ISC, ETSI TIPHON (www.etsi.org/tiphon), and MSF (Multi-Service Switching Forum, www.msforum.org).

Many softswitch implementations currently exist, but their state of development and refinement varies considerably, as does their degree of conformance to recognized standards. Competition in the marketplace remains keen because there are many potential vendors of software products. These vendors fall into four principal groups:

- *Major public network telecommunications vendors*: Most of the major public network telecommunications vendors have announced softswitch products, either as Class 4/5 replacements or as call servers for the packet-switched domain of third generation mobile networks. Where appropriate, these products are derived from earlier circuit-switched call control products, and therefore incorporate a lot of the architectures and feature sets covered earlier in this chapter.
- *Minor public network telecommunications vendors*: There are a number of smaller companies in the public network arena who have been producing the circuit-switched equivalent of a softswitch for some time, particularly for mobile systems. These are used to control

commercially available switches that support open control interfaces. Several have used their expertise in developing software that controls independent hardware to enter the IP softswitch market ahead of the major players.

- *IP hardware manufacturers*: A number of the leading IP hardware manufacturers have developed or acquired softswitch products. Their hardware will be the foundation of the media transport in VoIP, and the availability of a softswitch complements their portfolio. These suppliers do not have a legacy of circuit-switched call control, and are likely to adopt more distributed architectures based on servers that support each of the different functional components of the softswitch.
- *New manufacturers*: There are many new players in the softswitch market who have not previously been active in telecommunications. Some of these are developers of server software for IP networks, and others are new companies who see an opportunity to enter a space that has traditionally been dominated by a relatively small number of large companies. The approach taken by these new entrants is the same as the hardware vendors. That is, they adopt a distributed architecture that re-uses general purpose components wherever possible. For example, accounting is not specific to telephony. It is required by a variety of IP services and can be supported by specialist functions on dedicated servers. The distributed nature of the architecture provides niche opportunities for new players to compete for business in the softswitch market.

3.7 Future

Enterprise IP telephony, allowing the use of a corporate LAN or WAN for voice and data communications, is already quite widespread. For new businesses it saves on the cost of building parallel networks within their premises. For existing businesses the point of migration depends on the available lifetime of their current investment. Some expanding businesses have used it as a means of unifying diverse and distributed PBX systems within their constituent companies.

In public networks, a migration to IP has been slower due to their relative conservatism as well as the problems in the technology sector. Indeed, the initial focus for the public networks was for the support of voice services over ATM. However, a number of public network operators are now looking at VoIP as the basis for their next generation networks, and are starting customer trials to judge the maturity and capability of the technology.

In addition, the emergence of IP-based multimedia terminals will have an impact as intelligence increasingly moves from the core of the network to the edge of the network. Such terminals are likely to be based on the SIP standard.

Despite the ongoing evolution of the network towards IP and the migration of intelligence to the network edge, the continuing importance and need for a call processing capability should not be overlooked. There is still an important role for call processing to manage communication transactions between a myriad of endpoint types in the next generation IP-based network. Thus, call processing looks set to continue in some shape or form well into the future.

4

Advanced Intelligent Networks

Robert Pinheiro and Simon Tsang

Telcordia Technologies, USA

4.1 History of the Intelligent Network (IN/AIN)

IN (Intelligent Network) systems are an essential part of modern digital switch networks, and have been widely deployed worldwide since 1994. Prior to the introduction of intelligent networking, the logic necessary for the realization of advanced voice services was integrated into the functionality of the central office (exchange) switch using SPC (Stored Program Control). This meant that any modifications to these services, including the introduction of new services, required switch suppliers to change a switch's software. Moreover, the new software had to be loaded into *all* switches that would support the new or modified service. Because this process was dependent on the switch supplier's timetable for introducing these changes, telephone companies were hampered in their ability to introduce new services or to modify existing ones.

In particular, this approach introduced four major problems that hindered a network operator's ability to design and deploy new services quickly:

- *Switch downtime*: A large amount of switch downtime was required in order to install and test new service programs. In addition to the downtime necessary for simply upgrading a switch's software to incorporate the new service programs, additional downtime often resulted because the service programs were not 100 % error free on initial release and deployment. In addition, *every* SPC switch supporting the new service needed to be upgraded, resulting in additional downtime.
- *Interoperability*: There were no standardized platforms or architectures for providing services on switches. Programs were therefore vendor-specific and platform-specific. Programs written for one switch would often not work on another vendor's switch. Indeed, even programs written for different releases of the same switch would present interoperability difficulties.
- *Backwards compatibility*: Every time a new service or feature was introduced on the SPC switch, it was necessary to ensure that it was compatible with other software already running on the switch. Very often, this required adding software patches to the existing programs to make them interoperate with the new program. If there were many programs already

Service Provision – Technologies for Next Generation Communications. Edited by Kenneth J. Turner, Evan H. Magill and David J. Marples
 ISBN: 0-470-85066-3

provisioned on the switch, a significant amount of time and effort was required to write and provision these patches. For this reason, services were often introduced in groups ('service packs') rather than individually. This meant that service providers were restricted to yearly or twice-yearly release schedules.

- *Data distribution, provisioning, and maintenance*: Databases and other service data are integral parts of a service. With SPC switches, data has to be provisioned and maintained on every switch. If the amount of data is large this is a difficult and time-consuming task. Modifying or updating the service data is also difficult because it is necessary to make changes to *every* switch where the service is provisioned.

Although SPC switches offered network operators a way to provide digital services in their networks for the first time, the problems described meant that designing and deploying new services was a time-consuming and consequently expensive process, particularly for data-intensive applications.

In the mid-1970s, the introduction of CCS (Common Channel Signaling) provided a way out of this dilemma. With CCS, the network signaling required for establishing, managing, and ending calls could be carried out of band – outside of the actual voice path, and within a separate signaling network. This signaling network in North America uses SS7 (ANSI Signaling System Number 7) as the signaling protocol. By making use of CCS/SS7, service programs could be removed from switching system software, and placed in separate network elements called Service Control Points, or SCPs, designed specifically to host service programs. SPC switches evolved to become Service Switching Points, or SSPs, and were required to support a limited number of service triggers. These triggers caused the SSP to signal the SCP, via CCS/SS7, with a particular set of information when certain service-related events were detected by the SSP. For instance, such events included the dialing of a particular sequence of digits, or the arrival of an incoming telephone call. As a result the SCP would execute certain service logic and send back a message to the SSP with further directions for handling the call.

Thus IN/1 (Intelligent Network 1) was born. This new architecture resolved many of the problems associated with service design and deployment on SPC switches by separating the service logic from the switching elements. Service logic programs now needed to be provisioned at only one centralized point in the network, instead of at every switch. By removing service software from the switch and placing it in a programmable SCP, telephone companies would themselves be able to modify services and to introduce new services. In addition, the separation of service control and switching would provide for supplier independence and create open interfaces between network elements, opening up the possibility for building networks with different vendors' equipment.

The first IN/1 services, introduced in 1981, were 800 service and calling card services (also known as Alternate Billing Services). Following the introduction of IN/1, Telcordia Technologies (then known as Bellcore, or Bell Communications Research) introduced what it called the AIN (Advanced Intelligent Network) architecture. In addition to SSPs and SCPs, the AIN architecture introduced other network elements called IPs (Intelligent Peripherals) and SNs (Service Nodes). An IP is essentially a separate network node responsible for playing announcements and collecting user information – Personal Identification Numbers, for instance. A Service Node combines the functionality of an SCP and an IP.

The initial vision for AIN was defined by Bellcore as AIN Release 1 (Bellcore 1993a). However, the functionality specified by AIN Release 1 was extensive, and the supplier community decided that they could not realistically implement all of AIN Release 1 in a single product release. Hence, AIN Release 1 was replaced with three phased implementations: AIN 0.0, AIN 0.1, and AIN 0.2. The last of these phases, AIN 0.2, was specified in 1993 in Bellcore Generic Requirements document GR-1298-CORE (Bellcore 1993b). Following the publication of this document, however, additional AIN functionality has been gradually refined over the years. Beginning with Issue 2 of GR-1298-CORE in 1994 (Bellcore 1994), the designation '0.2' was dropped and the name became simply Advanced Intelligent Network. As of late 2002, the most recent specification of AIN functionality was given in Telcordia GR-1298-CORE (Issue 7), November 2001 (Telcordia 2001).

The GR-1298-CORE series of documents specifies the requirements for the AIN switching system, or SSP. The functionality specified in the document includes the AIN Call Model, triggers and requested events, details about the messages exchanged between the SSP and the SCP, and other requirements for SSP processing. A companion document, GR-1299-CORE (Bellcore 2001a), has also been released for each issue of GR-1298-CORE. The purpose of GR-1299-CORE is to specify requirements for the interface between the SSP and the SCP. Another series of Telcordia documents, the GR-1129-CORE series (Bellcore 2001b), concerns itself with the interface between an SSP and an IP.

On the international front, ITU-T (International Telecommunications Union – Telecommunications Sector) defined the CS 1 (Capability Set 1) series of recommendations in 1993. This was the first milestone in the international standardization of Intelligent Networks. The recommendation described the IN framework from the perspective of four different planes: Service Plane, Global Functional Plane, Distributed Functional Plane, and Physical Plane. ETSI (European Telecommunications Standard Institute) refined INAP (IN Application Part), which specifies the communications between the IN physical elements, and produced the first European INAP standard (ETSI 1994). ITU-T refined their IN recommendation in 1995 (ITU 1995a). IN CS 2 (ITU 1997b) introduced Call Party Handling in 1997. These recommendations and standards were an attempt by the industry to ensure that IN equipment from different vendors would successfully interwork. Outside of the USA, most IN deployments are based on CS 1 functionality and specifications.

4.2 Intelligent Network Architecture

4.2.1 AIN and ITU-T IN architectures

The simplified view of a generic AIN architecture is illustrated in Figure 4.1. In the figure, three SSPs (Service Switching Points) are shown: one each at the EO (End Office, exchange) serving the calling party and called party, and an intermediate SSP at an AT (Access Tandem) switch. The IP (Intelligent Peripheral) is associated with the SSP serving the calling party, and provides functions such as digit collection or announcements in support of a service. The SSPs communicate with the service programs at the SCP (Service Control Point) by means of SS7 (Signaling System Number 7) TCAP (Transaction Capability Application Part) signaling, which traverses an STP (Signaling Transfer Point) switch in the SS7 signaling network.

The ITU-T IN architecture is nearly identical. Figure 4.2 shows the physical elements (Points) and functional entities (Functions) that comprise the ITU-T IN CS 1 architecture (ITU 1995b).

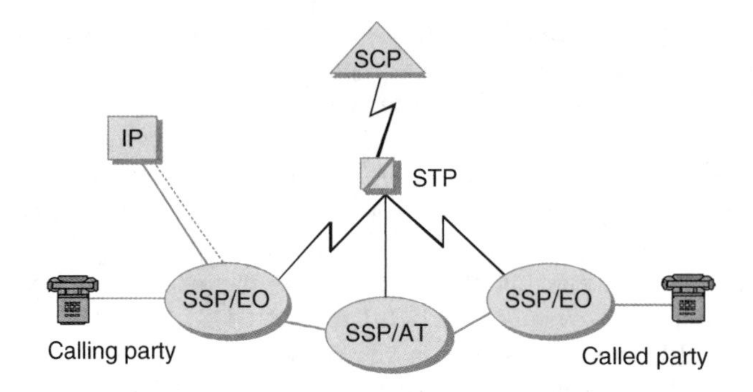

Figure 4.1 Generic AIN architecture.

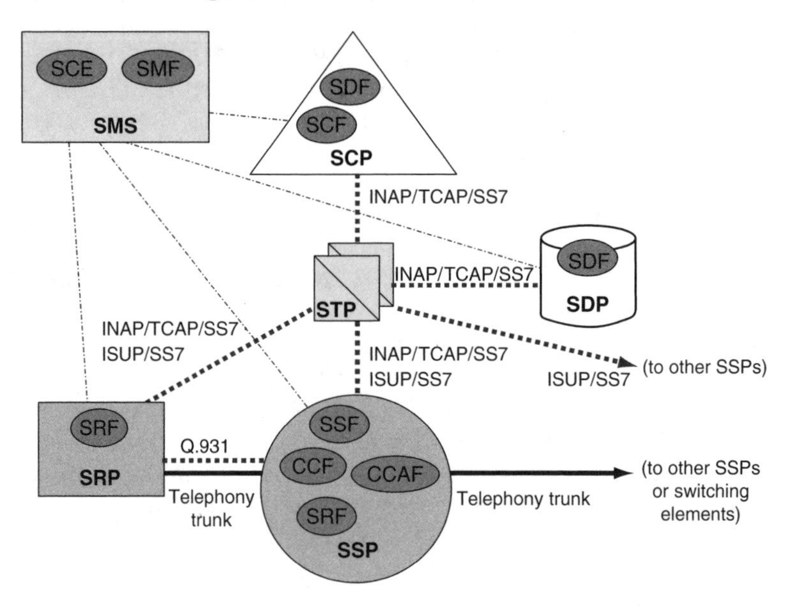

CCAF	Call Control Agent Function	SMF	Service Management Function
CCF	Call Control Function	SMS	Service Management System
SCE	Service Creation Environment	SRF	Specialized Resource Function
SCF	Service Control Function	SRP	Specialized Resource Point
SCP	Service Control Point	SSF	Service Switching Function
SDF	Service Data Function	SSP	Service Switching Point
SDP	Service Data Point	STP	Signal Transfer Point

Figure 4.2 IN architecture and functional elements.

The control protocol used between physical elements is INAP (Intelligent Network Application Part) that uses TCAP as the transaction and transport service. The physical elements and their associated functional entities are now described:

SCP (Service Control Point): This provides a centralized intelligence point within the IN architecture. It contains the SCF (Service Control Function) that provides SLPs (Service

Logic Programs) and an execution environment. The SCP may also contain an SDF (Service Data Function) that provides databases and directories for the SCF.

SSP (Service Switching Point): This is the true heart of the IN architecture and comprises numerous elements:

— *CCF (Call Control Function)*: This provides the IN call model abstraction called the BCSM (Basic Call State Machine). CCF to CCF intercommunication is via the ISUP (ISDN User Part) protocol (ITU 1993c).

— *SSF (Service Switching Function)*: This provides the interface to the SCF and is positioned between the CCF and SCF. It contains the DP (Detection Point) trigger mechanism, FIM (Feature Interaction Manager), and IN-SSM (IN Switching State Model) that provides the SCF with a view of call and connection view objects.

— *CCAF (Call Control Agent Function)*: This provides access for users. It is the interface between user and network call control functions.

— *SRF (Specialized Resource Function)*: This optional element provides user interaction capabilities such as announcements and digit collection.

SDP (Service Data Point): This provides databases and directories for the SCF. This is provided through the SDF (Service Data Function). The SDF interface is based on the X.500 protocol.

SRP (Specialized Resource Point): This provides user interaction capabilities such as announcements and digit collection. SRF (Specialized Resource Function) is the functional entity that provides the capabilities.

SMS (Service Management System): This manages services within the IN. The SCE (Service Creation Environment) provides a point for downloading new services onto the SMS. The SMF (Service Management Function) then provisions the new service onto the various

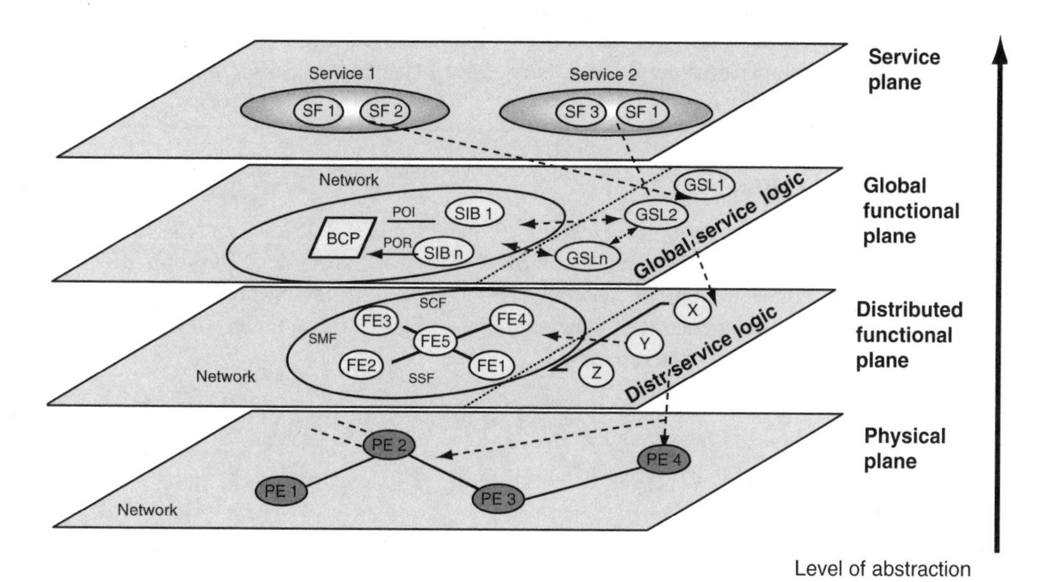

Figure 4.3 Service creation views of ITU-T Intelligent Networks.

elements within the IN architecture. The provisioning process includes arming or disarming DP triggers in the SSF, and downloading new announcements into the SRF. The interfaces and detailed functionality provided by the SMS are out of scope of the IN CS 1 and CS 2 recommendations.

It is possible to aggregate functional elements to create new physical elements. For example, an SN (Service Node) is a combination of CCF, SRF and SCF. An SSCP (Service Switching and Control Point) is a combination of CCF, SSF, SRF and SCF. These integrated components are frequently used in situations when it is uneconomical to build the entire IN architecture, or performance constraints demand that services are provided by integrated components rather than components distributed across the network.

The ITU-T Intelligent Network recommendation series provides the only existing standard for service creation. Since hybrid services will likely include elements of IN services, it is of interest to further examine the IN service creation model. The ITU-T IN model employs a view in which a service exists on four different planes, each plane providing a different level of abstraction from the underlying network details. The four planes are illustrated in Figure 4.3 and each is discussed below.

Service Plane

The SP (Service Plane) is the highest level in the ITU-T IN model. It is on this plane that SLPs (Service Logic Programs) are described and specified. Services are built up by composing one or more abstract Service Features (as described in the Global Functional Plane). Decoupling services from the underlying network implementation enhances re-use and network independence.

An example IN service is UPT (Universal Personal Telecommunications). UPT is composed of the following core service features: Authorization Code, Follow-me Diversion, Personal Numbering, and Split Charging. Optional service features for UPT are: Call Logging, Customer Profile Management, Customized Recorded Announcement, Destination User Prompter, Originating Call Screening, and Time Dependent Routing. Service logic is used to specify how the various service features are employed. At the Service Plane level, any distribution is implicit and hidden from the services.

Global Functional Plane

The GFP (Global Functional Plane (ITU 1995c)) is where the ITU-T IN concentrates its service creation efforts. It introduces the concept of SIBs (Service Independent Building Blocks). SIBs are abstract representations of network capabilities, and they enforce the concept of service and technology independence by decoupling services from the underlying technology used. SIBs have several key characteristics:

- SIBs are blocks of atomic functionality;
- SIBs are building blocks for service features;
- SIBs are generic and not network-specific.

The ITU-T Q.1213 recommendation (ITU 1995c) describes a variety of different SIBs that cover all the available functionality in an IN system. The SIBs use functionality provided from the Distributed Functional Plane. The documents also outline how SIBs can be used to build up higher-level SIBs or service features. Figure 4.4 illustrates how two simple SIBs – Queue

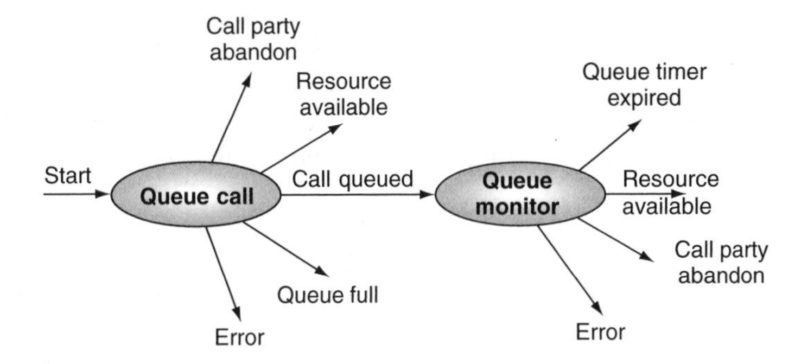

Figure 4.4 Example of queue call and queue monitor SIBs used to control call queues.

Call and Queue Monitor – can be combined to create a service feature for queuing, directing, and releasing incoming calls. The example shows a graphical composition, but the actual composition mechanics will depend on the specification and design method used in the service creation process.

At the Global Functional Plane, distribution is still implicit and hidden from the services and service features. Distribution aspects and the underlying network mechanics are revealed in the Distributed Functional Plane.

Distributed Functional Plane

The DFP (Distributed Functional Plane (ITU 1995b)) provides a view of the functional entities that comprise the IN model. For the first time, actual network elements and functions are exposed to the service designer. For ITU-T CS 1, Q.1214 defines the full set of functional elements including the SSF (Service Switching Function), the SCF (Service Control Function), and the FIM (Feature Interaction Manager). The information flows between the various functions are also defined. The Distributed Functional Plane uses functionality provided by the Physical Plane.

Many proprietary Service Creation Environments provide service creation at the Distributed Functional Plane level as it allows a designer greater control and visualization of how the service will behave when deployed on a real network. The Distributed Functional Plane provides this low-level control while still providing enough abstraction so that the service designer does not have to work directly with platform-specific code or signaling protocols. Distribution in this plane is explicit, so the network designer has to decide how and where to distribute service logic and data.

Physical Plane

The PP (Physical Plane (ITU 1995d)) is the lowest level defined by ITU-T. It represents the PEs (Physical Entities) in the Intelligent Network such as the SCP (Service Control Point) and the SSP (Service Switching Point). The definition outlines various ways to place the various functions within the physical elements. For example, an SCP can comprise an SCF (Service

Control Function) and an SDF (Service Data Function). Communication between the physical entities is by the TCAP-based INAP protocol (ITU 1995e).

Many proprietary Service Creation Environments allow service designers to create services at the Physical Plane. Service logic scripts determine how incoming INAP messages are to be dealt with, and use the INAP protocol to control various physical elements in the network.

4.2.2 Differences between AIN (USA) and ITU-T (International) Intelligent Network architectures

In general, the functionality provided by AIN is a subset of that provided by ITU-T. In those cases where the functionality is similar, terminology differences may exist. AIN uses TCAP only, while ITU-T uses TCAP as a transport for the INAP protocol. This leads to a key difference between AIN and the ITU-T architecture. AIN messages define a different TCAP message for each call event message, while ITU-T defines a generic application layer message that can carry different event messages. The AIN TCAP message set is therefore larger than the ITU-T INAP message set.

The functionality encompassed by an AIN SSP includes the following Functional Entities defined in the ITU-T DFP (Distributed Functional Plane):

- CCAF (Call Control Agent Function);
- SSF (Service Switching Function) or CCF (Call Control Function);
- SRF (Specialized Resource Function);
- SDF (Service Data Function).

The physical architecture supporting AIN is a subset of the ITU-T physical architecture, and does not include the following ITU-T interfaces:

- SSP – Service Node;
- SCP – IP (via SS7);
- SCF (Service Control Function) – SDF;
- SCF – SCF;
- SDF – SDF.

The functionality for OA&M (Operations, Administration and Management) specified in AIN is outside the scope of ITU-T's IN recommendations.

4.3 Components of IN Service Delivery

4.3.1 Service Switching Point

SSPs (Service Switching Points) provide two key functions in an Intelligent Network. They provide the switching functions for handling basic calls, and the IN-related service functions and triggers to support IN services provided through the SCP (Service Control Point). SSPs may also provide feature interaction management functions. SSPs communicate with SCPs via TCAP (AIN) or INAP (ITU-T) protocols. Inter-SSP and Intelligent Peripheral communication is via Q.931 or SS7 protocols. The SSP functions are now explored in more detail. See also Chapter 3 for more on call processing and call models.

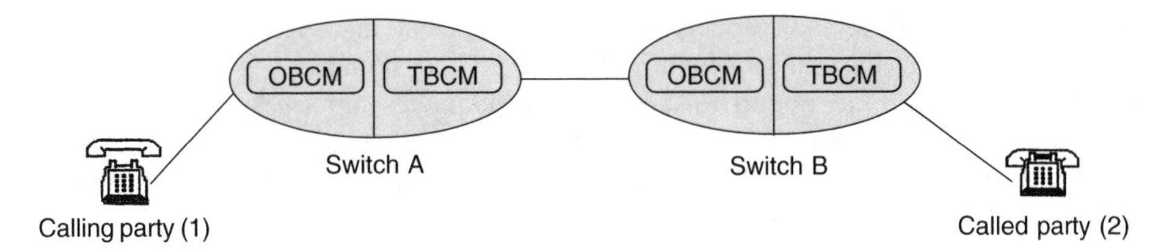

Figure 4.5 Two-party, interswitch call.

Call Model

The AIN call model describes the processing that is done by an SSP to establish a two-party call. The call model actually consists of two half call models: the OBCM (Originating Basic Call Model) and the TBCM (Terminating Basic Call Model). Every SSP that does AIN processing contains both call models. The OBCM is concerned with switching functions that pertain to the calling party's line, or the incoming trunk to an SSP, whereas the TBCM is concerned with switching functions that pertain to the called party's line, or the outgoing trunk at an SSP.

Figure 4.5 illustrates a simple, two-party interswitch call. When the calling party (1) calls the called party (2), the OBCM associated with Switch A communicates with Switch A's TBCM. This results in the call origination being terminated on an outgoing trunk at Switch A. The incoming trunk at Switch B has associated with it an OBCM, which communicates with the TBCM at Switch B. This results in the call on the incoming trunk being terminated on party 2's line.

Both of the half call models are composed of a series of PICs (Points in Call). Each PIC represents a different stage in call processing. Between these PICs are DPs (Detection Points) where events may be detected that result in a message being sent to an SCP. There are two types of such events: triggers and requested events. Events that are encountered as triggers are pre-provisioned as such, whereas the SCP 'arms' a particular DP for detection of a requested event. Hence, a DP may be either a TDP (Trigger Detection Point), an EDP (Event Detection Point), or both. A trigger is encountered when the SSP determines that it must query the SCP to get further instructions before call processing may continue. As a result of detecting a trigger, the SSP sends a TCAP query message to the SCP, and then suspends call processing while it awaits a TCAP response message. This response message may be returned to the SSP as part of a TCAP 'response package', meaning that the response message contains instructions for further call processing, and that the communication between the SSP and the SCP is terminated.

Alternatively, the SCP may return a TCAP response message to the SSP, together with another message that requests the SSP to monitor the occurrence of one or more requested events, called a Next Event List. This occurs by means of a TCAP 'conversation package' that creates an open transaction between the SSP and the SCP. During this open transaction, the SSP informs the SCP when a requested event is detected. Depending upon the type of requested event, the SSP either suspends call processing while awaiting a response from the SCP, or continues with call processing. At some point the open transaction is closed, either by detection of a requested event or by a specific message sent from the SCP.

The Originating Basic Call Models as defined by ITU-T and AIN are very similar. A simplified OBCM is illustrated in Figure 4.6. The PICs designated 1 to 10 describe three different stages of call processing. PICs 1–6 cover the processing necessary for call set-up. PICs 7–9 cover processing during a stable call, and PIC 10 is concerned with call clearing. Call processing begins with the O_NULL PIC and proceeds downward through subsequent PICs. Between each PIC is a TDP and/or EDP. When call processing at each PIC is completed, control passes to the DP that follows it. If the trigger criteria or event criteria are satisfied, the corresponding trigger or event will be detected. Following closure of the transaction between the SSP and the SCP that results from the trigger detection or requested event, call processing continues at the next PIC in the OBCM (unless the SCP directs call processing to continue at a different PIC).

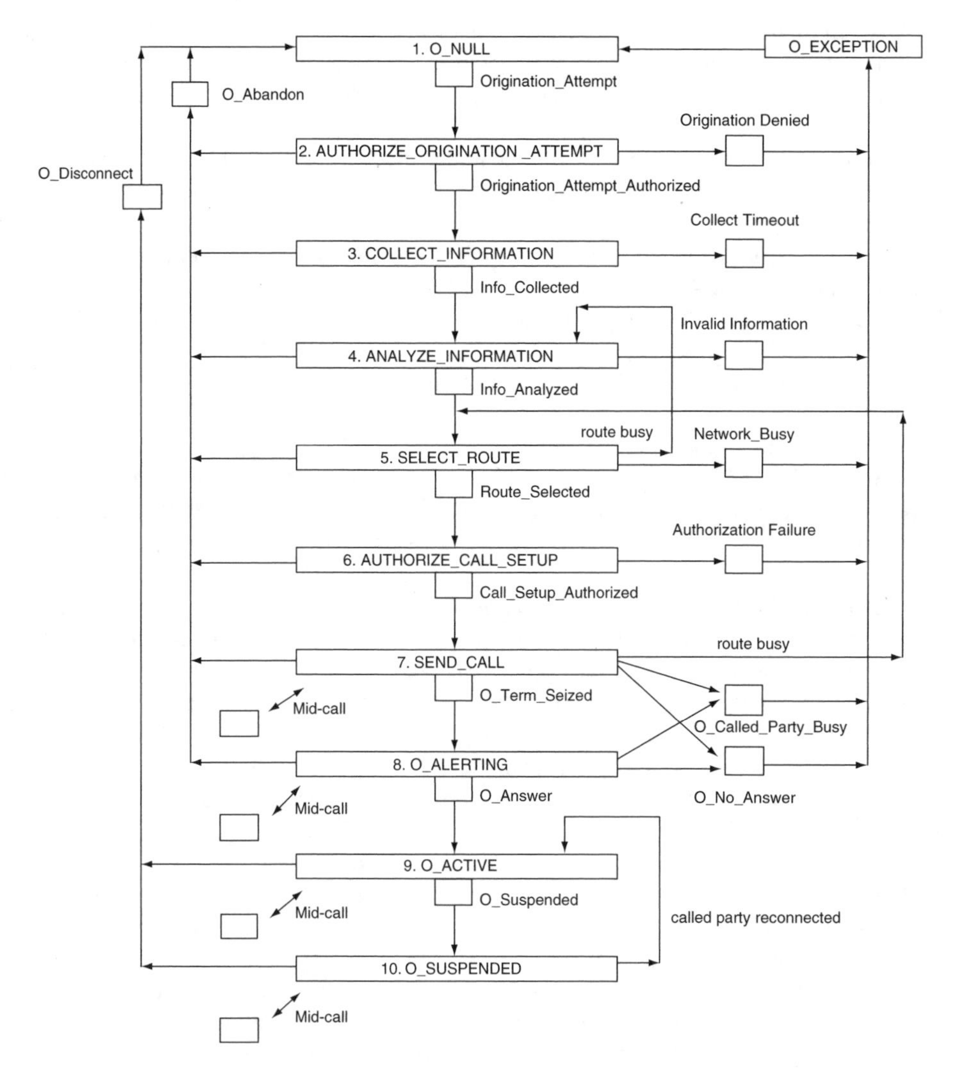

Figure 4.6 Originating Basic Call Model.

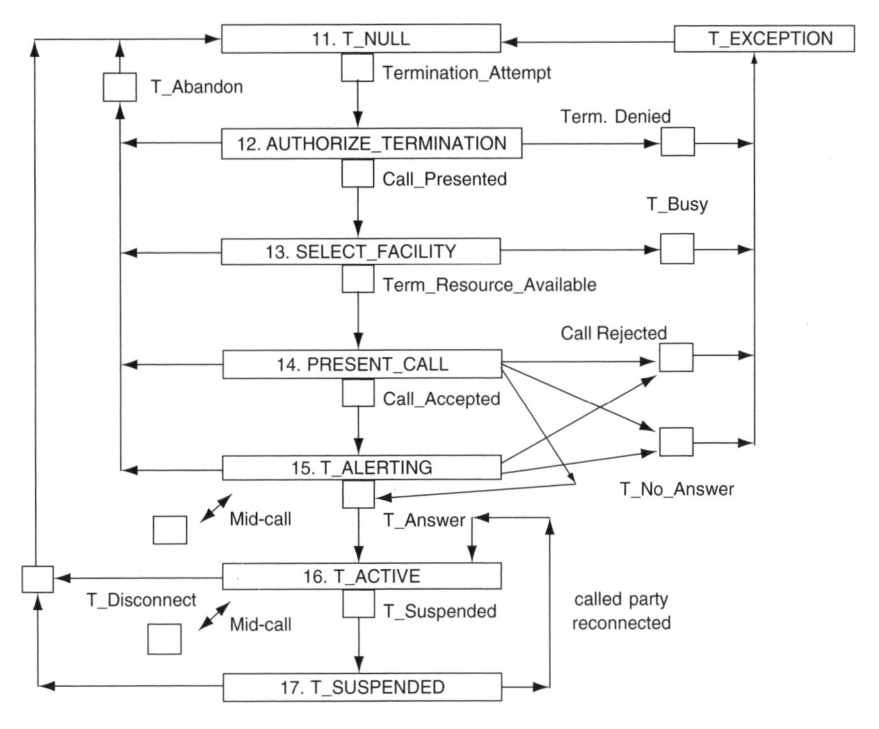

Figure 4.7 Terminating Basic Call Model.

The Terminating Basic Call Models as defined by ITU-T and AIN are also similar. A simplified TBCM is shown in Figure 4.7. Call processing begins at PIC 11 in the TBCM once a route is selected for the outgoing call. PICs 11–15 model the processing required for call set-up, while PIC 16 corresponds to processing during a stable call, and PIC 17 covers call clearing. As with the OBCM, TDPs and/or EDPs occur between the PICs, so that triggers and/or requested events may be detected.

Trigger Mechanism

AIN triggers may be classified as being subscribed, group-based, or office-based. Subscribed triggers are those associated with facilities that subscribe to the trigger, such as an individual subscriber's line. An example of a subscribed trigger in the OBCM is Off-Hook_Delay, which is encountered when the SSP receives enough information to process a call associated with a particular subscriber. For instance, Off-Hook_Delay would be encountered if the subscriber dials a particular string of digits such as a specific telephone number. Group-based triggers are associated with software-defined groups of users that share a customized dialing plan or routing pattern. If a trigger is group-based, then all calls originating at facilities associated with the group can encounter the trigger. An example of a group-based trigger in the OBCM is Customized_Dialing_Plan. This trigger is encountered when any member of the group dials a particular access code, or another series of one to seven digits that form part of a private customized dialing plan. An example of an office-based trigger in the OBCM is Office_Public_Feature_Code. This trigger is encountered by all calls originating at a particular central office from facilities associated with a public dialing plan. For instance, all callers served by a particular office who dial a public office feature code of the form *XY will encounter this trigger.

In the TBCM, all the triggers are subscribed. That is, the triggers in the TBCM are associated with individual subscriber's lines. Probably the most common triggers are Termination_Attempt, T_Busy, and T_No_Answer. Termination_Attempt is encountered when an incoming call is attempting to terminate on a subscriber's line. As their names imply, T_Busy is encountered when a called party's line is busy, and T_No_Answer is encountered when the called party's line does not answer (as determined by the expiration of a timer).

Feature Interaction Management

A unique feature of the ITU-T CS *n* series of recommendations is that an FIM (Feature Interaction Manager) is defined in the architecture. Feature interaction is covered in more detail in Chapter 13, so this section will cover only the IN realization of the FIM. The FIM component is defined in the Distributed Functional Plane (ITU 1995b) and resides in the SSF (Service Switching Function) or CCF (Call Control Function) as shown in Figure 4.8. It is therefore an integral part of the IN call processing and may be involved in every IN call.

Due to the FIM's location, all events to and from SLPs (Service Logic Programs) must be processed via the FIM. It may therefore control which events are passed between the SCP

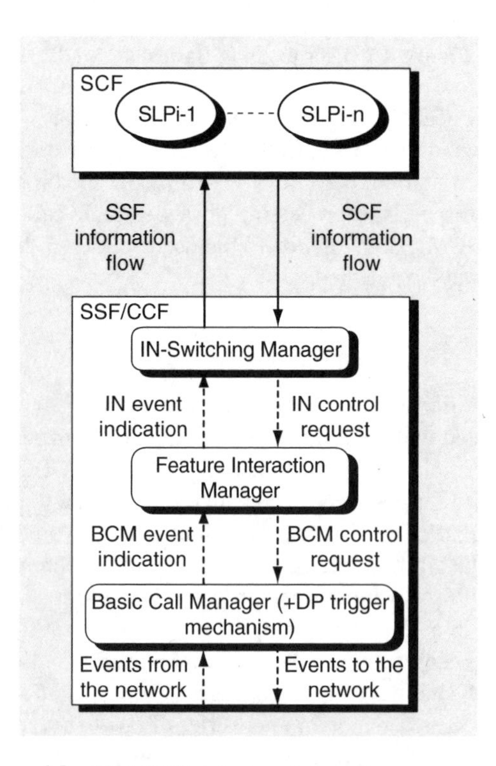

Figure 4.8 ITU-T CS 1 feature interaction management.

and SSP, how they are passed, and when they are passed. Indeed the FIM may even generate its own events in response to certain triggers. Although ITU-T defines the architecture for feature interaction management, it does not define the algorithms or mechanisms the FIM performs. This has been an area of much research in recent years, with some work focused specifically on the IN-FIM (Tsang and Magill 1998a, Tsang and Magill 1998b).

For more information on the feature interaction problem, refer to Chapter 13.

4.3.2 Service Control Point

The SCP (Service Control Point) is the centralized element in the Intelligent Network which provides services by executing stored SLPs (Service Logic Programs). The SCP is an element where standardization has proven to be difficult due to the variety of functions that can be provided. Therefore, this section describes the generic ITU-T SCP functions. Different SCP implementations will provide various subsets of the following functions. SCPs communicate with SSPs and Intelligent Peripherals via the TCAP (AIN) or INAP (ITU-T) protocols.

Service Control Function Model

Figure 4.9 shows the ITU-T service control function model. Several functions are shown, but the key functions are the SLEE (Service Logic Execution Environment) and Service Logic Program library. Services and service features are defined as SLPs (Service Logic Programs). SLPs are stored in the SLP library, and SLP instances are executed on the SLEE. It is common for SLPs to be implemented as scripts that are interpreted by the SLEE.

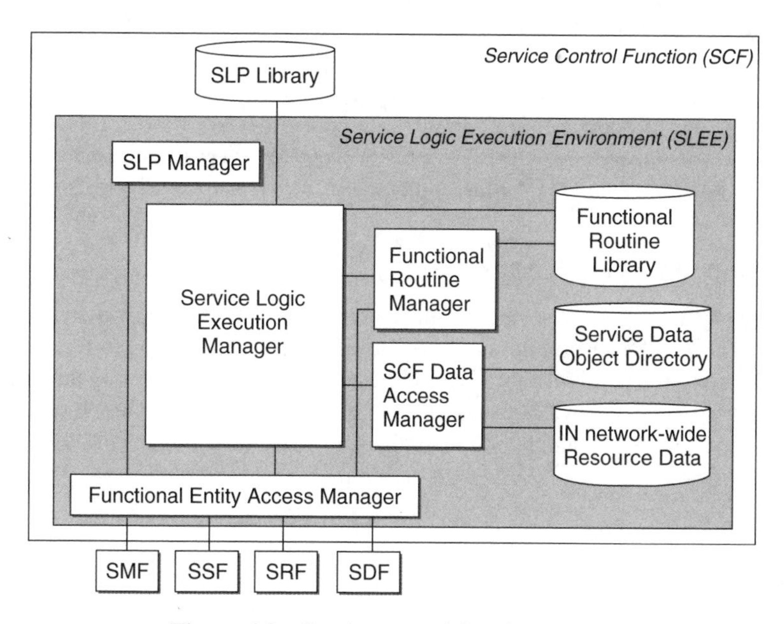

Figure 4.9 Service control function model.

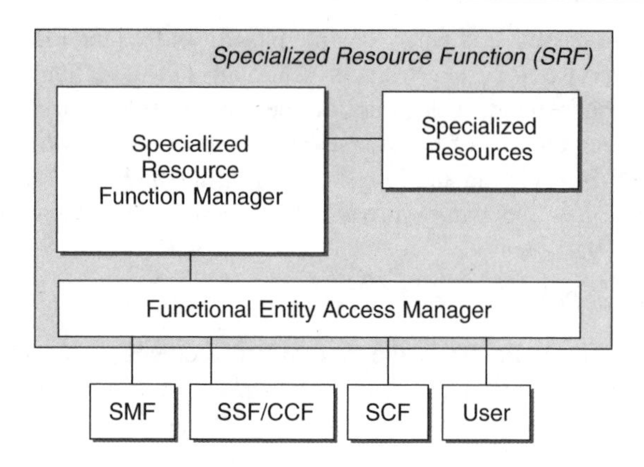

Figure 4.10 Intelligent Peripheral (Specialized Resource Function) model.

Service Independent Building Blocks

SIBs (Service Independent Building Blocks) are commonly used as building blocks for SLP scripts. SIBs are defined in ITU-T Q.1213 (ITU 1995c). They are discussed in Section 4.2.1 under the Global Functional Plane.

4.3.3 Intelligent Peripherals (Specialized Resource Points)

IPs (Intelligent Peripherals), also known as SRPs (Specialized Resource Points) in ITU-T, provide special resources necessary to provide specific IN services. Examples of specialized resources include: DTMF (Dual-Tone Multiple Frequency) receiver, tone generator, announcements, message sender and/or receiver, synthesized speech recognition devices with interactive prompting facilities, text to speech synthesis, protocol converters, and audio conference bridge. Intelligent Peripherals may be controlled by SCPs through the TCAP (AIN) or INAP (ITU-T) protocols. They can also communicate with other IPs or SSPs via Q.931 (ITU 1993d) or SS7 protocols.

Figure 4.10 illustrates the ITU-T functional architecture for an Intelligent Peripheral.

4.4 Intelligent Network Services

A wide range of AIN services has been defined and deployed by service providers throughout the world. Although AIN capabilities have been available since the early 1990s, various factors have contributed to the relatively slow deployment of AIN services. Among these factors are different implementations of AIN functionality by equipment suppliers, complexities involving the service creation process, and operational difficulties. Some examples of AIN services are summarized below.

4.4.1 Service examples

AWC (Area Wide Centrex): A service in the USA that allows organizations with multiple locations to provide abbreviated dialing services to their users. It also enables flexible call routing options for interlocation calls. The service utilizes a Customized Dialing Plan trigger that is encountered when a user dials an abbreviated four or five digit dialing code. The SSP queries the SCP for the full seven or ten digit directory number that corresponds to the abbreviated dialing code. In addition, the SCP can select different call routing options for handling interlocation calls, such as private versus public transport facilities, as well as selection of facilities based on time, day, or week.

Calling Name Delivery: An incoming call service that allows a subscriber to receive the name of the calling party, as well as date and time of the call. When a call arrives at the SSP that serves a subscriber, a Termination_Attempt trigger is encountered on the subscriber's line. A query message is sent to the SCP containing the directory number of the calling party. The SCP then requests information about the calling party's identity from a remote database, such as an LIDB (Line Information Database). When this information is returned to the SCP, the SCP provides the calling party name to the SSP via an appropriate response message. The SSP then provides this information to the subscriber's display device during the interval between the first and second rings of the subscriber's telephone.

LNP (Local Number Portability): A capability that is required in the United States by the Telecommunications Act of 1996. Essentially, LNP allows a subscriber to maintain a specific ten digit telephone number even if the subscriber changes service providers. To enable LNP, an office-based Local Number Portability trigger has been defined in the OBCM. This trigger may be encountered at an originating, intermediate, or terminating SSP by all calls originating from facilities associated with a public dialing plan. Although the trigger criteria for LNP may be complex, in the simplest case an LNP trigger is encountered when a 'ported' seven or ten digit telephone number has been dialed. The SCP that is queried by this event then responds by providing to the SSP new routing instructions for the call.

Internet Call Waiting: A service that combines the AIN architecture with the Internet. The service allows subscribers having only a single telephone line to be notified of incoming calls while the line is being used for dial-up Internet access. Once notified, subscribers can choose to take the incoming call, or they may choose other options such as routing the incoming call to voicemail. Although implementations may vary among service providers, Internet Call Waiting depends on a T_Busy trigger at the subscriber's SSP. When an incoming call activates this trigger, the queried SCP contacts an Intelligent Peripheral which determines whether the subscriber is online (by checking to see if an access number to an Internet Service Provider was dialed). If so, the SCP retrieves the calling party's name from LIDB information (as in the Calling Name Delivery service). This information is provided to the IP, which then signals a network node that provides interconnection between the Internet and the PSTN (Public Switched Telephone Network). This node responds by causing a pop-up window to appear on the subscriber's computer screen, informing the subscriber of the incoming call. If the subscriber chooses to take the call, the IP then waits for the subscriber to drop the call to the Internet Service Provider. When that occurs, the IP signals the SCP, which in turn directs the SSP to connect the incoming call to the subscriber.

In this version of the service, it is necessary for the subscriber to drop the Internet connection to take the incoming call. A more desirable solution would allow the subscriber to simply put the Internet session on hold while the incoming call is taken. Such scenarios may require the incoming call to be converted into Voice over IP, so that it can be routed to the subscriber via the dial-up IP connection without requiring that the connection be broken.

Table 4.1 ITU-T CS 1 service set

Abbreviated Dialing	ABD
Account Card Calling	ACC
Automatic Alternative Billing	AAB
Call Distribution	CD
Call Forwarding	CF
Call Rerouting Distribution	CRD
Completion of Call to Busy Subscriber	CCBS
Conference Calling	CON
Credit Card Calling	CCC
Destination Call Routing	DCR
Follow-Me Diversion	FMD
Freephone	FPH
Malicious Call Identification	MCI
Mass Calling	MAS
Originating Call Screening	OCS
Premium Rate	PRM
Security Screening	SEC
Selective Call Forward on Busy/Don't Answer	SCF
Split Charging	SPL
Televoting	VOT
Terminating Call Screening	TCS
Universal Access Number	UAN
Universal Personal Telecommunications	UPT
User-Defined Routing	UDR
Virtual Private Network	VPN

4.4.2 ITU-T CS 1 services

ITU-T services are defined in terms of service features. It is beyond the scope of this book to explore in detail each service, but for the reader's reference, Table 4.1 lists the target set of services for ITU-T Capability Set 1 as outlined in Q.1211 (ITU 1993e). Table 4.2 lists the target set of service features for ITU-T Capability Set 1. The reader is urged to read the ITU-T recommendation for more information on this topic.

4.5 Assessment of Intelligent Networks

Although intelligent networks have proven to be significantly more flexible and efficient for the introduction and modification of services, some problems remain. One of the goals of IN was to free the telephone companies from relying on switch suppliers for new and updated services. However, new IN services may still require suppliers to implement additional IN functionality (such as additional triggers) into the switches. Even though telephone companies are not dependent on switch suppliers to create new services, service creation has proved to be more difficult than initially thought. In addition, interoperability problems between the equipment of different suppliers have not been completely eliminated, since suppliers may

Table 4.2 ITU-T CS 1 service feature set

Abbreviated Dialing	ABD
Attendant	ATT
Authentication	AUTC
Authorization Code	AUTZ
Automatic Call Back	ACB
Call Distribution	CD
Call Forwarding	CF
Call Forwarding (Conditional)	CFC
Call Gapping	GAP
Call Hold with Announcement	CHA
Call Limiter	LIM
Call Logging	LOG
Call Queuing	QUE
Call Transfer	TRA
Call Waiting	CW
Closed User Group	CUG
Consultation Calling	COC
Customer Profile Management	CPM
Customized Recorded Announcement	CRA
Customized Ringing	CRG
Destination User Prompter	DUP
Follow-Me Diversion	FMD
Mass Calling	MAS
Meet-Me Conference	MMC
Multi-Way Calling	MWC
Off-Net Access	OFA
Off-Net Calling	ONC
One Number	ONE
Origin Dependent Routing	ODR
Originating Call Screening	OCS
Originating User Prompter	OUP
Personal Numbering	PN
Premium Charging	PRMC
Private Numbering Plan	PNP
Reverse Charging	REVC
Split Charging	SPLC
Terminating Call Screening	TCS
Time Dependent Routing	TDR

implement the same function in slightly different ways. Also, as discussed elsewhere, feature interactions between growing numbers of services are also a problem.

In concept, IN provides the answer for service and network providers who wish to design and deploy services with the least effort and in the least amount of time. In reality, this has happened only to a limited extent. Services such as Freephone, Premium Rate, Mass Calling, and VPN (Virtual Private Network) have been successfully deployed to provide a valuable source of revenue in many countries worldwide. However, there have also been drawbacks:

- IN service functionality is restricted to what is provided in the application protocols (e.g. ETSI INAP) that define the interfaces between IN network elements. These interfaces have been designed to meet the needs of only a certain set of telecommunications services. Therefore, although it is possible to customize services or build similar new services, it is not possible to build radically new or innovative ones. Ultimately, this restricts service differentiation.
- In reality, the INAP interface functionality depends wholly on the capability of the SSP (switch). For example, the SSP may not be able to perform certain functions such as creation of temporary connections to external Intelligent Peripherals. Apart from limiting the functionality a service can provide, this issue also poses an interoperability problem in a network that has SSPs (possibly from different vendors) with various levels of functionality. SCPs from one vendor may not work with SSPs from a different vendor.
- Generally, the INAP interface is vendor-specific to varying degrees because of the close coupling with the SSP. This occurs even when ETSI compliance is claimed. Even if INAP message names are the same, the message parameters may not match exactly. This leads to serious difficulties when interworking different vendors' equipment. If this problem occurs, one solution is to upgrade the SCPs with the INAP interface of other vendor SSPs.
- Another problem with INAP development is that the process to standardize INAP is slow. Standards bodies generally work in two to four year cycles, and this severely limits innovation and time between INAP developments. As a result of this, vendors often differentiate themselves from their competitors by developing proprietary INAP interfaces, since this process is much faster. Although this approach offers more features, it also means that interworking between different vendors' equipment becomes increasingly difficult. One good example is CPH (Call Party Handling) in INAP CS 2. Because vendors have disagreed on how CPH should be implemented, this has generally been implemented in a proprietary manner and does not follow the ETSI or ITU-T interface recommendations.
- IN Service Management Systems (SMSs) and interfaces have never been standardized by ITU-T or ETSI. This means that in a multi-vendor environment, even if the INAP interface permits interworking, service provisioning may not be possible from one vendor's SMS to another vendor's SSP. This restricts service creation to a single-vendor environment.

Although there are drawbacks, IN has enabled services such as Freephone, Premium Rate, One Number and others to be deployed and used on the huge scale they are today in the PSTN. Networked call centers and VPN services are also based primarily on IN architecture and systems. Because there is a massive installed base of IN network equipment, systems, and services, this cannot be easily discarded. For the foreseeable future it will need to interwork with NGNs (Next Generation Networks).

4.6 Future of Intelligent Networks

As the PSTN gradually gives way to NGNs based on the Internet Protocol, voice services carried on the circuit-switched PSTN will become packetized as VoIP (Voice over IP) services in the NGN. The ability to provide advanced voice services in this new environment will rely on service platforms such as SCPs or 'applications servers'. Value-added voice services that have previously been deployed on IN SCPs may still be usable within NGNs. However, the IN triggers and messages previously associated with SSPs will now be associated with different elements in the NGN from which these services will be invoked.

Prior to the complete replacement of the PSTN by an all-IP NGN, there will be a need to transport voice calls between the PSTN and the NGN. Media gateways under the control of a Media Gateway Controller, or softswitch, will perform the conversion between circuit-based and packet-based voice. It is at the softswitch that these triggering functions will occur. The softswitch will encounter triggers when certain events occur, causing service logic in the SCP to be invoked. For IN triggers to be encountered at a softswitch, a mapping must occur between the IN call model and the MGCP (Media Gateway Control Protocol) that specifies the interface between the softswitch and the media gateway. Because triggers are encountered when certain events are detected in the IN call model, a mapping must exist between such events and the information available to the Media Gateway Controller via the MGCP. See Chapter 3 for more information on softswitches.

Even though the future telecommunications will most likely be Internet Protocol based, it is likely that the demand for IN services will continue. To facilitate this, IN components (such as the SCP) will continue to provide advanced service support and play an important role in the world's telecommunications infrastructure.

5

Basic Internet Technology in Support of Communication Services

Marcus Brunner

NEC Europe Ltd, Germany

5.1 Introduction

5.1.1 Overview

The impact of Internet Technology on the communication industry has been large. However, it is not really clear what the Internet means. Originally the Internet was assumed to be the worldwide IP (Internet Protocol) network. Recently the meaning has, at least in popular media, changed towards everything running on top of IP, such as email or the WWW (World Wide Web). What this change in meaning indicates is the interest of people in the service and in what they get from the Internet compared to how the technology is implemented.

This chapter mainly addresses service issues in IP-based environments. The term service is one of those terms saying everything and nothing. In the case of Internet Services we can basically differentiate between transport services, such as the Integrated Service Architecture or the Differentiated Service Architecture, and end-system services such as Web services, email services, and voice services. Both types of service are covered in this chapter.

Figure 5.1 shows the area covered in the rest of the chapter. The figure shows an example scenario covering enterprise, core, and home networks running Voice over IP calls between all these parties.

The IP network provides the basic transport service, including various types of transport service guarantee. The technologies are meant to operate in core networks and in larger enterprise networks.

Voice over IP can be regarded as a service running on top of the IP-based infrastructure. It runs a specific signaling protocol for setting up voice calls. The Voice over IP signaling protocol

Service Provision – Technologies for Next Generation Communications. Edited by Kenneth J. Turner, Evan H. Magill and David J. Marples
 ISBN: 0-470-85066-3

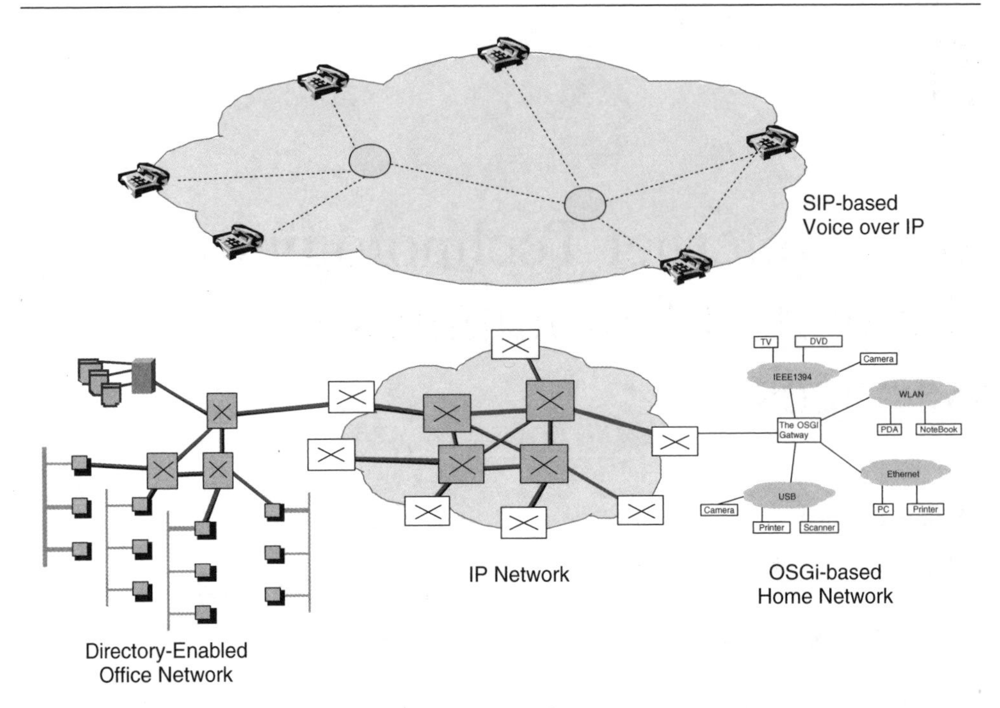

Figure 5.1 Overview of Internet areas.

is independent of network-level signaling. It just uses the IP infrastructure for transporting the signaling messages and for transporting the encoded voice.

A directory-enabled network runs several enterprise internal services including databases, email, and other distributed applications. It is still based on IP network technologies together with some QoS (Quality of Service) technologies. The enterprise services are controlled and managed via several independent applications, but using the same directory to read and write management data.

The Home Network on the other hand has the requirement that it must run without sophisticated management tools and applications, but in a heterogeneous networking environment.

Finally, the last section in the chapter will give an outlook on networking research. This part, not shown in Figure 5.1, tries to integrate the different requirements in one very flexible approach. These new developments are termed active networking technologies.

5.1.2 The Internet standardization process

The premier standardization body for Internet technology is the IETF (Internet Engineering Task Force, (Bradner 1996)). It operates in an open fashion allowing everybody to participate. It consists of a large set of working groups, each having an open mailing list to which anyone can subscribe and contribute. Additionally, three times a year, IETF face-to-face meetings are organized to sort out disagreement and solve problems quickly.

IETF documents include Internet Drafts, which are proposals and working documents without formal significance. Then there are RFCs (Requests for Comments), which are documents agreed upon by the working group as well as by the IESG (Internet Engineering Steering Group). As soon as a document achieves the status of an RFC, it means it is ready for implementation. RFCs are classified according to whether the content is really a standard (standards track), informational, or experimental. If an RFC is on the standards track, it passes through three stages. First it is called a Proposed Standard, which means the proposal can be implemented. The second state, Draft Standard, is entered as soon as two interoperable independent implementations are available and tested. Finally, the document eventually becomes a Full Standard if, and only if, the technology is widely deployed.

5.2 Transport Service Quality in the Internet

In this section several standardized approaches are covered in order to guarantee the quality of a transport service, often called QoS (Quality of Service). The difficulty starts when trying to define what quality is. Does quality lie in the guaranteed rate of bytes transported, or more in the time the service is available and accessible? Quality of Service is discussed from a Service Management perspective in Chapter 7. The IETF standardized approaches follow more the notion of short-term quality such as bandwidth guarantees.

5.2.1 Plain old IP

The plain old IP service is called a best-effort service since no bandwidth or delivery guarantees are provided. However, on closer inspection the plain old IP service does have some quality attributes as well. The service guarantees fairness in the sense that no packets are preferred over others. Additionally, the TCP (Transmission Control Protocol) provides fairness in terms of bandwidth used by TCP sessions. All TCP sessions running over the same link share the link capacity in a fair way. This type of fairness has been undermined by some applications using several TCP sessions, e.g. for file download. Other types of transport protocol, such as UDP (User Datagram Protocol), do not provide this fairness.

Another feature of the IP architecture is its relative robustness against node and link failures. At least in the networks running dynamic routing protocols the transport service recovers from failures by learning the new network topology and forwarding traffic according to the new calculated routes.

Concerning bandwidth guarantees, many people believe that they are not needed since there is enough capacity available anyway. The argument for the so-called over-provisioning of capacity is as follows. It is much cheaper to add more capacity to the network compared to adding management and control complexity in order to avoid bottlenecks and to provide guaranteed services. So, sound network planning on a long timeframe provides QoS by increasing the capacity at bottleneck links in advance. Additionally, a sound topology can optimize packet latency mainly by a more meshed topology for the network.

Nevertheless, this approach is not practical for certain environments, and might be a short-term solution only. Environments where plain old IP does not solve the problems are cases where the link technology is expensive, such as wireless networks. Also in environments where stringent delay, loss, jitter, and bandwidth guarantees are required, plain old IP has limitations. Finally, there might be business opportunities providing a whole range of different customized transport services.

5.2.2 Integrated Services

The first approach that tried to include Quality of Service features in IP networks was called IntServ (Integrated Services, Braden *et al.* 1994). The idea was to operate only one single integrated network instead of several dedicated ones. The additional feature required is the ability to run real-time applications such as multimedia applications over the Internet.

The Integrated Service Architecture was built for individual application instances to request resources from the network. This implies that per-flow handling is proposed, so every router along an application's traffic path needs per-flow handling of traffic. An IntServ flow is defined as a classifiable set of packets from a source to a destination for which common QoS treatment has been requested.

Signaling

In order to let an application communicate with the network and to configure the network to provide a certain service, some sort of signaling is required. First the application needs to request a transport service from the network; secondly the service needs to be provisioned. The signaling protocol for the IntServ architecture (Wroclawski 1997a) is RSVP (Resource Reservation Protocol, (Braden *et al.* 1997). It provides both functions at the same time. However, note that the IntServ architecture is too often directly associated with RSVP. RSVP can also be used for other types of QoS and non-QoS signaling.

In the following we provide a short summary of RSVP as the only IP-based signaling protocol. RSVP establishes resource reservations for a unidirectional data flow. RSVP is classified as a hop-by-hop protocol, which means that RSVP communicates with all routers on the data path in order to set up a particular service.

A feature that makes the protocol simpler, but also limits its applicability in certain environments, is that it is independent of the routing protocol. So no QoS-based routing can be implemented.

In RSVP the data sender advertises QoS requirements, because the sending application most likely knows the data flow requirement best. This advertisement message searches the path through the network to the destination. All entities along that path then know the sender's capability and QoS requirement, and they can change this specification if they are not able to provide the required service.

Eventually, the destination is made aware of the requirement and can adapt it to its capabilities. This is the point where the real reservation starts. The reservation is performed starting at the data receiver, so RSVP is often called a receiver-oriented signaling protocol.

One of the biggest problems for signaling protocols is state management. This denotes the function of deleting unused states; in this case reservations are no longer valid. Since the Internet is an inherently unreliable network, RSVP has chosen a soft-state approach. This means that a state has significance for a certain time only; afterwards it needs to be refreshed in order to stay alive. This approach allows RSVP gracefully to handle error situations, which happen often in unreliable networks. It also handles routing changes in the same way.

Figure 5.2 shows the basic RSVP operation. The PATH message contains the QoS requirements and searches the path through the network. The RESV message is what really reserves the resource on each node for that particular service. After that operation, the sender can start sending data and obtain the guaranteed service. Since RSVP is a soft-state

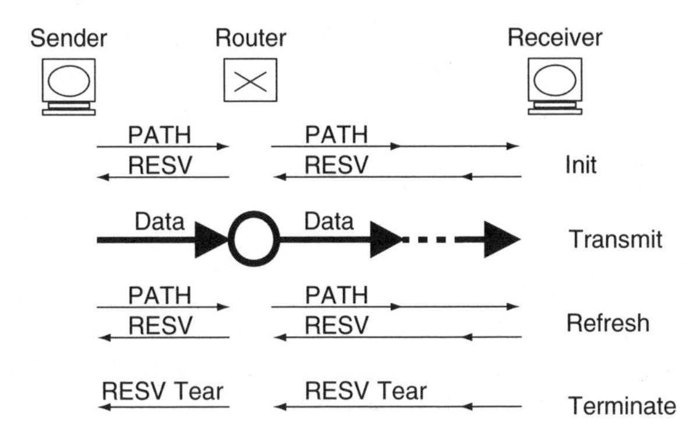

Figure 5.2 Basic RSVP operation.

protocol, periodic refreshes are used to keep the reservation valid. Finally, RSVP tear-down closes the session.

RSVP also has its drawbacks. It is regarded as fairly complex, because both one-to-one and multicast applications are supported. Additionally, the design of RSVP has scalability problems for a large number of flows, which appear in the core of the network.

Provided services

From a transport service point of view two services, the Controlled Load and the Guaranteed Service, have been defined. It is possible to define other services, but none has been standardized so far.

The Guaranteed Service provides firm (mathematically provable) bounds on end-to-end packet queuing delays. This service makes it possible to provide a service that guarantees both delay and bandwidth (Shenker *et al.* 1997).

The Guaranteed Service is more suitable for intolerant real-time applications. These applications implement a play-out buffer whose size is bounded by the worst-case jitter, which in turn is bounded by the worst-case delay. The service does not guarantee any average or minimum latency, average jitter, etc. The application must provide the characteristics of its expected traffic. The network calculates and returns an indication of the resulting end-to-end latency and whether the service can be guaranteed. The delay is computed from the link delay plus the queuing delay, where the queuing delay again depends on the burst size.

The service parameters include the sending rate and the burst size, together with the requested delay bound. The network guarantees timely delivery of packets conforming to the specified traffic (rate and burst size). From a service provider perspective this service allows for almost no multiplexing, which makes it relatively expensive.

The Controlled Load Service basically provides a service similar to that provided when the links are unloaded (Wroclawski 1997b). Unloaded conditions are meant to be lightly loaded networks or not congested networks. The service requires the packet loss to approximate the basic packet error rate of the underlying transmission system. The transit delay of most of the packets should not exceed the minimum transit delay (sum of transmission delay plus router processing delay).

The Controlled Load Service allows applications to get a good service, but not a fully guaranteed service. From a service provider point of view, this allows for a higher degree of multiplexing compared to the Guaranteed Service.

Discussion

One characteristic of the Integrated Services model is that routes are selected based on traditional routing. This means that packets follow the normal path. Reservations are made on that path, or they are rejected if no resources are available on that path. Therefore IntServ does not provide QoS routing capabilities. QoS routing is a mechanism where the path is chosen according to the required service characteristics. The reasons for this decision are that it is in general not possible to force packets to follow non-shortest paths (see later for a discussion on how MPLS provides that function), and that QoS routing algorithms are very complex and still not well understood.

The most severe problem of the Integrated Service approach is that it is not suitable for backbone networks. The reason is that each flow is handled separately on the signaling (control) plane and on the data plane, which implies millions of flows in backbone routers at the same time. The algorithm to handle flow separation does not scale well to these high numbers of flows. Also RSVP has some scalability problems for a large number of flows, mainly in terms of processing capacity and memory usage.

5.2.3 Differentiated Services

As an answer to the previously mentioned scalability problems for the Integrated Service approach, DiffServ (Differentiated Service) was developed. Compared to IntServ it breaks with end-to-end significance, is more incremental to the existing Internet, and is simpler. Scalability is achieved because DiffServ does not deal with individual flows, but with CoS (Classes of Service). All traffic is mapped to a small set of traffic classes when entering the DiffServ network.

The basic design goal of DiffServ is to perform the expensive data plane functions at the edge of the network, and to keep the core router functionality very simple. The core routers basically maintain a small number of service classes with defined per-router behavior called Per-Hop Behavior (PHB). The edge routers classify incoming packets into one of these service classes, and perform policing and traffic shaping. Policing refers to the function of checking whether a traffic stream conforms to a negotiated rate, and if not, then dropping packets of that stream. Traffic shaping refers to the function of making the traffic conformant by buffering. Basically, it equalizes short-term variations within a flow.

Figure 5.3 shows a conceptual model of the functionality on an edge router. An edge router classifies packets based on several IP packet header fields such as source and destination address, source and destination port number, and protocol type. It then marks the IP packet with a so-called DSCP (DiffServ Code Point), which is a replacement of the original IP header field ToS (Type-of-Service). The traffic may then be policed or shaped. For instance, a token bucket policer would drop all packets not conforming to the token bucket parameters.

In the core routers the packets are then classified based only on the DSCP and handled according to the defined Per-Hop Behavior. Figure 5.4 shows an example core router configuration. Packets are classified based on the DSCP value in the ToS field. Depending on the

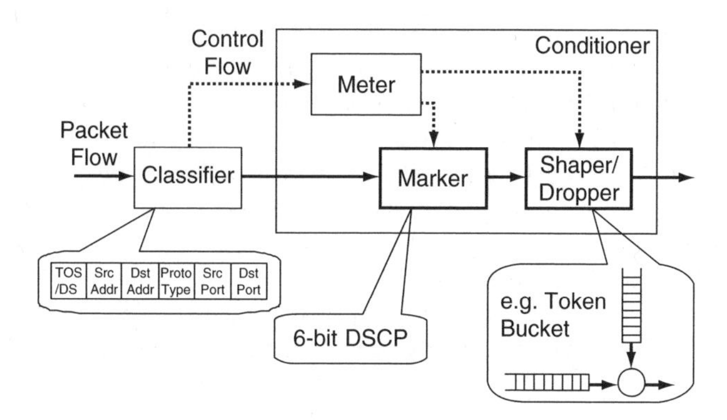

Figure 5.3 Example DiffServ edge router.

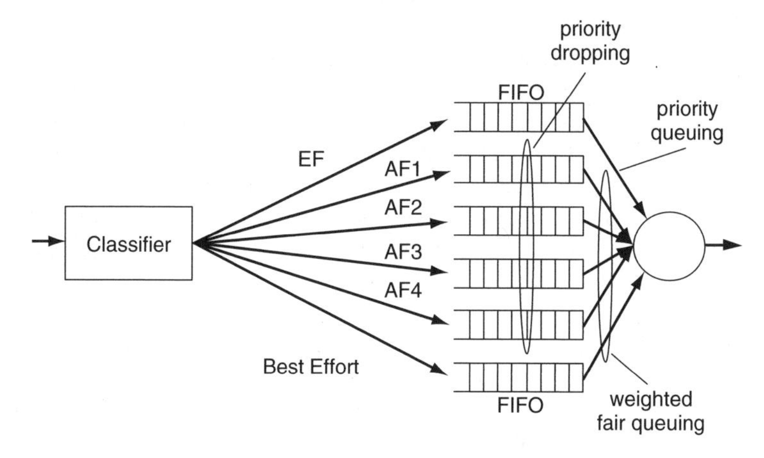

Figure 5.4 Example DiffServ core router configuration.

local DSCP to PHB mapping, the packets are placed in separate queues. This is where the differentiation of packets takes place.

The example configuration runs a priority queue, where all EF packets (Expedited Forwarding, see below) are handled with explicit priority over the others. Then five queues share the rest of the bandwidth using, for example, weighted round-robin scheduling. Within the queues for AF traffic (Assured Forwarding), algorithms take care of the drop precedence if the queues are filling up. See below for a description of Per-Hop Behaviors.

This example shows that in core routers no policing and shaping happen. The classification based on a few possible values of the DSCP is much simpler. Finally, the packet scheduler with only six queues in the example is far more efficient compared to one with millions of queues.

Standardized Per-Hop Behaviors

The relationship between the PHB and DSCPs is mainly of operational interest, because a network administrator can choose the DSCP to PHB mapping freely. However, all PHBs have default/recommended DSCP values.

There are currently three PHBs standardized: the Class Selector PHB, the Expedited Forwarding PHB, and the Assured Forwarding PHB. Additionally, there is the best-effort PHB, but this is the normal IP behavior and is not further discussed here.

The Class Selector PHB is partially backward compatible with the definition of the ToS field in the IP header. Nichols defines the behavior with the following rules (Nichols *et al.* 1998). Routers should give packets the same or higher probability of timely forwarding compared to packets with a lower order Class Selector Code Point.

The EF PHB for Expedited Forwarding (Davie *et al.* 2002) is simple high-priority packet handling. It specifies that routers must service packets at least as fast as the rate at which EF PHB packets arrive. The definition is pretty open, but in reality a priority scheduler processing packets from the EF queue with highest priority can easily achieve this PHB. Other scheduling disciplines are also useful. The handling rate must be equal to or greater than the expected arrival rate of EF marked packets.

The AF PHB for Assured Forwarding (Heinanen *et al.* 1999) is actually a whole group of PHBs. The group can have a number of PHB classes. Within a class a number of drop-precedence levels are defined. The number of classes is normally four, and the number of levels three. The class really defines a certain service class into which packets at the edge of a DiffServ node are classified. The drop-precedence level defines which packets should be dropped first compared to other packets of the same class in the case of overloaded links. More intuitively we can talk about colored packets and mean the drop precedence. For example, packets can be marked green, yellow, and red. Packets marked red have a higher drop probability compared to packets marked with yellow, and these in turn have a higher drop probability than the packets marked green.

Providing transport services

Given the basics of DiffServ, it is possible to consider the types of transport service that can be provided. The services can be classified along several dimensions. First, based on duration, a service can be regarded as dynamically provisioned or statically provisioned. Secondly, a service can be guaranteed quantitative (constant or variable), qualitative, or relative. Thirdly, the traffic scope may include point-to-point, one-to-any, or one-to-many (multicast). For one-to-any the ingress node for the DiffServ domain is known, but the egress node is not known and can be any node. It also means that traffic with different destinations at a particular point in time is present in the network.

In order to provide a guaranteed quantitative transport service, an admission control function must be added to the network. Normally this is handled by a QoS management system that is responsible for the network configuration and for further QoS management functions. This kind of QoS management system is normally called a QoS server or a bandwidth broker.

The QoS server performs connection admission control to the traffic class. It is aware of current routing tables and may modify them (QoS routing). It is also aware of all reservations concerning the particular traffic class. Furthermore, the QoS server is responsible for the proper configuration of the QoS-related components on all routers of the network, including traffic shapers, traffic classifiers, and schedulers.

This QoS management system is normally thought of as a centralized entity, but it does not need to be. The system is able to deal with only a limited number of reservations, which means reservations are more based on aggregated traffic than on single small flows.

Nevertheless, the granularity of reservation is mainly a decision of the network provider, and depends also on the maximum number of service requests a QoS management system can handle. The question of how dynamic service requests can be also impacts upon the QoS server performance.

There was an effort in the IETF DiffServ Working Group to standardize services that use Per-Domain Behavior (PDB) in their terminology. Compared to the term Per-Hop Behavior, which defines the behavior of packets at a router, the PDB defines the behavior of packets or flow passing through a complete DiffServ domain. The idea behind this was to define services using the DiffServ components available. However, the standardization effort was stopped because it was very difficult to find many different services with predictable behavior.

As described above, there are also services with qualitative and relative guarantees. These are typically provided in a static way and are normally bound to a service contract. Additionally, they often do not need even admission control.

Example services

The two most frequently cited services are the Premium Service and the Assured Service. A third one presented here is called Better than Best-Effort service.

The Premium Service is a service with bandwidth guarantee, bounded delay, limited jitter, and no packet loss. It is provided in a DiffServ network with the EF PHB, which gets preferred handling in the routers. Access to the EF traffic class must be controlled, and must be checked for each new connection. Strict policing of flows at the DiffServ edge mainly performs this. A strict policer means a policer that drops all packets not conforming to the negotiated bandwidth. Additionally, admission control needs to be performed. Admission control functions check whether there is still enough capacity available to accommodate a new service request. This requires a global view of the network, including routing tables, network resources, and all current reservations. Only if access to the network is controlled can a guaranteed Premium Service be provided.

The Assured Service is less stringent in the guarantee that it provides. It mainly defines a service with assured bandwidth guarantees and with near-zero loss for packets conforming to the committed rate. Packets that exceed the committed rate are not lost if enough capacity is available.

This service can be implemented with the Assured Forwarding PHB and a so-called two-color marker at the edge router. This entity marks packets conforming to the committed rate with green; packets above that rate are marked with yellow. If a three-color marker is used, a second maximum rate can be a service parameter; all packets above the maximum rate are marked red. Coloring is implemented with the drop precedence feature of the Assured Forwarding PHB.

In the admission control function, the committed rate is checked against availability in order to guarantee near-zero loss. If a link is congested, the red packets are dropped. If the link becomes more congested, the yellow packets are also dropped to give green packets preferred handling within an Assured Service class.

The Better than Best-Effort service can be classified as a typical relative guaranteed service. The easiest implementation uses the Class Selector PHB, which defines the priority of classes. So all traffic having a static Better than Best-Effort service gets a higher priority than that of the Best-Effort class. There are no quantitative guarantees with such a contract, only the provision that traffic gets better service compared to Best-Effort service traffic.

Other services are also possible, combining the different components freely to obtain different behavior.

Discussion

Basic DiffServ scales very well, but it does not provide sufficient means for providing certain types of QoS. It often needs to be coupled with a QoS management system. Typically, this has a negative impact on the scalability of the DiffServ approach. The challenge is designing it in a way that keeps scalability as high as possible.

We also need to consider the time required for all these configuration actions when the QoS server receives per-flow requests from individual applications. This solution has a significantly lower scalability than basic DiffServ and it would not be feasible for a core network. In order to make such a system reasonably scalable, some aggregation of requests is necessary.

In general, the Differentiated Services Architecture is useful for a backbone network, giving guarantees for traffic aggregates instead of single application flows. In other environments, DiffServ makes sense from a business point of view. It allows service providers to differentiate individual customers and various service offerings, where a more expensive service obtains better treatment.

From a management point of view, the open definition of DiffServ provides problems. It allows for too many very flexible solutions. It is therefore difficult to achieve an interoperable solution with equipment and software from different vendors.

5.2.4 Multi-Protocol Label Switching (MPLS)

MPLS (Multi-Protocol Label Switching) is not directly a QoS technology, but it can support QoS provisioning by extending IP routing. With MPLS, IP packets are labeled when entering the MPLS network and they then follow a fixed LSP (Label Switched Path).

The major features of MPLS are as follows. IP packets belonging to an LSP are forwarded based on a short label (label switching) instead of being based on the longest-prefix address look-up (IP routing). This simplifies and speeds up forwarding. The label itself has only per-link significance, and can be changed on each node along a path.

IP packets can be routed explicitly by fixing the route of an LSP. An LSR (Label Switching Router) examines the label on each arriving packet and handles the packet accordingly.

Figure 5.5 shows the basic architecture of MPLS. At edge nodes packets are classified and labeled. Then they are sent according to the local LSR configuration towards the next hop based on the label. In the core, labels are then the only way to find the way through the network.

With packet labeling based on packet classification, packets can be assigned to LSPs on a per-flow basis. This is particularly interesting because there can be several LSPs between a pair of edge nodes, with different QoS characteristics or just for load balancing.

MPLS can be used for providing QoS when combined with other QoS technologies. For example, when run over ATM (Asynchronous Transfer Mode), MPLS can map LSPs to ATM virtual connections and provide IP QoS per LSP based on ATM QoS. Another example is combining MPLS with DiffServ. Using different LSPs for different classes of service allows routing the classes individually. Additionally, routing traffic along less used paths and around bottlenecks provides better Quality of Service.

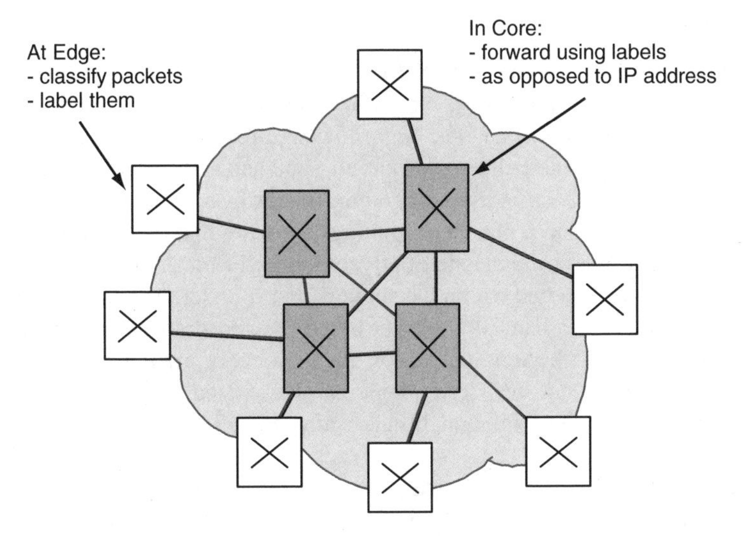

Figure 5.5 MPLS architecture.

A label distribution protocol is used for setting up Label Switched Paths. Label distribution protocols are a set of procedures by which one LSR informs another of the label and/or FEC (Forward Equivalence Class) bindings it has made. Note that FEC binding denotes the traffic carried by a certain label. The label distribution protocol also encompasses any negotiations in which two neighboring LSRs need to engage in order to learn of each other's MPLS capabilities.

Discussion

MPLS is not a QoS technology *per se*, but provides functionality that makes QoS networking easier. Basically, QoS can be more easily provided through fixing paths. Monitoring the network and directing traffic around bottlenecks achieve some quality as well.

On the other hand, it is questionable whether adding another layer below IP is a good choice. It adds management complexity because an additional technology needs to be managed. It also adds some overhead for the networking layer. Finally, it relies on IP because the complete control plane (signaling for label distribution and routing protocols) is based on IP standards.

5.3 Internet Telephony

Several different architectures can be considered under the heading of Internet telephony. Some people regard an IP transport network between central offices as Internet telephony. The other extreme case is people using IP-capable telephones for voice with IP protocols only. In this section we mainly cover the second case. Different technologies for IP telephony will be discussed, mainly those that are based on the SIP (Session Initiation Protocol).

5.3.1 PSTN versus Internet

The PSTN (Public Switched Telephone Network) makes use of two very distinct functional layers: the circuit-switched transport layer and the control layer. The circuit-switched layer consists of different types of switch. The control layer consists of computers, databases, and service nodes that control the behavior of circuit switches and provide all the services of the PSTN. All the signaling traffic used for control travels over a separate signaling network. The main characteristics from a high-level point of view are that the network contains most of the intelligence and the terminals are relatively dumb. From a service point of view, the service logic needs to be handled within the network and is under the control of the provider.

The Internet on the other hand is a packet-based network, where the network has only minimal functionality and the terminal nodes are reasonably intelligent. From the service point of view this means that the service logic can be located on end systems. This is, in principle, a very competitive situation because anybody can easily provide any kind of IP-based service.

Internet telephony is meant to be cheaper than conventional telephony. However, in many respects it is cheaper only because the quality and reliability are lower than in traditional telephony. Another oft-stated benefit is the integration of telephony into other types of computer communication such as Web services, email, CTI (Computer–Telephony Integration), etc. There are also some challenges for Internet telephony. Quality of Service must be established to guarantee certain voice quality. Internet telephony must also be integrated with the existing PSTN architecture, including call signaling.

5.3.2 Session Initiation Protocol (SIP)

The SIP (Session Initiation Protocol, Rosenberg *et al.* 2002) is a signaling protocol developed by the IETF to control the negotiation, set-up, tear-down, and modification of multimedia sessions over the Internet. It is, together with H.323 (ITU 1996b), a candidate for becoming the standard signaling protocol for IP telephony. However, SIP is not restricted to IP telephony, as it was designed for general multimedia connections.

SIP entities include user agents at terminals and proxy servers. A SIP user agent consists of a user agent client and a user agent server, sending requests to other SIP entities or replying to received requests respectively. SIP proxy servers route SIP messages between different domains and resolve SIP addresses. Figure 5.6 shows a simple example SIP message flow.

In this example a caller's user agent sends a SIP INVITE request to a proxy server. The proxy server (and potentially several further proxy servers) locates the callee by looking up call forwarding tables or other routing information, and forwards the message to the callee's SIP user agent. If this one accepts the call, it starts ringing while waiting for the callee to pick up the handset. This is indicated to the caller by the 180 RINGING reply, which generates a ringing tone in the caller's terminal. A 200 OK reply indicates to the caller that the callee accepted the call. Now the caller has to confirm the call by sending a SIP ACK message.

Starting with this message, the user agents can communicate directly with each other. This bypasses the proxy server because the INVITE message and the 200 OK replies contain sufficient address information about both parties. However, the SIP user agents

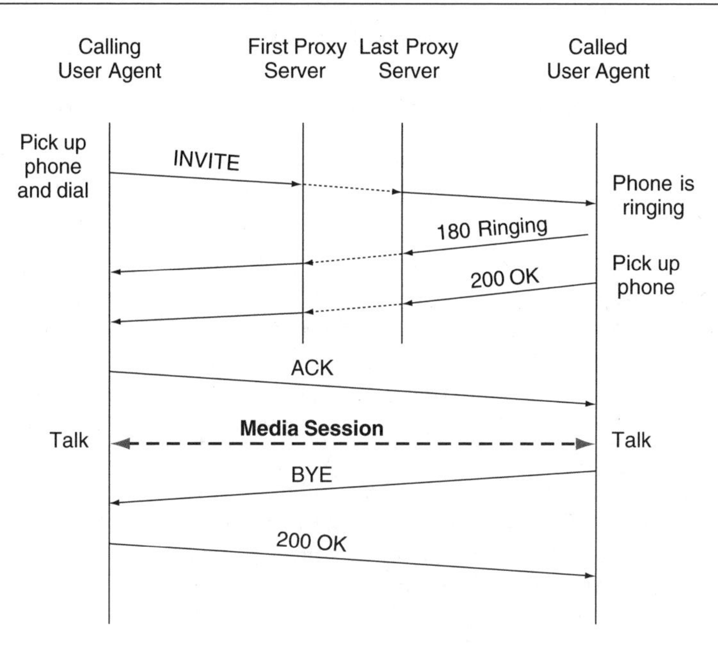

Figure 5.6 Example SIP message flow.

may decide to continue signaling using proxy servers. The BYE request signals session termination. Either party can send it, and the other party must acknowledge it with a 200 OK reply.

For setting up the media session, the INVITE request and its replies may carry a payload specifying voice coder–decoder, transport protocol, port numbers, and further technical session information. Typically, the SDP (Session Description Protocol, Handley and Jacobson 1998) is used to encode this media information.

SIP incorporates elements from two widely deployed Internet protocols, namely HTTP (HyperText Transport Protocol, Fielding *et al.* 1999) used for the Web, and SMTP (Simple Mail Transport Protocol, Klensin *et al.* 2001) used for sending email. From HTTP, SIP borrows the client-server design and the use of URIs (Uniform Resource Indicators). From SMTP, SIP borrows the text encoding scheme and the header style. A user is addressed with a URI similar to an email address. In most cases it consists of a user name and domain name (e.g. sip:user@foo.org).

Discussion

SIP functionality concentrates on just signaling. Its very simple design, restricted functionality, restricted usage of servers, and independence from other components of sessions make it highly scalable. SIP integrates well into the Internet architecture.

Another protocol suite for performing signaling has been developed by the ITU-T, namely H.323. Comparing SIP with H.323 might go too far and might raise 'religious' concerns. In general the functionality is very similar. SIP is considered simpler to implement mainly because of its text-based nature. On the other hand it has some performance problems, especially for SIP proxies since these need to handle a high number of call set-ups. Additionally, the text representation is not so compact and might be a problem for wireless networks.

5.3.3 Media transport protocols (RTP, RTCP)

Given that a signaling protocol negotiates and sets up a multimedia session, the next step is sending the media data over the network. Designed for this task are RTP (Real-time Transport Protocol, Schulzrinne 1996) and RTCP (Real-time Transport Control Protocol, Schulzrinne 1996). RTP runs on top of UDP (User Datagram Protocol, Postel 1980). RTP is the data transport protocol providing synchronization through time stamps, ordering through sequence numbers, and indication of payload type. The payload type defines the encoding of audio and/or video information.

The purpose of RTCP is to provide feedback to all participants in a session about the quality of the data transmission. It periodically sends reports containing reception statistics. Senders can use this information to control adaptive encoding algorithms, or it might be used for fault diagnosis and Quality of Service monitoring.

5.3.4 SIP services

SIP capabilities beyond the example above include call transfer, multi-party sessions, and other scenarios containing further SIP entity types and using further SIP message types. Section 5.3.2 describes the basic SIP functionality for session establishment and tear-down. In this section more advanced features of SIP are discussed. All of the features use a server in the network to provide the service, whereas SIP itself also works without server support.

User mobility

SIP allows a user to be at different places, which means that the user can have different IP addresses. The user is still reachable by an incoming call. Since a SIP address is a URI, the domain name can be resolved into a SIP server handling the session initiation (INVITE message). A SIP server (SIP proxy) allows a user to register the current location (IP address) where the user can be reached. The SIP server receiving an INVITE message looks up the user's registration and forwards the call to the registered place.

Redirecting calls

SIP also allows redirection of calls. This means a server handling an incoming INVITE could reply that the person is currently not reachable via that server. It can give one or more possible other addresses that can be tried to reach the person. Such an address could be a different URI for the user, or a different server where the user potentially can be found. Concerning user mobility, in this case the server does not tell where the person really is. Rather the server indicates somewhere else that the person could be, or where more information about the location might be known. A SIP user agent of a caller then tries to contact the list of potential servers hosting that particular user.

The same feature can be used to implement group addresses. Calling a group address will redirect the call to a person available to take the call. This is one of the main functions needed for call center solutions.

Discussion

SIP can be seen as a basic platform for service creation in many regards. The SIP servers for redirection, call forwarding, etc. can be extended to implement several other features still using the basic SIP protocol and SIP handling engine on the server. The service logic on SIP servers can influence the decision in various situations. Also, the user can influence these decisions by providing the SIP server with user-specific call preferences.

Through the flexible nature of SIP and SDP, it is easy to extend SIP to provide completely different services. For example, a SIP server could be used as a game server bringing users together for distributed games. SIP then establishes a gaming session between several parties. This would mean SDP must be extended to contain game-specific information.

Since SIP was inspired by several other protocols such as HTTP and SMTP, it is easy to integrate with these systems to create combined services.

From a business point of view, the SIP server need not be part of an ISP (Internet Service Provider) domain. Any company can provide SIP services by running a SIP server somewhere in the world. This means it is possible to decouple Voice over IP from base IP connectivity.

5.4 Directory-Enabled Networks (DEN)

DEN (Directory-Enabled Networks) is a standardization activity by the DMTF (Distributed Management Task Force). The main goal of DEN is to achieve better interoperability on the management data level. Additionally, it provides a standardized way for sharing management data among applications and service-providing elements.

DEN tackles the problem of several islands of technologies, each of them maintaining its own management solution and system. It tries to overcome the traditional element-focused and technology domain-centric management, which hinders the integration of the technologies towards fully integrated networking solutions.

The key concept of the DEN solution to management is that a service-enabled network is viewed as one entity. So everything, including networking devices, computers, services, locations, organizational entities, and users are managed under a single framework. The key component in this concept is a central directory, where all management data is stored in a standardized way. Service Management applications are now able to relate the management data and can operate on it.

In order to standardize this vision, two different components are used. First, it is necessary to standardize a remote directory access protocol. Secondly, the data model of the directory needs to be standardized.

The remote directory access protocol allows any management application component to remotely access the directory from any place, to read and to write management data. Naturally, the access needs to be secured. As access protocol, DEN has chosen the LDAP (Lightweight Directory Access Protocol, Wahl *et al.* 1997), currently in its third version. Meanwhile, extensions and additions have been defined to overcome some of the deficiencies.

The Lightweight Directory Access Protocol was developed to simplify the complex X.500 directory system (ITU 1993f). Initially, LDAP was meant to access X.500 servers in an easier way. Subsequently, LDAP servers have also been built.

The second component in DEN is the standardization of the data model represented in the directory. A standard data model is used because different entities need to access the same data and need to understand what these data mean. Therefore the naming of data as well as the semantics of data must be standardized. DEN uses the CIM (Common Information Model, DMTF 1999) as the basis for its standardized LDAP schemas.

The Distributed Management Task Force (DMTF) standardizes the Common Information Model (CIM). The CIM is a database and directory-independent information model, defining only the pure data but no particular mapping to any implementation. The CIM defines a whole set of classes together with their attributes. It defines associations and aggregation between the classes. A core information model pre-defines the overall structure of the model. From that core model several technology-specific models have been derived including physical components, network elements, applications, and services. The CIM is regularly updated and extended.

Given the standard CIM, LDAP schemas can be derived. Mappings to any data model representations are also possible. There needs to be a mapping function between the CIM and an LDAP schema, because LDAP uses a more specific representation of data. The main mapping problems include the naming of objects and the representation of relationships modeled as associations in the CIM.

DEN has a very broad scope and tries to unify different worlds, which is always a very difficult task – even more when people from different companies need to agree on such a standard. For the CIM the problems lie in the standardization of a single information model. It is again difficult to agree on a high level of abstraction. In many cases a sound information model is the most valuable part in management software, so companies hesitate to participate in the standardization effort because they lose a competitive advantage.

5.5 Open Services Gateway Initiative

The Open Services Gateway Initiative (OSGi 2002) aims to specify a service gateway in order to let small devices coordinate and cooperate. The target environments are mainly home networks. The major challenge is that there are many small devices such as smart phones, Web tablets and PDAs (Personal Digital Assistants), but also television sets, video recorders, camcorders, PCs, etc. in a home environment. Many of them are equipped with increasing networking capability. However, the communication facilities are heterogeneous and range from small wireless networks to high-speed wired networks.

To exploit the capabilities of these residential networks, the diversity of networks and devices must be coordinated. However, home users want to use the services without dealing with the control and management of the technology itself.

The approach of OSGi lies in providing a centralized OSGi gateway in the home as a coordination entity as well as a gateway towards the Internet (Figure 5.7). The specifications of the OSGi consortium define APIs (Application Programming Interfaces) that enable service providers and application developers to deploy and manage services in home networks.

The OSGi framework contains an embedded application server for dynamically loading and managing software components. These components can be instantiated dynamically by

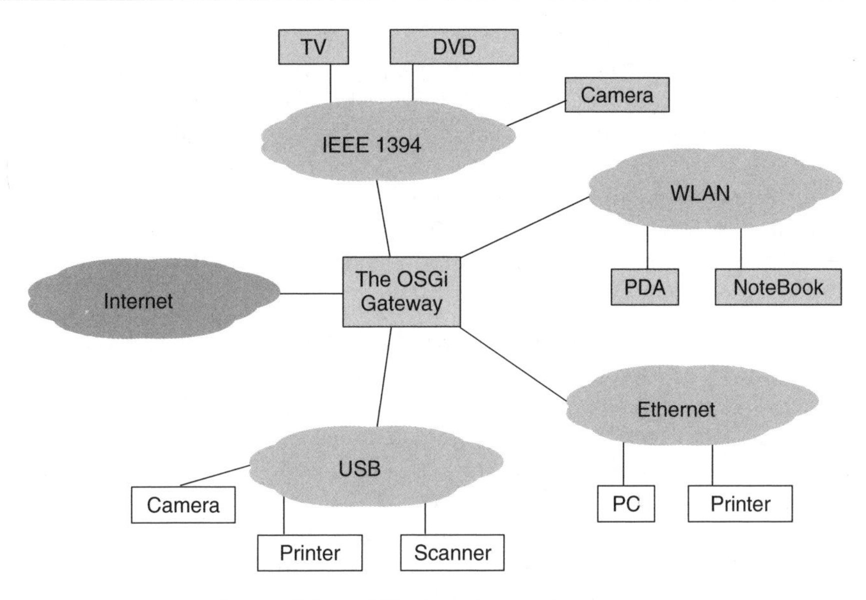

Figure 5.7 OSGi Home Network scenario.

the framework to implement particular services. Additionally, the framework provides HTTP support services, logging functionality, security, and service collaboration.

One of the most important features of the OSGi framework includes device discovery, service discovery, and self-management capabilities. Some of the underlying networking technologies already provide these functions. This is where OSGi can serve as a platform to interconnect these subnetworks of different technologies.

All these features aim at ease of operation for home networks, interoperation of several services, devices and applications in home networks, and their access to the Internet.

5.6 Active Networks

The following briefly discusses a new type of paradigm in networking, namely Active Networks. The technology is still in the research phase.

Assuming the future will be very service-centric in many regards, networks need to adapt to this paradigm. Note that the term 'service' is quite overloaded with different meanings. Given this, new kinds of network need to focus on network services and the ability to provide these services easily. The following lists various observations known from other businesses that are, or will become, important in the networking area.

- *Fast time to market*: New services must be implemented and deployed in a very short time. Not only is it necessary to have a good idea for a service, but also even more important is the short time it takes to bring the service to customers.
- *Differentiation*: Service providers on various levels must differentiate their services from their competitors in order to gain competitive advantage.

- *Customization*: Customers are eager to get what they want. Providers meeting the customers' requirements best will potentially win over others. Easy customization of services will allow a service provider to acquire customers with a broad range of different requirements.
- *More features*: It is very important to have as many features as possible to be attractive to customers.
- *Flexibility*: A high degree of flexibility is needed in order to keep up to date. The faster the innovation process proceeds, the more important becomes the flexibility of the infrastructure to adapt to new upcoming customer requirements. Furthermore, flexibility supports heterogeneous infrastructures to connect customers.

Assuming the service-centric business model is valid for the networking business as well, active networking may be a solution to the above requirements for new networks.

End systems attached to a network are open in the sense that they can be programmed with appropriate languages and tools. In contrast, nodes of a traditional network, such as ATM switches, IP routers, and Control Servers are closed, integrated systems whose functions and interfaces are determined through (possibly lengthy) standardization processes. These nodes transport messages between end systems. The processing of these messages inside the network is limited to operations on the message headers, primarily for routing purposes. Specifically, network nodes neither interpret nor modify the message payloads. Furthermore, the functionality of a node is fixed and is changeable only with major effort.

Active Networks break with this tradition by letting the network perform customized computation on entire messages, including their payloads. As a consequence, the Active Network approach opens up the possibilities of (1) computation on user/control/management data messages inside the network and (2) tailoring of the message processing functions in network nodes according to service-specific or user-specific requirements.

The paradigm of active networking can be applied on different planes. Figure 5.8 shows a three-plane network model including a data, a control, and a management plane. The terminology in the area of active technologies is not yet uniform. Therefore, many people talk about mobile code or mobile agents if the technology is applied for management purposes.

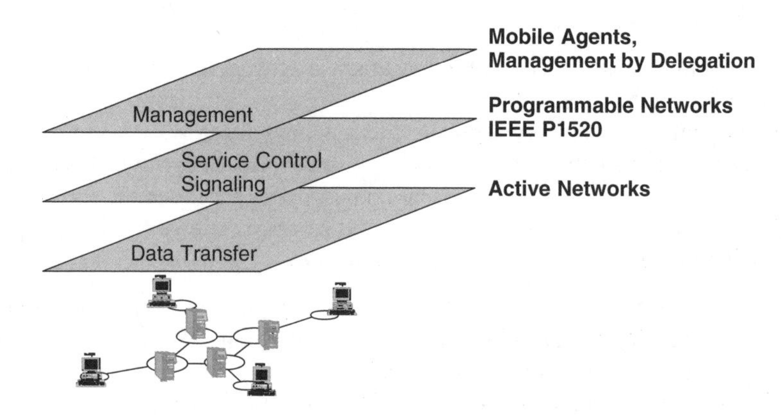

Figure 5.8 Programmability in different functional planes.

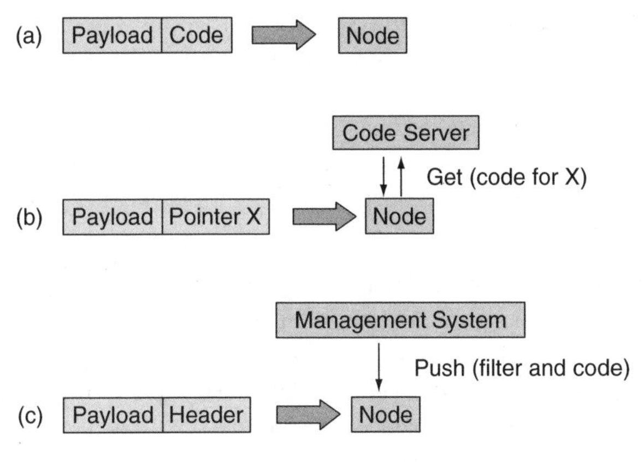

Figure 5.9 Code loading mechanisms.

Networks using active technologies in the control plane are often referred to as programmable networks.

The mechanism to add new service functionality dynamically within the network is loading software code to the nodes. Three distinct methods exist, together with variations of loading the code into the network nodes. Figure 5.9(a) shows the most extreme approach of adding processing code to each message. This is also referred to as the capsule-based, mobile agent, or integrated approach.

Figure 5.9(b) shows the second approach of adding just a pointer to code into the message header at a standardized place. The code is then fetched from a defined code server if it is not already available on the node.

Figure 5.9(c) shows the third approach of loading the code via a management action, so the network administrator is responsible for the specific code installed on the network nodes.

Different mechanisms for loading code have been introduced above. However, the administrative entity installing the code was not specified. Note that the code loading mechanism is independent of the installer of the code, so the characteristics are orthogonal to each other. However, different combinations are superior to others in terms of implementation performance or complexity. Therefore they are often referred to as only one characteristic of Active Network architecture. There are three approaches: installation by a user/application, by a customer, or by the owner of the infrastructure. A user/application basically refers to the case where the code is installed by an end-system into the network, whereas a customer refers to a corporate network connected to a network provider.

A network user or a distributed application may install new message handling code into the network, typically on the path to its destination. This will allow an application vendor to include application-specific message handling code in the application. This will tailor the application for efficiency by including the networking part of the distributed application. This case is very easy to implement by adding the code to the messages. With other code loading mechanisms in place, there needs to be management interaction with the network or code server. One problem is the number of different pieces of code used in a network. This problem is very severe with this approach, because potentially every user is able to install new code into the network nodes. On the other hand, it allows for great flexibility because every user/application can flexibly change the network behavior.

In the literature of active networking, many application-aware network services are proposed that can profit from active networking technology. The following lists the basic network service scenarios used to enhance distributed application performance or reduce network resource consumption.

- Local node condition access enables network services to react quickly to information collected on the nodes in the network, aiming to enhance the end-to-end service performance. The most obvious case is where active messages access node conditions on the node they reside on, e.g. adapt to resource shortage on a node. The enhancements are feasible only if it is possible to optimize the global service performance with local information.
- Caching allows network services to store information on the network node, aiming to reduce network traffic and latency.
- Sensor data fusion, where information is gathered from different points in a network, has the goal of reducing network traffic by aggregating information in the network node.
- Filtering also reduces the network traffic by throwing away less important packets in a controlled way, e.g. early packet discarding, or it enhances security by not allowing packets to pass through the node (firewall functionality).
- Applications in heterogeneous environments benefit from application-specific computation within the network to cope with heterogeneity. For example, transcoding of images in a multi-party videoconference may provide each party with the best possible quality.
- The concept of Active Networks allows a provider or even a customer to deploy new networking services very quickly. It is just a matter of installing new software on a networking infrastructure.

5.7 Conclusion

This chapter has outlined technologies used in IP-based environments to support services of various kinds. The technologies are very different, and are difficult to bring into perspective without a clear application scenario. All of them should be regarded as building blocks or a toolset for creating a service-enabled networking environment. Specific solutions have not been discussed, so the examples given only sketch a potential roadmap for their deployment.

6

Wireless Technology

James M. Irvine

University of Strathclyde, UK

6.1 Introduction

Wireless systems, by their very definition, have no fixed physical connection between the transmitter and the receiver. The wireless link can be provided by infrared or inductive loop systems, but by far the most popular implementation is a radio system.

The untethered nature of wireless systems gives them two key advantages – no physical connection to constrain movement and reduced infrastructure requirements due to the lack of a fixed connection.

Until relatively recently most wireless terminals were constrained by size and battery technology, which in themselves limited mobility. The traditional application for wireless has therefore been to reduce infrastructure, and so it made an obvious choice for broadcast entertainment systems. This application has continued for satellite systems, where very large areas can be covered with relatively small, albeit expensive, infrastructure. Other recent examples of this driver for wireless technology have been WLL (Wireless Local Loop) systems, where a new operator can use wireless systems to allow them to compete with an existing operator without having to replicate their infrastructure. This is a major factor in developing countries without comprehensive fixed networks, but the ability of wireless to allow the entry of new operators to challenge incumbent operators in countries with highly developed fixed networks is now less important as regulators require local loop unbundling and access to the existing (wired) local loop.

However, wireless systems are inherently more limited in terms of transmission capacity than wired systems. In a wireless system the available spectrum is shared between all users, and subject to other interference and environmental disruptions. In a wired system the physical medium, such as a cable or fiber, is a dedicated resource for each user, so additional capacity can be obtained by adding additional media. Manufacturing new radio spectrum is not an available option! More use can be made of the radio spectrum if the links are short, such as for an indoor system.

Service Provision – Technologies for Next Generation Communications. Edited by Kenneth J. Turner, Evan H. Magill and David J. Marples
 ISBN: 0-470-85066-3

While more efficient use can be made of the radio interface to improve capacity, the same can be done for wired systems, so that wired systems retain an advantage. After a number of years of 9.6 kbit/s data service rates for mobile phones, GPRS (General Packet Radio Service) and 3G (Third Generation) systems are beginning to allow 64 kbit/s, which has been available for many years to ISDN (Integrated Services Digital Network) subscribers. However, fixed line operators are deploying DSL (Digital Subscriber Line) technologies allowing hundreds of kilobits per second, retaining their advantage.

While services to fixed terminals can be provided by wired connections, wireless is a true enabling technology for mobile devices. In developed countries the focus of wireless has moved away from broadcast applications to fixed terminals, towards point-to-point communications to movable terminals. This causes a major increase in the complexity of the network, but also many more possibilities in terms of the services that can be offered. The network must be able to stay in contact with a terminal as it moves, and must be able to direct any incoming connection to its location, which implies that a two-way signaling flow is always present. This always-on connection means that users can be offered services with a permanent connection wherever they happen to be.

There are a very large number of different systems in use. The main types are summarized in Table 6.1.

The spectrum available for radio systems is limited, and different types of system have different requirements. Lower frequencies diffract better and therefore travel greater distances. It is also possible to reflect such signals off the ionosphere, allowing reception beyond the horizon. High frequencies suffer higher attenuation and much less diffraction, limiting their range, but that can be an advantage since it limits interference with other systems. It is therefore more efficient if short range devices use higher frequencies. Another factor is that as technologies have improved, it is possible to extend the range of usable frequencies upwards so that newer systems tend to use higher frequencies. On the other hand, older systems jealously

Table 6.1 Types of wireless system

	Personal area	Local area	Suburban	Rural	Global
Area	On the person or in a single room	A building, street or campus	Town	Country	Continent
Cell range	<10 m	<500 m	<5 km	<30 km	>30 km
Data rate	Various	<2 Mbit/s	<384 kbit/s	<144 kbit/s	<64 kbit/s
System types	Personal Area Network, Wire replacement	WLAN, Cordless, Cellular	Cellular, Local Broadcast	Cellular, Broadcast	Broadcast, Cellular
Movement	Stationary	Stationary or low mobility	<120 km/h	<500 km/h	Aircraft speeds
Systems	Bluetooth, IrDA	802.11x, Wi-Fi, WLL, 3G Picocell	2/2.5G or 3G Microcell, DAB	2/2.5G or 3G Macrocell, DAB, DVB-T	2/2.5G Satellite, DVB-S

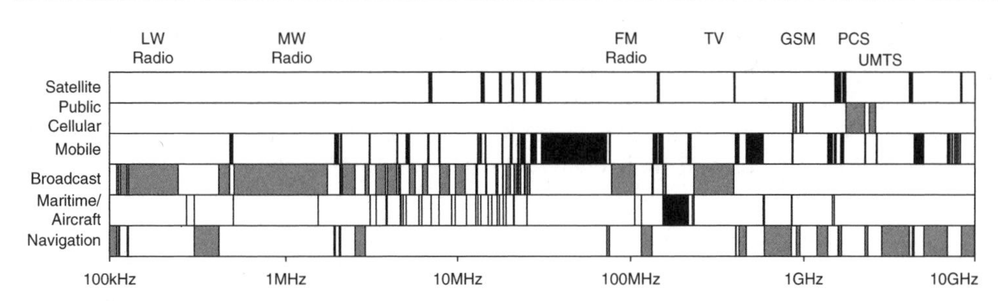

Figure 6.1 Frequency allocation in the UK.

guard the frequency allocations they already have – possession is nine points of the law. Figure 6.1 gives an overview of current allocations, using the UK as an example.

Most radio spectrum is licensed by governments, although international agreements set by World Radio Conferences every three years attempt to agree standardized bands. Some spectrum is set aside for unlicensed use, for such items as cordless phones. Manufacturers must produce such equipment to agreed specifications and limit their power to prevent interference.

6.2 Cellular Systems

6.2.1 Introduction – the cellular principle

The first personal portable sets predate the Second World War, the 'walkie talkie' being patented in 1938. Technological limitations restricted the frequencies available, and so military uses predominated. However, from as early as 1947 some radio channels were opened up to allow a mobile telephone service to be operated in cities. These systems were more like broadcast systems, covering hundreds of square miles. For example, New York had a system which allowed about 40 simultaneous calls.

The radio spectrum is a finite resource. If one person is using a given frequency in a given location it is not available to another user unless there is some means of sharing it, in which case each user gets a smaller capacity. The key innovation which makes mobile telephony practical is the cellular principle. The area is divided into a number of cells, and different radio resources are used in each cell such that the same resource is not used in neighboring cells. The resource may be the carrier frequency, time or code depending on the multiplexing method used. FDMA (Frequency Division Multiple Access) assigns each user in the cell a different carrier frequency for their communication. TDMA (Time Division Multiple Access) breaks each carrier into different time slots and assigns a different one to each user. CDMA (Code Division Multiple Access) assigns a different spreading code or modulation method to each user. These latter two techniques are almost always used in combination with FDMA so that the radio spectrum used by the system is divided in both frequency and time, for example.

Dividing the system into cells complicates the mobile radio network, since the system has to keep track of the cell the user is currently in, and switch the call between cells as the user moves using a process called handover. However, the cell size can be reduced to match the density of users, so rather than having 40 simultaneous calls in a city we can have that many, or more, calls in a single building if the cell sizes are small enough.

Although the principle of cellular operation was known as long ago as 1947, the lack of usable frequencies meant that it was 1979 before the first commercial cellular network was launched in Tokyo. Europe saw its first network (in Scandinavia) in 1981, with the USA waiting until 1983 for its AMPS (Advanced Mobile Phone Service) system. These were all analogue FDMA systems, and are known as first generation (1G).

6.2.2 2G systems

Digital processing of signals greatly increases flexibility and makes more complex multiplexing systems like TDMA and CDMA practical. Digital systems – termed 2G (second generation) started to roll out in the early 1990s. By far the most popular of these, with almost 70 % of users worldwide, is GSM (Global System for Mobile Communications) based on TDMA. Japan has its own PDC (Personal Digital Cellular) system, while the USA has a number of different systems, the main ones being IS-54 and IS-95 (or cdmaOne) based on CDMA.

GSM (Global System for Mobile Communications): The *de facto* worldwide standard, GSM is a TDMA system using a 200 kHz carrier divided into eight time-slots. Frequencies around 900 MHz were used originally, with a further band near 1.8 GHz to expand capacity. The 1.9 GHz band for PCS (Personal Communication System) is used by GSM operators in the USA. An extension to a 450 MHz band has also been defined to replace lower frequency 1G systems in Scandinavia, since lower frequencies propagate more easily, allowing larger cells and more economic coverage of sparse areas. GSM is a very comprehensive standard covering the whole system, with defined interfaces that allow networks to be sourced from different vendors.

Almost as an afterthought, the GSM standard included SMS (Short Message Service). The fact that this was included in the standard meant that it was available to all users by default, giving a very large user base. Billions of SMS messages are now sent each month. SMS contributes a significant proportion of many operators' revenues.

GSM has been extended to include phase 2 enhancements such as half-rate speech (allowing two speech users per slot), high-speed circuit-switched data, and cell broadcast SMS messages.

D-AMPS or IS-54: This is a digital update to the North American AMPS 1G system, and was first specified as a standard in EIA/TIA Interim Standard 54 (IS-54). Each of the 30 kHz AMPS carriers is divided into three TDMA slots, allowing the support of three users where the AMPS system supported only one. In North America, the system is often referred to simply as 'TDMA' after its multiplexing method.

IS-95 or cdmaOne: This is a CDMA system, with each user's signal spread by the modulating code over the entire 1.25 MHz of the carrier bandwidth. It is used in North America and South Korea.

6.2.3 Early data services

The various 2G standards all support circuit-switched data services, but take-up has not been very large. GSM allows Group 3 fax transmission and data transport at up to 9.6 kbit/s, as does PDC, while cdmaOne supports data transfer speeds of up to 14.4 kbit/s. However these speeds are eclipsed by even narrowband fixed line modems. Attempts were made to squeeze Internet browsing over the 9.6 kbit/s GSM data service using WAP (Wireless Application

Protocol), a Web transmission system optimized for a wireless environment. WAP proved a flop, with slow, unreliable connections and little dedicated content. GSM phase 2 included HSCSD (High-Speed Circuit-Switched Data), giving circuit-switched operation at up to 64 kbit/s, but operator support has been limited, with many waiting for the 2.5G GPRS system.

Packet-based data services have been much more successful. One of the earliest was CDPD (Cellular Digital Packet Data), which is a packet-based data overlay for the 1G AMPS network in the USA. The system carries data at up to 19.2 kbit/s using vacant intervals in voice channels.

The most successful data service so far has been the GSM SMS service, which is message-switched. Messages are limited to 160 characters, but the system's simplicity has proved a benefit where more complex WAP services have suffered from compatibility problems between handsets. SMS is being extended, first with cell broadcast capabilities to send slightly shorter messages to everyone within a particular area (for traffic updates, for example), and now to MMS (Multimedia Message Service), giving longer messages including pictures and sound.

Another very successful packet data service has been *i*-mode, which is carried over PDC networks. Although limited to 9.6 kbit/s, short emails can be sent and specially designed websites can be browsed. NTT DoCoMo who designed the system have been much more open to third party developers than some of the operators using WAP, and the cut down version of HTML needed for the website was easier to work with than the WML (Wireless Markup Language) used for WAP. Having a single operator also meant fewer compatibility problems, avoiding website designers having to have several versions of their sites for different terminals. *i*-mode is spreading with NTT DoCoMo's involvement in other markets, and it is likely that in the long term WAP and *i*-mode will converge within XML.

6.2.4 2.5G systems

2.5G systems are enhanced packet-based data systems which extend the capability of 2G systems towards the higher data rates promised by 3G systems. They use the same or compatible extensions of the air interface, but the core mobile network does require quite major extensions to accommodate them. However these extensions have been designed with 3G in mind, offering operators an upgrade path. The main extensions are as follows.

GPRS (General Packet Radio Service): This is an extension to GSM which transmits packet data in one or more slots in the frame. Multiple coding schemes are used depending on the level of interference. If all eight slots are used for a single user, and there is little interference, data transfer rates of up to 170 kbit/s are theoretically possible. However, most terminals can use only a smaller number of slots. This and capacity restrictions on the network mean data rates of 20–40 kbit/s are more likely in practice, at least initially.

cdma2000 1×RTT: This is an extension to cdmaOne networks. The theoretical maximum data rate is 144 kbit/s, but something less than 64 kbit/s is more likely in practice. cdma2000 1×RTT offers a first step to full cdma2000 3G systems.

6.2.5 3G services

Second generation systems provide quite simple services, dominated by voice. Messaging is also usually available as an application, but other types of data transport are fairly rudimentary.

Table 6.2 3G transport requirements in different environments

Environment	Indoor/ Short range outdoor	Suburban	Rural satellite
Bit rate	2 Mbit/s	384 kbit/s	144 kbit/s
Mobility	<10 kph	<120 kph	<500 kph

The focus changes with 3G to stress the importance of data (Ahonen and Barrett 2002). 3G was originally conceived as a universal (i.e. worldwide) network capable of carrying video, which imposed a requirement of being able to carry 2 Mbit/s to an individual user. Video compression technology has improved to the stage where video phones can operate over 64 kbit/s, and indeed the first commercial 3G WCDMA implementation, NTT Do-CoMo's FOMA (Freedom of Mobile Multimedia Access), has done just that. However, the 2 Mbit/s target has remained, although this only applies in local areas, as shown in Table 6.2.

As well as increasing bit rates, 3G also offers support for emerging IP-based services, both real-time and non real-time. It is optimized for packet-switched operation, and has QoS support which is essential for multimedia services.

The much richer suite of services supported by 3G can be broken into four classes depending on delay characteristics: *Conversational*, with a delay very much less than a second, *Interactive* (delay about a second), *Streaming* (delay <10 s) and *Background* (delay >10 s).

3G also envisages a more complex business model. In 2G systems, the network operator was the main player, providing the service and usually also providing the terminal. In 3G, the richer range of services allows a distinction between the network operator providing transport and the service provider supporting the service itself, be it a voice call, Web browsing, video, email, etc. Application providers, providing traffic information for example, are also envisaged (Figure 6.2).

6.2.6 *3G technologies*

The original vision for 3G was for a single standard which would be deployed worldwide and therefore could be used by users wherever they went. However, a number of competing technical and business requirements made that impossible (Harte *et al.* 2001).

In Europe, and most of the rest of the world, spectrum has been set aside for 3G. This is the radio equivalent of virgin territory, allowing the most technically efficient solution to be

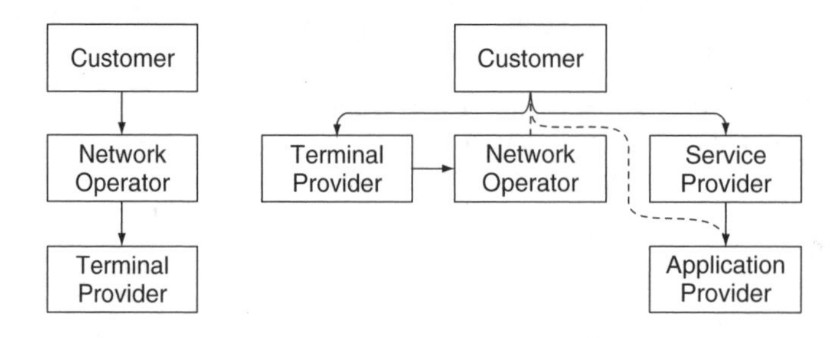

Figure 6.2 Business relationships in 2G (left) and 3G (right) systems.

designed. CDMA has advantages over TDMA in terms of the flexibility with which radio resources can be assigned. After much debate over the relative merits of TDMA and CDMA, Europe decided to go for CDMA for its system, originally known as UMTS (Universal Mobile Telecommunication System). For high data rates – up to the 2 Mbit/s envisaged for 3G – these advantages apply only when the carrier bandwidth is large, so Europe and Japan defined the WCDMA (Wideband CDMA) system as their 3G proposal (Holma and Toskala 2002).

WCDMA is designed for symmetrical operation, where the mobile to network (uplink) and network to mobile (downlink) transmissions are equal. It uses FDD (Frequency Division Duplex), where uplink and downlink transmission use different frequencies in so-called paired spectrum – matching blocks of spectrum far enough apart so that transmission and reception will not interfere within the confines of a handset. However, transmitting asymmetric services, such as Web download, on paired spectrum is inefficient, since one direction will be underutilized. The solution is TDD (Time Division Duplex), where uplink and downlink share the same carrier, but at different times. The relative uplink and downlink time can be changed to accommodate the service characteristics. Since TDD uses only one carrier, it can use unpaired spectrum. The Europeans therefore proposed a second system, TD-CDMA, which uses WCDMA with TDD. The large degree of commonality between WCDMA and TD-CDMA means that dual-mode terminals are possible. TD-CDMA has been extended into a TD-SCDMA (synchronous) system currently under study in China.

The Americans had a problem with WCDMA on two fronts: they had already allocated the 3G spectrum to 2G systems, meaning that the 5 MHz bandwidth of WCDMA was difficult to accommodate, and WCDMA was not compatible with existing CDMA systems being deployed in the US and elsewhere. This led to cdma2000, and a suite of standards either using the existing 1.25 MHz carrier of cdmaOne or, for cdma2000 3x, grouping three carriers into one wideband carrier for higher data rates and more flexibility (Harte *et al.* 2001). WCDMA and cdma2000 are similar in principle, but detailed differences make them incompatible. cdma2000 1×RTT added higher data rates to a cdmaOne system, while cdma2000 1×DV (data voice) and cdma2000 1×DO (data only) are further extensions to the 3G data rates.

EDGE (Enhanced Data rates for GSM (or Global) Evolution) introduces higher order modulation to the GSM system to increase data rates when the signal quality is good. It allows data rates of up to 384 kbit/s, while requiring little modification to a GPRS network. Europe saw EDGE as an enhancement to GSM, and both EDGE and GSM as being replaced by 3G. However, EDGE attracted a lot of interest from GSM operators fearful of not getting a 3G license, and from US operators seeking an upgrade from TDMA systems. With their backing, EDGE was also put forward as a 3G technology.

The ITU met in early 1999 to agree a 3G standard from 13 proposals, all of which could deliver the basic requirements the ITU laid down for 3G. In the event, agreement on a single proposal proved impossible, and five proposals were chosen – WCDMA, TD-CDMA, cdma2000, EDGE and DECT, an extension to the DECT cordless standard (see Section 6.5.3). Figure 6.3 shows the development path for these technologies.

As can be expected when different technologies are in competition, the relationship between technologies and phases shown in Figure 6.3 is subject to some debate. Many commentators consider EDGE to be a 2.5G technology, as it was originally designed as an intermediate technology to extend GSM networks towards the requirements of 3G. However, EDGE was supported by operators in the USA as a 3G technology in its own right, and has

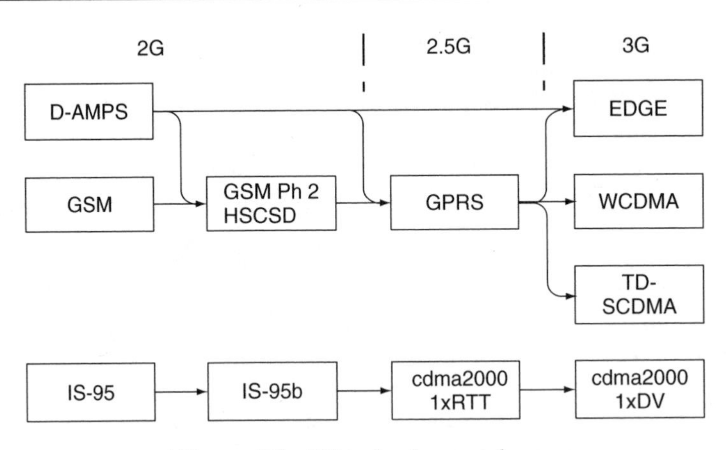

Figure 6.3 3G technology roadmap.

been accepted by the ITU as a 3G technology. On the other hand, some operators have taken to referring to cdma2000 1×RTT as a 3G technology, allowing claims to be made about operating the 'first 3G' network in a particular country. While it is true to say that it does meet some of the data rate requirements of 3G, cdma2000 1×RTT does not meet all the ITU requirements of a 3G technology, and is not considered by the ITU as 3G.

6.2.7 Handover and location management

In a cellular system, the radio connection to the user is provided by a series of base stations. The area covered by a base station is known as its cell. As a user nears the edge of a cell, communication with the base station will become more difficult. The mobile network has to recognize this, and arrange for communication from a neighboring base station to maintain the call. The call is said to handover to the new cell.

In Figure 6.4, a user starts in cell 3 and moves towards cell 1. At point A, the user can pick up the signal from cell 1. Handover must occur before point C, where the signal from cell 3 is lost, and ideally would occur at the boundary between cells, point B. In view of different radio propagation, due to the terrain for example, cells will not form regular hexagons. Some overlap is required to allow the network to switch the user to the next cell before the signal from the current cell is lost or 'dropped'.

The distance between areas where the same resource is used is called the re-use distance. The smaller the re-use distance, the more users can be accommodated with a given set of resources, but the greater will be the interference from other users. For FDMA, cells are formed into clusters, with each carrier frequency being used once in each cluster. The smaller the cluster size, the nearer the interfering cells, but the more carrier frequencies for each cell. CDMA systems usually operate with a cluster size of one, so that neighboring cells use the same carrier frequency. Users are distinguished by the code they use for transmission.

On TDMA systems, communication with the existing base station is usually terminated when the call is handed over. This is known as hard handover. On CDMA systems with

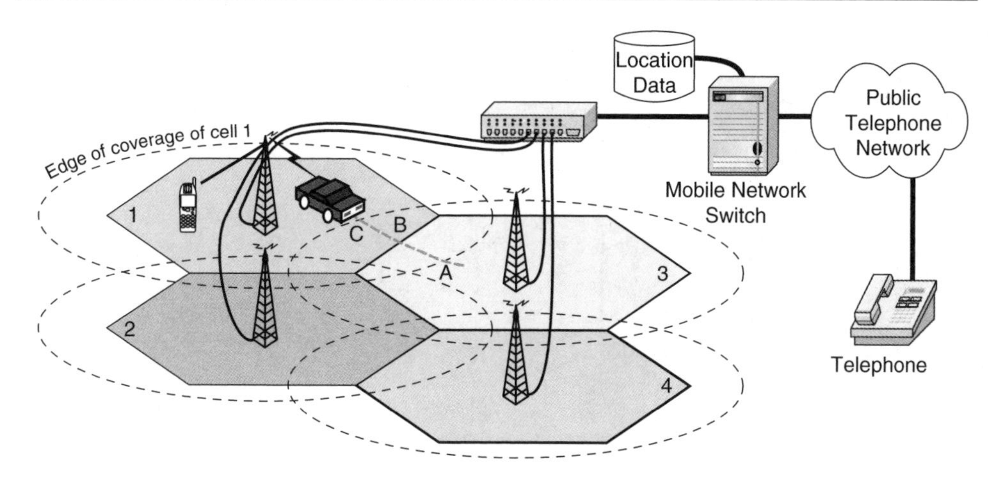

Figure 6.4 Simple mobile network.

neighboring cells on the same carrier frequency, the user may be able to communicate with both base stations at the same time. This is known as soft handover, and has the advantage that when the mobile is far from the base station and vulnerable to interference, then at least one of the signals is likely to be received well enough. When the user moves nearer one base station and the signal strength exceeds a threshold, the second link is dropped.

While handover works to maintain a call once it is in progress, there is the initial problem of finding a mobile to connect a call to it in the first place. For this purpose, groups of cells are formed into LAs (Location Areas). A cell's LA is transmitted along with other information, such as which carriers are available, in a broadcast control channel in each cell. Whenever a mobile detects that it is in a different LA, it sends a signal to the network informing it of its new LA, and the network updates its information accordingly. This includes the point when the mobile is first switched on, the previous location area in this case being stored on the SIM card. When the network needs to contact a mobile – a process called *paging* – it need only contact the cells in the LA, rather than every cell in the system.

There is a trade-off in LA design. Large LAs mean that many cells have to be contacted to page a mobile, which takes a significant amount of signaling traffic. Small LAs, on the other hand, reduce paging traffic but increase the amount of signaling for LA updates. Unlike a page, which at least has the promise of a call which can be charged for, LA updates cost the network operator with no direct benefit which they can charge the user for. The optimal size of an LA therefore depends on the calling pattern and the movement of users, with operators wanting busy highways and railways to cross as few LA boundaries as possible.

6.2.8 Mobile networks

The mobile network must route a call to the correct base station, knowing the location of the user. Second generation cellular networks are circuit switched. Figure 6.5 shows a GSM network, which is typical of 2G systems.

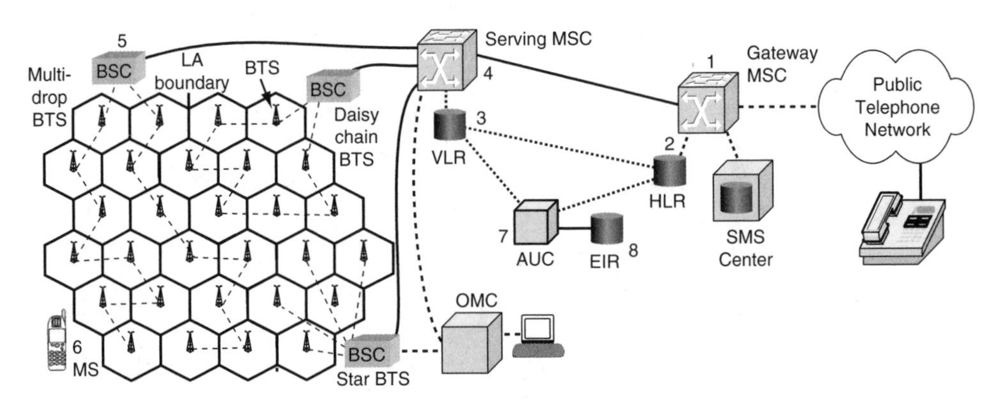

Figure 6.5 GSM network structure.

The Base Transceiver Station (BTS) provides the physical transmission to the user terminal. Base stations usually have little intelligence, simply transmitting the signal under instruction from their base station controller.

The Base Station Controller (BSC) provides radio resource management for the BTSs it controls, for example power management. It also controls handover along with the MSC, and provides information to the Operation and Management Center on network performance.

The Mobile Switching Center (MSC) is the heart of the mobile network. Its functions include routing, call management, location registration, interworking with different networks (for example, transcoding ADPCM (Adaptive Dynamic Pulse Code Modulation) voice transmission to the vector quantization codecs used in GSM) and billing. In GSM, it also provides a gateway to the SMS center, which handles SMS messages.

There are two types of location register. The HLR (Home Location Register) contains all the permanent information for a given user, including the user's subscriber number, authentication and encryption keys for security purposes, billing information, services the user is allowed to use, and so on. The HLR also contains a pointer to the user's current location by including the address of the MSC where the mobile can be located (i.e. its serving MSC) and its VLR (Visited Location Register). The VLR contains the address of the location area where the mobile currently is.

Other network elements are an OMC (Operations and Management Center) to control the network, and an AuC (Authentication Center) which stores the authentication keys for all SIM cards, and which is used to establish if the terminal is valid. It also sets up encryption for each call, and an EIR (Equipment Identity Register) containing details of valid IMEI (International Mobile Equipment Identity) numbers. Each terminal has a unique IMEI, allowing stolen or faulty terminals to be blocked.

The function of these different elements can be illustrated by considering the process of receiving a call, as illustrated in Figure 6.5. A connection request will arrive at a GMSC (Gateway MSC, 1), which acts as an MSC connected to another network such as the PSTN. The MSC translates the called number into the subscriber identity and then contacts the relevant HLR (2), checking that the mobile exists and is not blocked, and then contacts the VLR (3) addressed by the HLR to get its current serving MSC. The GMSC contacts the SMSC (Serving MSC, 4), which in turn contacts the VLR to get location information and availability. If the mobile is available, the SMSC initiates a paging call to the Location Area (5). If the mobile responds, it is authenticated with the VLR using the Authentication Center (7) and Equipment Identity Register (8). Only if this is successful are resources assigned to the call and it is connected. If the mobile does not respond, a signal to that effect is passed back to the Gateway MSC, for the calling party to be informed or a supplementary service such as voicemail or call divert to be activated.

A mobile-initiated call is simpler, as there is no need to locate the mobile first. The mobile will have registered and authenticated when it was switched on. It makes a call request which is forwarded to the MSC. The MSC checks the destination, authorization to make a call with the VLR, and resource availability to the destination network. If this is successful, the MSC allocates resources and sends a message back to the mobile alerting it to a connection set-up.

2.5G networks add packet data transmission capabilities to circuit-switched 2G networks. As shown in Figure 6.6, two new elements are added to the network to allow this, mirroring the serving and gateway MSC functions. The radio access network of BTS and BSC remains the same, but data packets are sent to the first entity which controls packet data transmission to the mobile. This is called the SGSN (Serving GPRS Support Node) in GPRS and UMTS networks, and PCF (Packet Control Function) in CDMA networks. The second entity handles roaming and access to other networks, and is called the GGSN (Gateway GPRS Support Node) in GPRS and UMTS networks, and the PDSN (Packet Data Serving Node) in CDMA systems. Both these entities connect to existing network registers such as the HLR.

In the first version of a UMTS network (Release 99), Node B replaces BTS (the name was a temporary one in the standardization process and no replacement was agreed, so it stuck). The RNC (Radio Network Controller) performs a similar job to the BSC, only taking on some of the lower level functions of an MSC. The packet-switched side of the network is the same as the GPRS packet-switched network. Future versions of the UMTS network (Release 4 and Release 5) (Kaaranen *et al.* 2001) first envisage an increased use of IP for transport, differentiating transport from service. Release 4 breaks the MSC into a control entity and a series of Media Gateways to control specific services. Release 5 foresees the complete removal of the circuit-switched core network, so that circuit-switched services are

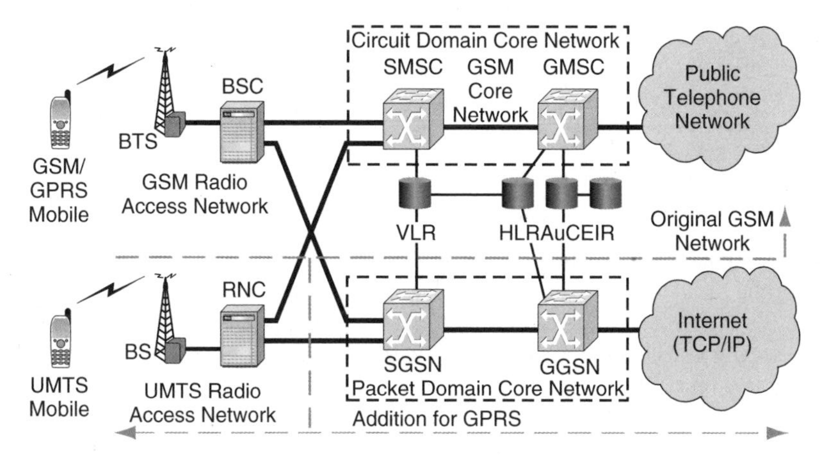

Figure 6.6 Network extensions for GPRS.

carried over the IP-based packet core network. This is called an 'All IP' network, and has the advantage of simplifying network construction and deployment. However, this is still some way off, and is used by some as the definition of 4G (fourth generation).

6.2.9 *Location technologies*

The fact that mobile users move provides an opportunity to sell services to them based on their location. While it is a trivial matter to find which cell a user is located in, more detailed information is more difficult, and therefore costly, to obtain. However, the US FCC (Federal Communications Commission) set a requirement that by October 2001 operators be able to locate mobiles making emergency calls to within 30 m. This is because about 30 % of callers making emergency calls do not know their location. While this is not a problem with fixed lines (whose location can be traced), it becomes so as the proportion of mobile calls increases. Although many US operators have been allowed to delay implementation, the fact that such a large market requires the technology means that it is general within 3G systems, and so can be used for location-based services.

There are a number of ways to ascertain location with the required accuracy. One of the simplest is to use GPS (Global Positioning System). However, while this is very accurate, it requires a GPS receiver in each handset, which is costly. It also cannot provide a solution on its own, as it requires a signal from at least three satellites, which may not be possible indoors or in street canyons.

The time difference between when a mobile user is picked up at one cell site and when it is received at the next can be used to find the user's position. This is called the TDOA (Time Difference of Arrival) method. A more sophisticated variation, called E-OTD (Enhanced Observed Time Difference), has been incorporated in the GSM standard.

Rather than the time, the angle of arrival of the user's signal to different base stations can also be used to calculate position. However, this method can be affected by reflections from large objects like buildings, which means that the path to the base station is indirect.

An interesting aspect of location-based services is the almost total lack of debate amongst users of the privacy issues concerning their operator knowing their location at all times. This is in spite of some governments introducing legislation requiring operators to log this data so it can be supplied to the authorities later. Even current cell-based location systems can provide much information on user's movements, especially in areas of dense coverage such as cities. It remains to be seen whether this view changes once location-based services become more visible to the user.

6.2.10 *Mobile satellite systems*

As well as being used for broadcast systems, satellites can be used for mobile radio systems too, and for the same basic reason – very large coverage for little infrastructure. The earliest system was by Inmarsat, a large consortium involving major telecommunications companies and operators from around the world. From 1979 it offered an analogue system, Inmarsat-A, from geostationary satellites. Initially targeted at maritime applications, services included voice, fax and data at up to 64 kbit/s. This was developed to the Inmarsat-B digital system in 1993, offering similar services but with increased efficiency and lower cost. The development of enhanced video compression schemes has recently enabled reasonable quality video transmission over this system, and journalists can often now be seen giving live front-line reports using videophones on this system. Four satellites cover most of the globe, the exceptions being near the poles (Antarctica and the Arctic Ocean). Their coverage overlaps in Europe and North America to increase capacity in these regions, and the system has a number of spare satellites available for lease to other organizations that can replace the four operational satellites if required.

In order to maintain geostationary orbit and so maintain a fixed point in the sky, satellites have to have an altitude of 35 750 km. This has significant problems in terms of delay, which is about 0.25 seconds for a round trip to the satellite. Such delays are just about acceptable for a voice call, but a mobile to mobile call, involving two such delays, is not. For this reason, mobile satellite systems with smaller altitudes, either low earth orbit (LEO, about 200 km to 3000 km in altitude) or medium earth orbit (MEO, about 6000 km to 12 000 km) have been deployed. These have the advantage of smaller handsets and delays, as well as giving coverage near the poles, but at the cost of much higher complexity and larger constellations of satellites. Being nearer the atmosphere, the orbits decay more quickly, requiring the satellites to be replaced. Systems include the following. Inmarsat/ico is a MEO TDMA/FDMA based system with 12 satellites. Iridium was a very large and very expensive TDMA LEO system that went bankrupt, forcing its main backer to request permission to bring down the 66 satellites; the system was given a reprieve with new backers. Globalstar was a CDMA LEO system with 48 satellites which also went into receivership. Odyssey is a CDMA MEO system with 12 satellites. Teledesic is backed by Microsoft. Boeing has proposed a large TDMA based LEO system with up to 840 satellites for wireless Internet access.

6.2.11 Market trends

The growth in the mobile market has been very strong in recent years. While the exponential growth of the late 1990s was replaced by linear growth in 2001, that still represented a rise of over 250 million subscribers. One of the reasons for the slowing rate of growth is the fact that some markets in Europe, the US, and Asia have become saturated, with penetration rates over 70 % (see Figure 6.7). Further rapid growth in these markets is unlikely, with growth coming from new services rather than basic connectivity. On the other hand, in the rest of the world penetration rates are still low, and many of these countries have poorly developed fixed line infrastructure, so in these areas rapid growth is still possible and indeed likely. China in particular has a subscriber base growing at a rate of two million each month, and many less developed nations have more mobile subscribers than fixed line users.

The rapid growth of the Chinese market has meant that the US has lost its top spot in terms of subscribers (having about 130 million to China's 144 million at the end of 2001). South East Asia as a whole is overtaking Europe as the largest mobile market (see Figure 6.8).

While in developing markets connectivity is key, in the mature markets growth depends on new services delivered by 2.5G or 3G systems. An early example is MMS (Multimedia Messaging Service), allowing pictures and sound clips to be sent as enhanced SMS-type messages. Many operators are placing their faith in location-based services, while video phone services have not proved as popular as was hoped in the initial 3G networks. However, until there are more 3G networks in operation and more services are developed, it is difficult to foresee the popularity of applications. The GSM SMS service was never expected to be as popular as it is when it was developed.

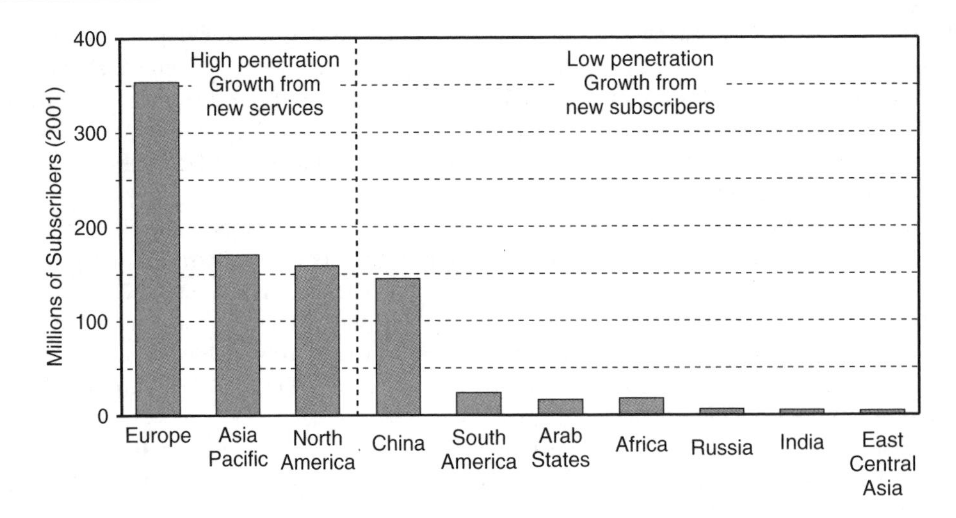

Figure 6.7 Mobile market worldwide in 2001 (figures from Vodafone).

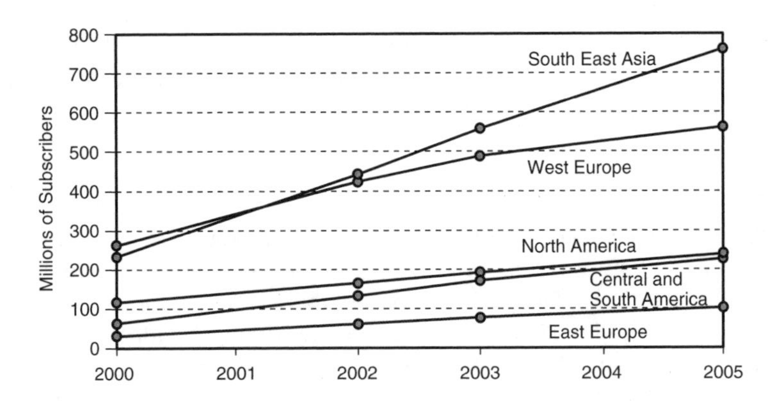

Figure 6.8 Trends in mobile market development.

A key issue is whether new services will generate new revenue or simply displace existing revenue sources. If a customer's expenditure on SMS simply moves to MMS, operators will be faced with the problem of transmitting more data, requiring more network investment, to receive the same Average Revenue Per User (ARPU). Regulatory pressure on prices has kept ARPU roughly constant in the UK in spite of increased calling times. In some markets, particularly the teenage market, mobile phone expenditure already consumes such a proportion of disposable income that significant increases seem unlikely.

There are a number of open questions with regard to 3G take-up. A major one is the popularity of data services, for if data services are not very popular, 2.5G will suffice in most markets. A related question is the willingness of operators to invest in 3G networks. 2.5G provides a stopgap to build revenues and ride out the current nervousness of investors in new technologies. However, against this must be set the willingness of regulators to wait, as many licenses were written with deadlines by which networks had to be operational. Market conditions have encouraged regulators to be more flexible, but some operators who were unsuccessful in getting a license are now complaining that their competitors obtained the licenses by making extravagant claims they are not being held to.

A final question relating to the success of 3G will be the impact of other technologies such as Wi-Fi and broadband fixed access. This might give users a cheaper, albeit less convenient, option for high data rate services. For most current applications requiring high data rates, mobility is not very important.

6.3 Private Mobile Radio

PMR (Private Mobile Radio) systems were the first form of mobile radio system. Unlike public cellular systems, where a network provider builds a network and offers access to that network charged by the call, PMR systems are usually built and operated by the company which is using the communication service, for example a taxi company. Capacities tend to be lower, and bandwidth small (less than or equal to 25 kHz) to allow a large number of networks in a relatively small amount of spectrum. Dunlop *et al.* (1999) provide a full discussion of the PMR market, and the differences between PMR and cellular.

Since they may well be run by companies outside the telecommunications sector, PMR networks must be relatively simple to deploy and operate. In fact, some 'systems' are just an open channel, usually repeated by a base station, so that any transmitting user can be heard by any other user. 'Call control' is simply waiting for other users to stop talking before starting to talk oneself. Slightly more complex is dispatcher operation, where a dispatcher operates from a central point. Transmitting users are heard only by the dispatcher, who can transmit to all other users. Such systems are not cellular in the true sense as they consist of only a single cell. More complex forms of addressing are possible, right up to systems like MPT1327 and TETRA (Terrestrial Trunked Radio) which have all the features of a cellular system.

PMR is mostly still analogue, equating to 1G cellular systems. The capacity driver which forced the public cellular operators to move to digital to accommodate more users is not really present in PMR where call volumes are much lower. The MPT1327 system has become a *de facto* standard for analogue PMR outside the North American market, where proprietary systems by Motorola and Ericsson dominate. MPT1327 can operate in a number of configurations, including a full cellular system with voice and group calls as well as some data capability.

TETRA (Dunlop *et al.* 1999) is a digital PMR system, equivalent to a 2G public cellular system. It has 25 kHz carriers, and uses TDMA so that four voice users can share a carrier. It has more advanced data services up to 28.8 kbit/s Since it was developed after GSM, it has most of its features plus a few extras. While TETRA is extremely flexible, this flexibility has a cost disadvantage in a market where voice and simple data services predominate.

The traditional advantage of PMR systems was cost and the ability to provide tailored services, for example data services to a company database. Now that public cellular service is ubiquitous, the cost advantage has reversed, particularly since public cellular operators are now providing PMR-type services with unlimited calls to a specific user group for a fixed monthly fee over their networks. With the more advanced data services of 2.5G and 3G, public cellular operators may make further inroads into the PMR market, leaving only the emergency services as the major users of PMR since they require very quick call set-up and call prioritization.

6.4 Broadcast

Broadcast entertainment was one of the first commercial applications of wireless. Its key advantage – the ability to reach a large number of subscribers with minimum infrastructure – remains relevant today. By getting in early, broadcast radio and television have managed to obtain large sections of prime spectrum in most countries, but the requirement of backward compatibility with older equipment has meant that it has been used relatively inefficiently.

Traditional analogue systems are quite limited, providing only entertainment with very limited additional services in the form of Teletext or closed captioning (for television), or RDS (Radio Data System) for radio. However, as with mobile radio, there is a movement towards digital technology. Digital systems allow the transmission of data, and make it easier to target services to specific groups of users by addressing or encryption. Digital transmission also allows much more efficient compression techniques, so more channels can be offered in a given piece of spectrum. However, take-up of the new systems has been relatively slow.

There are a number of different digital standards for television and radio services (Tvede *et al.* 2001). The main one for radio is DAB (Digital Audio Broadcasting), although a number of proprietary standards exist, particularly for satellite broadcast. The situation for video is more open, with DVB (Digital Video Broadcasting) being promoted by Europe and a number of other countries, HDTV (High Definition TV) being promoted by the US, and ISDB (Integrated Services Digital Broadcasting) by Japan. However, all these systems share common features, and are not limited to video but are able to transport many different types of data.

One of the reasons for the slow take-up of digital services has been the popularity of analogue systems. While a customer may have only one phone, most have several radios, and many have several televisions. Switching to digital would require the replacement of a number of devices, and digital systems are more expensive. Cheaper receivers are becoming available, but most other equipment, like video recorders, is still designed to operate on an analogue system, so it will be some time before the consumer can easily have a fully digital solution.

In the analogue systems, a single radio or television channel is transmitted on each carrier. Both DAB and DVB use a multiplexing system, where a single wideband carrier carries a number of channels. The advantage of this approach is that different capacities can be assigned to the different channels, corresponding to the different levels of compression that are used. This means, for example, that a talk-based radio channel could be allocated less capacity than a high-quality music channel, or a television news channel could be allocated less capacity than a movie channel.

Broadcast networks have a very simple structure. The different content streams are brought together at the multiplexer, and the multiplexed stream is then sent to the broadcast infrastructure. This coding and multiplexing takes time, so that for live broadcasts there is a noticeable delay in the order of a few seconds. This multiplexed data stream is highly suited to satellite systems with geostationary satellites, since very many channels can be received with a fixed dish pointing at one or more co-located satellites.

Although they are similar in principle, the actual transmission and modulation schemes used by DAB and DVB are quite different. DAB uses a very robust modulation scheme, and is designed to be received in moving vehicles, and by portable terminals indoors at a height of

1.5 m. This limits the transmission rate which can be accommodated. DVB can be transmitted over a number of underlying media. There are three types: DVB-T for terrestrial broadcast, DVB-S for satellite broadcast, and DVB-C for cable broadcast. DVB-T itself has two different modulation schemes, offering a compromise between coverage and transmission rate. DVB-T is designed to be received through fixed, directional aerials high up on buildings.

Both DAB and DVB can be used to transmit data. However, regulators often limit the amount of data that licensees can transmit. For example in the UK, 20 % of DAB data transmission can be data, of which half must be program-related but half can be any other data.

6.5 Local Wireless

6.5.1 Introduction

An important application of wireless is for short range links between devices. In this environment it is not so much the possibility of movement that is important (although it is still a useful benefit), but rather the simplification by removal of interconnecting wires. This has significant benefits for consumer equipment and for portable equipment which may be used in a number of locations.

Unlike the cellular systems considered in Section 6.2 where standards cover the entire system, local wireless standards are more limited, usually covering only those parts of the system necessary to specify the wireless link. They are designed to be a wireless component of a larger system.

Applications range from very short range wire replacement to longer range local distribution systems. An early example of the former is the IrDA infrared system for connecting devices. Although the system is cheap, incompatibilities between implementations have meant that take-up has been poor. Bluetooth is aimed at the same market, while HomeRF was designed as a consumer answer to higher data rate applications like video transfer up to 10 Mbit/s. Wireless 1394 now looks more promising for short range links, with WLAN (Wireless Local Area Network) for longer range links up to 50 Mbit/s, as key HomeRF backers defected to this technology.

Systems for local distribution include cordless technologies and WLAN. WLAN promises a cheap and reasonably standard method of transferring high data rates over tens of meters. The principal short range wireless technologies are discussed in more detail below.

6.5.2 Bluetooth and ZigBee

Bluetooth is designed as a low-cost, low-power cable replacement technology, for example between a phone and a laptop, or a hands-free unit. The system uses frequency hopping spread spectrum in the unlicensed 2.4 GHz band, giving speeds of up to about 720 kbit/s at ranges up to 10 meters (Stallings 2002). This is much lower than WLAN systems, but Bluetooth's cost and power consumption are much lower too. Low power is particularly important for small, light, battery-operated devices.

Bluetooth uses *ad hoc* networking, meaning that it has discovery protocols to ascertain automatically what devices are available without the user having to configure them. One proposed application of Bluetooth is a so-called PAN (Personal Area Network) of devices within a few meters of the user, so that, for example, a PDA (Personal Digital Assistant) could surf the Web by using a Bluetooth link to a WLAN-enabled laptop which may be sitting in a briefcase. Phones, printers, and other devices could also be connected to the network whenever they are switched on and within range.

Bluetooth has been around since 1999, but products have taken a long time to appear. One of the early problems was compatibility, but the controlling body for Bluetooth, the Bluetooth SIG (Special Interest Group) which now has over 2000 members, has used the time to ensure extensive interoperability tests were carried out. This should ensure the technology does not go the same way as IrDA.

For devices with more demanding power constraints and low data rates, ZigBee, promoted by the ZigBee Alliance, provides a solution. While similar in concept to Bluetooth, ZigBee has a lower bit rate (tens to a couple of hundred kilobits per second, depending on the frequency band) but requires a much smaller protocol stack, making it suitable for very small, low-power devices. Applications include telemetry and alarms.

6.5.3 Cordless technologies

Cordless phones are short range wireless handsets designed to connect to a base unit which is connected to the PSTN. Their popularity started in the US. The number of people importing them illegally into Europe caused authorities to set aside frequencies for an analogue FM system.

Cordless phone systems have to cater for large numbers of users in small areas and be self managing – it would not make for a popular consumer product if users had to check with neighbors before buying a phone. In addition, the security of analogue phones, particularly the earlier ones, left a lot to be desired, with the problem of eavesdropping and of making calls through the wrong base unit. Digital systems have several advantages for cordless phones, including greater spectral efficiency, and support for much more complex security and resource management functions. Popular digital cordless systems include the European DECT system (Digital Enhanced Cordless Telecommunications) and the Japanese PHS (Personal Handyphone System). Both have the capability of being much more than a cordless voice telephone. DECT has support for a wide range of data services, PBX functionality, and the ability to integrate into GSM networks. Public access services using DECT were also envisaged, but this has not proved to be generally successful. More successful in this regard has been the PHS system, with large areas of coverage in Japan.

Enhancements to DECT in the form of new coding and modulation raise the possible data rates above 2 Mbit/s, and in this form DECT has been agreed as one of the ITU's five 3G standards.

6.5.4 WLL

Wireless Local Loop is where the final connection to the subscriber is replaced by a wireless connection. This can be in the form of a multiple user cordless system, and indeed DECT is a popular WLL technology. The advantage of this system is that users have a cordless phone which they can use in the local area (although handover is not supported). The disadvantage is the limited range of cordless technologies (only a few hundred meters) so that a relatively large amount of infrastructure is required on the part of the operator.

Range can be enhanced considerably through the use of a fixed directional antenna fitted on the roof of the customer's premises. This is costly to fit, and in developed countries with good infrastructure it is generally uneconomic for voice only services.

6.5.5 *Fixed Broadband Wireless Access or WiMax*

It is possible to offer broadband services through WLL in a system called LMDS (Local Multipoint Distribution Services) or BWA (Broadband Wireless Access, Stallings 2002). The IEEE 802.16 standard (IEEE 2003), trademarked by the IEEE as WirelessMAN, defines a range of air interfaces. The system allows point-to-multipoint links over distances of many kilometers, initially for frequency bands between 2 GHz to 11 GHz, though plans are in place to take this up to 66 GHz and bit rates of 286 Mbit/s.

The WiMax Forum has been formed to market the technology and ensure compatibility between products. Many initial devices use the unlicensed 5 GHz band, simplifying deployment.

6.5.6 *WLAN*

The lack of infrastructure required by wireless systems has great advantages in an office environment, as communication requirements are high and fitting buildings with cabling is disruptive and expensive. WLANs (Wireless Local Area Networks) provide a solution (Stallings 2002). Although Europe has developed the Hiperlan 2 standard, IEEE standards predominate as for wired LANs with the main ones being:

- *802.11b*: This operates at 2.4 GHz in an unlicensed band and offers up to 11 Mbit/s shared between all users (Ohrtman and Roeder 2003).
- *802.11g*: This is an upgrade to 802.11b, using the same band and offering up to 54 Mbit/s shared among all users, falling back to 11 Mbit/s if 802.11b devices are present. It is backward compatible with 802.11b.
- *802.11a*: This is incompatible with 802.11b or g, operating in the 5 GHz band which has fewer users (2.4 GHz is also used by Bluetooth and some cordless phones). It offers a capacity of 54 Mbit/s shared between users.

6.5.7 *Public access WLAN or Wi-Fi*

WLAN systems were designed for private business environments, but with a growing user base of laptops with WLAN capabilities, many companies are now providing public access WLAN access as unwired Internet cafes in places like airport terminals or coffee shops. These are cheap to set up and operate – unlike traditional Internet cafes, no premises are required. They do not require a license, and give customers broadband access with no further infrastructure to buy.

While Wi-Fi does not provide true mobility, it is questionable how much that is actually required for data services. Users normally choose where and when to access a data network, rather than receiving calls at arbitrary times, and not having to provide ubiquitous coverage significantly reduces costs.

Current Wi-Fi coverage is slightly anarchic, with many different systems without interoperation agreements. However, some 'national' networks are being proposed, and operators are increasingly cooperating.

6.5.8 Ultra-Wide Band

UWB (Ultra-Wide Band) is not necessarily a local wireless technology – in fact, early radio transmission used this technology, including the first trans-Atlantic transmission – but current proposals focus on local applications. UWB transmits over a very wide band at very low power, making it highly efficient. However, wide bandwidths make regulators and other operators very nervous. Although the low power makes interference negligible, spectrum licenses are usually granted on an exclusive basis. In February 2002 the US regulator, the FCC, granted a limited license for use in the 3.1 GHz and 10.6 GHz range indoors or in handheld peer-to-peer applications. However, these rules may well be relaxed with experience, making UWB a technology to watch in the future.

6.6 The Future of Wireless

This is currently a time of great change for wireless technologies. The big change from analogue to digital has been achieved in some areas (cellular) but is ongoing – slowly – in others (broadcast). The telecommunications crash has taken some of the luster off many business plans, but many companies, particularly in South East Asia, see a future for 4G services offering 100 Mbit/s within cities.

Meanwhile in Europe, with 3G services just starting to be deployed, most companies do not want to talk about 4G – preferring the euphemism 'beyond 3G'. A huge question is the popularity of 3G services, and how much customers will be willing to pay for them.

Wi-Fi does look as if it could be a major threat to enhanced 'cordless' technologies like TD-SCDMA and DECT, due to its higher data rates and head start in the market. Whether it will affect other 3G systems by creaming off lucrative high data rate users is less certain, because these systems target ubiquitous coverage. However, Wi-Fi still has the edge in terms of data rates, especially as new WLAN standards with higher data rates roll out. Combined with a WiMax backhaul, low-cost broadband wireless networks can be deployed easily.

The switch to digital in the broadcast world is taking longer than was expected, much to the disappointment of regulators who would like to see capacity freed for new mobile services by the more efficient digital broadcasting. An initial boost has been given bydigital satellite services, but customer reaction to digital television has been lukewarm. This may change with the increasing levels of interactivity that digital television makes possible.

A major open question is the effect of wireless systems on health or, almost more importantly for the industry, its perceived effects. Systems with high data rates require higher capacities, which in turn require more transmitter sites. However, there is an increasing public backlash against base stations (although in fact exposure is very much less than from a mobile phone since the latter is held close to the body). HAPS (High Altitude Platform System) may provide an answer by replacing terrestrial base stations with directed beams from balloons or aircraft operating high above cities.

Part II

Building and Analyzing Services

This part of the book builds upon the technologies described in Part I. Here the focus is on the building of services and the range of concerns impacting the service designer. Hence, in addition to looking at service creation simply from a functional point of view, issues such as Quality of Service, security, and compatibility are considered. Service creation and appropriate tools and techniques for creating services on a range of underlying networks are described. One chapter examines the advantages of the rigor that formal methods give to the analysis of services. In addition, service architectures and service capability APIs are explored for providing the framework for building services.

Chapter 7 (Service Management and Quality of Service) describes QoS (Quality of Service) requirements for voice, video, and data services in the context of Service Management. A brief introduction to Service Management opens the chapter. This highlights the challenge of providing the infrastructure to administer and maintain services over a communications network. This high level overview of Service Management and then Service Level Agreements leads into a detailed discussion of the attributes of QoS. An end-user's perspective, taken from outside the network, is used to present the QoS requirements. Human factors are used to justify and organize the presentation of service requirements because they are invariant and independent of technology. Implications and alternatives are presented. The chapter also presents techniques to handle QoS in abnormal or unusual circumstances such as overload or failure conditions. This informs service designers as to what choices they might have if QoS targets cannot be met. The interface alternatives are of particular use in cellular and Internet-based systems, where meeting all QoS needs can pose significant challenges.

Chapter 8 (Securing communication systems) describes approaches to making communications systems secure. Although security is often pursued as an academic discipline, here it is presented in a manner that encourages proper communications design. The chapter views secure communications as the proper controlled access to a range of objects such as data, communication channels, and executable programs. Before describing techniques to control access, the chapter introduces cryptography and forms of cipher, encryption, and hash functions. This explanation is complemented by a description of forms of attack on communications systems. Underpinning access control is the need to determine and verify the identification of user objects. Authentication mechanisms such as passwords, biometrics, and digital signatures are described; non-repudiation and anonymity are also covered. Finally, the chapter considers issues for access control and highlights the use of access techniques through an example application type. This digital cash example reinforces the role of solid design practice.

Chapter 9 (Service creation) describes service creation across a broad variety of service environments. It starts by describing telephony services, in particular within the Advanced Intelligent Network. The Telcordia service creation tool SPACE is described, leading into a discussion of validation, deployment, and monitoring of services. Following on from these Service Management issues, the chapter broadens out to consider Internet services, and hence integrated services. While service creation for Intelligent Network services has been addressed with graphical Service Creation Environments, service creation in the Internet is supported with scripting tools, protocol stacks, and defined interfaces. These are explained in the context of integrated services and Web services. Looking forward, the chapter highlights the innovation required to create a suite of new tools for service creation. These should make service creation for integrated voice and data networks faster, easier, and more reliable. In addition, these tools must be extended to support more of the stages involved in provisioning and managing emerging services.

Chapter 10 (Service architectures) considers the underlying architectures required to support communication services. The chapter surveys a succession of approaches, beginning with the limitations of the Open Systems Interconnection layered model. Its lack of a programming view led to a range of middleware technologies. The chapter then explains the gradual evolution of middleware technology. Specifically, client-server technologies such as the Distributed Computing Environment are presented, followed by distributed object technologies such as the Common Object Request Broker Architecture and Java Remote Method Invocation. Component-based technologies such as the CORBA Component Model, Enterprise Java Beans and .NET are explained. Crucially, the chapter examines existing practice in applying such middleware technologies within the communications industry. Although there are many developments (e.g. JAIN and Parlay) the use of these technologies is not yet widespread. The existing architectures tend to be too complex and heavyweight for communications, and there is little support in the key areas of mobile computing, ubiquitous computing, or multimedia services. The chapter concludes that these challenges demand a new approach to service architecture. It points to interesting work in the field of reflective component-based middleware that offers one promising way forward to meeting the middleware needs of the future.

Chapter 11 (Service capability APIs) focuses on approaches to make the capabilities of networks available to service providers in a secure, assured, and billable manner. The chapter introduces open standards that insulate the service provider from a mix of underlying network technologies. The chapter begins by discussing the limitations of the public telephone network and addresses the shortcomings of the Intelligent Network. However, much of the description assumes converged networks and so must manage both Internet and public telephone network service creation. The chapter provides an overview of standardization activities for the Java Programming Environment. Here there is joint work being done by standards bodies on a set of comprehensive and powerful programming languages, along with service architecture independent APIs. The joint activities are being performed by working groups of the Parlay Group, 3GPP, 3GPP2, and ETSI. These efforts have much in common with TINA (Telecommunications Information Network Architecture) described early in the chapter. Of particular note is the TINA business model, so in this context the chapter highlights common themes, with the environment being specified by JAIN and the activities of Parlay. More recent work on Web services is described as a specific service architecture for embedding abstract and easy to use service capability APIs.

Chapter 12 (Formal methods for services) explains how formal methods provide the means for a rigorous approach to the creation of services. Formal methods are precise languages and techniques for specifying and analyzing systems. They have often been applied to communication protocols. Work on formal methods tends to be rather specialized, however the chapter gives a technical introduction suitable for communications engineers. Formal methods are mainly used in the early phases of development, such as requirements specification and high-level design. These allow some automation of the testing process, but it is rarely practicable to test a service implementation completely. It is possible to demonstrate the presence of errors, but usually infeasible to show that no further errors exist. The chapter describes a range of formal methods: model-based, logic-based, state-based, algebraic, and structural. To reinforce these descriptions, Open Systems Interconnection and Open Distributed Processing are used in examples. This is followed by a review of how different types of formal method have been applied to services, permitting an evaluation of the various approaches. Looking forward, the chapter concludes by considering the prospects for formal methods in the broader range of services expected from converging network technologies.

Chapter 13 (Feature interaction: old hat or deadly new menace?) poses a rhetorical question. Feature interaction is a familiar problem to many in telephony, yet may prove a serious hindrance to the complex services expected from converging network technologies. Feature interaction occurs when the behavior of one service or feature alters the behavior of another. The term originated with telephony features, although the name has remained for modern communication services. The chapter begins by explaining how feature interaction can affect leading-edge services. Feature interactions and their classification are then described in some detail. The response of the communications community to feature interaction is then explored, dealing more specifically with the response of vendors, operators, and researchers. However, with changing technology and a changing regulatory environment, feature interaction is an increasingly critical problem. The chapter concludes by categorizing the nature of feature interactions in future services.

7

Service Management and Quality of Service

Pierre C. Johnson

Consultant, Canada

7.1 Overview

This chapter is primarily about QoS (Quality of Service) requirements for voice, video, and data services. QoS needs are a key ingredient used in Service Level Agreements. To set the context for Service Level Agreements and QoS, a brief introduction to Service Management opens the chapter. Service Management is the challenge of providing the infrastructure to administer and maintain services over a telecommunications network. A high-level overview of the direction in which Service Management and Service Level Agreements are headed is provided.

The basic attributes of QoS are explained in detail. The end-user's perspective, from the outside of the network, is used to present QoS requirements. Human factors are used to justify and organize the presentation of service requirements because they are invariant and independent of technology. Specific targets for all media types are described individually and listed in a table for easy reference. Implications and alternatives are presented as well as techniques to handle QoS in abnormal or unusual circumstances such as overload or failure conditions. The detailed explanation of QoS requirements, along with the description of mitigating alternatives, will help service designers to know what choices they might have in the interface design if QoS targets cannot always be met. These latter interface alternatives are of particular use in cellular and Internet-based systems where meeting all QoS needs can pose significant challenges.

This chapter does not cover technical alternatives within network design to help achieve or guarantee QoS. It is mainly concerned with what QoS needs are. Chapter 5 on Internet technology, for example, provides information on technical approaches to be used within networks to address QoS requirements.

Consistently meeting QoS needs on newer technology remains difficult. This in turn helps to keep the Service Level Agreement and Service Management environments evolving.

Service Provision – Technologies for Next Generation Communications. Edited by Kenneth J. Turner, Evan H. Magill and David J. Marples
 ISBN: 0-470-85066-3

7.2 What is Service Management?

Service Management comprises the functions required to ensure a telecommunications service can be administered and maintained. These functions can be broken down into the following tasks:

- *Service assignment*: establishing information necessary to assign a service to a customer (e.g. billing address, location(s), type of service provided).
- *Connection management*: the making or changing of connections in the network required as part of the service.
- *Fault management*: the detection, diagnosis and correction of hardware or software failures, which could interfere with the delivery of the service.
- *Accounting*: the ability to monitor service availability and usage in order to ensure appropriate revenue is collected from the customer.
- *Performance monitoring*: usage statistics on services, on quality metrics, and on critical resources required to ensure commitments are being kept and to forecast required network growth or equipment upgrades.
- *Security*: technology and procedures required to ensure customer privacy as well as the safety and integrity of the network.

It is desirable to automate as many of the aforementioned Service Management tasks as possible in order to minimize costs and improve the delivery of services. Rapid identification and resolution of faults relies on automation of Service Management.

7.2.1 Service Management historically

In the 1970s, when telecommunications services were primarily POTS (Plain Old Telephone Service), most of the tasks of Service Management were handled by the central office switch in conjunction with computer database systems that handled customer records and billing data. This was also a time before deregulation of the industry. The equipment and networks that a service provider had to interface with were entirely within their control. A few proprietary solutions enabled central office equipment to exchange relevant information with clerical staff, with craftspeople, and with database systems. This was sufficient to meet the Service Management needs.

7.2.2 Service Management evolution

With the advent of deregulation, which began in the 1980s, service providers had to interact with other service providers or re-sellers. This meant that the handling of tasks such as connection management, fault management, accounting, performance monitoring, and security had to enable service providers to manage their part in conjunction with another service provider. If a customer complained of a broken broadband connection which spanned more than one service provider's network, the ability to do fault or connection diagnosis now needed to take into account that the entire network was no longer guaranteed to be within a single service provider's domain. This implied a need for standard interfaces and protocols where one service provider's network bridged into another and that those protocols could handle the needs of Service Management.

The introduction of (non-POTS) services such as mobile systems (cellular telephony), video conferencing, and data services also increased the complexity of the Service Management challenge because many of these services also employed different networks and equipment. The result of the evolution of services and the deregulation can be summarized as a very significant increase in the number of different types of equipment, from different vendors, that needed to share information to automate the task of Service Management.

Families of standards and protocols such as TMN (Telecommunications Management Network) or SNMP (Simple Network Management Protocol) are being used to overcome the inherent complexity of having so many different types of equipment owned by different providers, and to ensure that they conform in the manner in which they receive and report relevant Service Management information. Evolution of Service Management can be thought of as the process of migrating from proprietary systems to these more open standards. This remains an ongoing process and is by no means complete.

See the websites listed at the end of this chapter for more information on TMN and SNMP.

7.3 Service Level Agreements

The deregulation of the industry and the evolution of services have also affected the definition of what a service is. The requirements that basic telephony as a service had to meet were prescribed by regulation, as were the tariffs that a telco could charge its customers. Services today may not be strictly defined and so customers and service providers use contracts to spell out the requirements (e.g. the performance of the service) and what will be paid.

An SLA (Service Level Agreement) is in essence a contract between a service provider and a customer, or between two service providers. SLAs are used to specify service constraints and what QoS (Quality of Service) will be provided as well as the cost of a service. They are particularly useful in helping nail down the specifics of a service where the service is not governed by any regulation. For instance, SLAs would be used to specify the costs, the types of connection (e.g. voice, video, data channels, and protocols), size (e.g. number of channels or bits/second), data reliability (e.g. percentage bit error rate tolerable), responsiveness (e.g. connection set-up time or server response time) and availability (e.g. 24 hours seven days a week with no more than x seconds of downtime in a year) required. New technology has made it possible to combine media types in more permutations and combinations than were possible before, and these new service type offerings create the need for new SLAs to govern them.

Having an SLA between service providers can simplify the Service Management task by partitioning the problem into separate regions, thus enabling each service provider to focus only on Service Management within their domain. This can remove much of the need to share Service Management information between service providers, but at the expense of having to monitor the connection points between service providers to ensure conformance to the SLA.

In addition to the value in partitioning the Service Management problem, SLAs are fundamentally driven by the need to ensure all services are viable. Customers want to know that the services they are purchasing will meet their needs and that the services will be usable. Poor audio quality can render a voice service undesirable or unusable. High bit error rates on data channels effectively reduce the channels' throughput if retransmission is required, or can make the application less usable if retransmission is not an option, as in some video streaming services.

The ultimate justifications for what is specified in an SLA are the QoS requirements for the services as seen from the outside of the network.

There are currently instances where guaranteeing QoS is not technically feasible. Many existing Internet-based networks can only offer 'best-effort' handling of packets. One way to deal with lack of concrete QoS guarantees is to have relative classes of service in SLAs. A premium class would only be guaranteed better service than a less premium class (e.g. 'gold' class packets get priority over 'silver', which get priority over 'bronze', and so on). This approach, which can be included in an SLA, avoids making any absolute guarantees and relies on relative value of being better or worse than another grade of service. It can ignore the question of whether a grade of service meets end-user QoS requirements.

7.4 Quality of Service

Unless otherwise stated in this chapter, Quality of Service refers to the end-to-end QoS as measured or perceived by customers on the outside of the network. QoS can be specified for intermediate points in a network or at a juncture between two providers' networks (e.g. in an SLA) but QoS at these points is a subset of end-to-end requirements. It is most useful to start with and fully understand the end-to-end constraints, which are the foundations for all other QoS targets.

7.4.1 What is Quality of Service?

QoS (Quality of Service) can be defined as the quantitative and qualitative characteristics that are necessary to achieve a level of functionality and end-user satisfaction with a service.

The characteristics we begin with are the ones that are manifest from outside the network. These include the characteristics of the following media types: audio, video, still image reproduction, and digital data. Digital data includes software applications, documents, databases, and files. All telecommunications services as manifested externally are comprised of one or more of these media types. These media types do not have the same QoS needs; therefore the QoS targets for services are a function of the media types they carry.

Two media types which may not be externally visible to end-users but which are most critical to the functioning of services and networks are analogue and digital signaling or control information. These internal media types are usually treated as higher priority than external media because the integrity of the network relies on them. QoS targets for media types within the network, which are not intended for end-users, are not discussed in this chapter.

The end-user is also concerned with the responsiveness and reliability of a service. In other words, the user wants a service to respond in a timely fashion when requested, and once in use, wants the likelihood of failure to be slim.

QoS can be thought of as providing a measure of how faithfully the various media types are reproduced, as well as how reliably and responsively the reproduction can be counted upon.

7.4.2 QoS attributes

QoS attributes tend to fit into two categories: quality and timing. In the quality category the key attributes are:

- *Fidelity*: how faithfully the source content is reproduced. This is usually a function of bandwidth available, sampling granularity, and encoding schemes.
- *Loss*: missing packets in a digital stream resulting in missing portions of the source content.
- *Corruption*: having bits or packets changed resulting in incorrect or modified source content.
- *Security*: ensuring the source content is protected from being received by unintended recipients.

In the timing category the key attributes are:

- *Delay*: also known as latency, is the average amount of time elapsed from the time source material is sent until it is presented at the receiving end.
- *Jitter*: also known as delay variability, is the extent to which actual delays deviate from the average. Jitter represents a measure of how much the minimum and maximum delays differ for a single media stream.
- *Synchronization*: the difference in delay between more than one media stream, which need to be sent and received together (e.g. sound and video).
- *Set-up time*: how long it takes to establish access to the service. This is also known as start-up time or start-up delay.
- *Tear-down time*: how long it takes to discontinue access to the service and free resources to allow another set-up to be initiated.

Set-up and tear-down times are not generally given as much attention as the other QoS attributes because they tend to be less frequent events. In the extreme case of an 'always on' service, set-up should only happen once and tear-down should never happen. Of the two, set-up time is usually given more attention because it is more observable to end-users, e.g. dial tone delay LSSGR (Local Switching Systems General Requirements) in the POTS environment.

7.4.3 Typical QoS trade-offs

One QoS attribute can be improved with no impact to other attributes so long as we have access to more bandwidth or processing power. In the absence of more bandwidth or processing power, improving one or more QoS attributes often has consequences for another. Improved fidelity is usually achieved by employing more sophisticated encoding or compression algorithms, which usually increases delay at the endpoints but can also save time via reduced bandwidth. Reducing loss or corruption is achieved by using protocols that retransmit required packets, which usually increases delay. Jitter and synchronization can be reduced by buffering the incoming data streams, which usually increases delay. Conversely, for each of the examples just mentioned one can reduce delay but at the expense of one or more of the other QoS attributes.

In general, it is no coincidence that delay seems to be the favorite QoS attribute to suffer. The trend towards IP (Internet Protocol)-based packet networks reinforces the fact that delay is the QoS attribute most at risk. Most of the specific requirements about QoS targets in this chapter refer to timing related targets because delay is the most vulnerable.

In the next couple of sections, delay is the QoS attribute receiving most attention because if its importance. In section 7.4.6, specific targets for each QoS attribute are listed.

7.4.4 Human factors and QoS

Human factors provide the best framework to derive QoS targets by describing the limits and constraints of the end-user. Most human factors such as the limits of the sensory or cognitive systems are well understood and are unchanging, unlike technology. They provide a solid target against which a service can be determined to meet, fail, or surpass the end-user's need. How these limits apply across all demographics of users is well understood and their relative invariance makes them ideal as a starting point. They provide a foil against the argument that QoS targets will always be tightening up because users always want it better or faster. For instance, in the film industry the number of frames per second (24 to 30) has been chosen to take advantage of 'flicker fusion'. A rate of any fewer frames per second prevents users from perceiving smooth motion. Many more frames per second are wasted because people cannot perceive any difference. This standard has been in place for nearly a hundred years, regardless of the particulars of film technology, because it is based on fundamental human factors.

7.4.5 Service types as a function of QoS

The following description of QoS categories and target zones is derived from the same model as was used to create the ITU (International Telecommunications Union) recommendation: ITU-T Recommendation G.1010 (ITU 2001). The description which follows provides background as well as interface alternatives to mitigate poor QoS, which are not covered by the ITU recommendation.

There emerge four categories of service in terms of delay, when considering QoS from a human factors point of view. In order to keep track of which human factors each category corresponds to, and hence their justification, they are named here (from most to least strict) using terms that correspond to the human system to which they relate:

- *Perceptual*: limits based on the perceptual limits of the human sensory systems (e.g. auditory or visual). These limits are the shortest in terms of delay, typically within 200 milliseconds.
- *Cognitive*: limits based on limits such as short-term memory and natural attention span which range from 0.25 to 3 seconds.
- *Social*: limits based on social expectations of what a reasonable response time is when a question or request is posed. The user's understanding of how complicated the original request was can temper these limits. Typically these are delays of up to 10 seconds.
- *Postal*: limits based on expectations of delivery to another person of things like mail or fax. Expectations range from tens of seconds to several minutes and, in some cases, hours. In contrast with the *Social* category the response for these service types is generally perceived by a person other than the sender, which is one of the reasons for more relaxed timing needs.

In human terms, the reproduction of the source material must be either precise (i.e. error free) or can be forgiving (i.e. a low number of errors may be of no consequence). Precise usually corresponds to digital source. Forgiving usually corresponds to analogue source.

Both the *Perceptual* and *Cognitive* categories are neurologically based and are therefore valid across all demographics. The *Social* and *Postal* categories may be subject to cultural and experience-related variation. Fortunately for service and network designers, the strictest

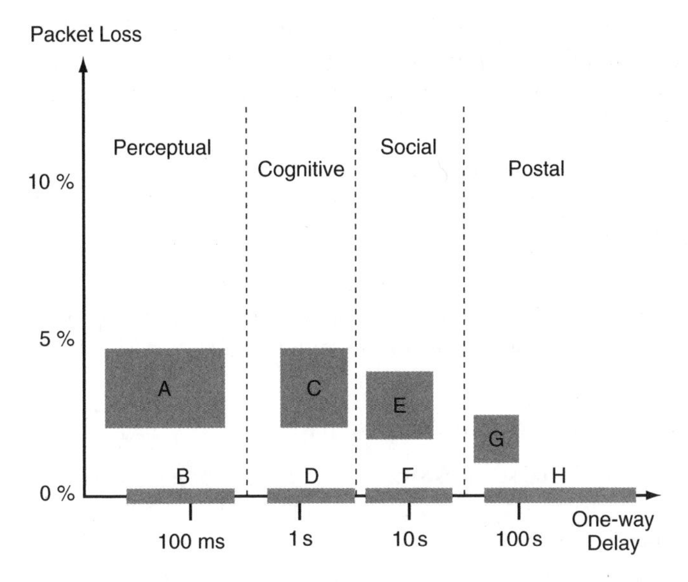

Figure 7.1 Human factors based QoS targets.

categories are the most invariant. This avoids the problem of having the toughest targets shifting over time.

The four QoS delay categories along with the two types of reproduction (i.e. precise and forgiving) yield eight target zones in terms of delay and loss. These eight target zones can be approximately represented as in Figure 7.1 with the horizontal axis representing delay timing on a logarithmic scale and the vertical axis representing loss (percentage of packets lost). The target zones are shaded gray. Each one has been arbitrarily labeled with a letter so they can be referred to in the following explanation. Target areas A, C, E, and G are for source media that can tolerate some loss (e.g. analogue) and where some trade-off of loss versus delay may be possible. Targets B, D, F, and H on the horizontal axis are for source media requiring 0 % loss (e.g. digital) and where there can be no compromise on loss versus delay. Delay is the only QoS attribute that can be allowed to vary.

Each of the target zones (A through H) is explained below with examples and options on how to mitigate delays that are beyond the target:

Zone A: QoS for perceptual and forgiving media

Two-way conversational voice and/or video are the typical service types for this target. The source content is analogue in nature and provides a continuous stream of information. Some lost information is tolerated because human hearing and visual systems can compensate for some noise. The delays must typically be within 200 milliseconds. Delays in excess of the target introduce noticeable pauses and/or synchronization skew in the conversation (e.g. satellite communications). These effects interfere with emotional cues, lead to user frustration, and undermine trust. As a result, these effects render the service unacceptable. Interface designers cannot mitigate this effect to lessen its impact. In some cases users can mitigate some of the effects by adopting a Ham Radio protocol in their conversation, in which only one person speaks at a time and they provide a verbal cue to the other side indicating they have finished speaking so the other person can begin, but this is rarely entirely satisfactory.

Zone B: QoS for perceptual and precise media

Services based on digital media such as telnet sessions and interactive or immersive computer games use this target. The source material content is digital. Loss is not acceptable and hence 0 % is the target. The delays must typically be within 200 milliseconds. Delays outside the target reduce the usability of the service by not enabling the user to remain in step with the other end. There is nothing that can be done to mitigate delay in excess of the target.

Zone C: QoS for cognitive and forgiving media

This is for one-way analogue services such as voice messaging systems. It is similar to Zone A with respect to loss but the one-way distinction means that the source content can be delayed more from the source to the destination without the user noticing. The delay is only apparent at the beginning of the stream. Delays can be in the range of one second or so. If the delay in receiving the stream is greater than the target, it can be mitigated by giving the user some feedback within a second to ensure they know the material is coming (e.g. a message or tone confirming the message about to be played back).

Zone D: QoS for cognitive and precise media

Interactive digital services such as Internet browsing and E-commerce on the Web use this target. Loss is not acceptable. As with computer interfaces in general, delays need to be within a few seconds. As response is delayed beyond the target, users' short-term memory and attention spans are stressed. The task the user is trying to complete becomes more challenging. We know, from the studies done on human computer interfaces dating as far back as the 1970s, that user error increases and satisfaction drops. Delays beyond the target zone can, in some cases, be mitigated somewhat by providing feedback to the user within the target period that their request is pending. This feedback can help reduce frustration but it cannot lengthen short-term memory nor the fundamental interruption in the flow of the task for the user.

Zone E: QoS for social and forgiving media

Services involving one-way streaming analogue source material such as audio and video use this target. The distinction with Zone C is that the content is more voluminous or continuous in nature (e.g. Internet radio) and hence more difficult to re-start or playback. Start-up delay can be up to ten or so seconds given that the duration of the stream is likely to be orders of magnitude longer. Unlike the QoS targets mentioned previously, the stresses for the user are not a function of neurology but of expectation based on experience. The role of feedback about progress to the user while the stream is starting up can significantly mitigate the usability of the service. This progress feedback needs to begin within the target zone. If the delay is roughly an order of magnitude beyond the target then users will suspect the service is not working, even with most common types of progress feedback.

Zone F: QoS for social and precise media

Similar to Zone E except that the source is either digital or of a static (non-streaming and persistent) nature such as still image downloading or FTP downloads (e.g. software downloading). Unlike Zone E, loss is not acceptable. The start-up delay is similar to Zone E in the ten second range. Start-up delay outside the target zone can be handled identically to Zone E services. The transmission of content in Zone F is usually a finite task. This means the transfer of material will come to an end because the data file is finite in size. Forecasting completion of content transfer provides an opportunity for a progress indicator to the end-user to show how close to termination the request is (e.g. count down indicators or completion bars).

Zone G: QoS for postal and forgiving media

Non-digital content such as fax is the typical service. The end product, unlike the other forgiving media types, is static and persistent in nature. This makes errors more noticeable which is why the acceptable loss is lower. The acceptable delay is much larger (i.e. anywhere from 20 seconds to a minute and a half). This is partly due to expectation and because the user who sends the content is not the same as the user who receives the content. Unless the sender makes immediate contact with the receiver the delay is not perceived at all. Mitigating delays well beyond the target zone may be achieved by providing feedback to the sender.

Zone H: QoS for postal and precise media

Digital services such as email and Usenet have this target. Loss must be 0 %. Acceptable delay is hugely variable and provides a very broad target ranging from a few minutes to hours. Mitigating delays beyond expectation is something that cannot typically be done in real time but rather by enabling users to query the status or progress of their mail message if they wish.

The documents by Conn (1995) and by Nielson (1997) provide additional reading on how appropriately to represent delay and why it needs to be minimized for Web-based services.

7.4.6 QoS targets summarized

This method of categorizing QoS targets into eight target zones on the basis of end-user needs was included in ITU-T Recommendation G.1010 (ITU 2001). More details and precision about these targets and examples of how they apply to different services can be found in the ITU document.

Table 7.1 lists the targets for each media type that services may employ. For further background on QoS targets see Ang and Nadarajan (1997), Kitawaki and Itoh (1991), Kwok (1995), Nielson (1994) and Vogel *et al.* (1995).

All QoS targets must be met under normal operating conditions. Normal in this case means a network system is running within its maximum capacity and all the equipment and links in service. Having traffic offered beyond the capacity of the network system results in one or more components in the network being in overload. Equipment failures (e.g. cable cuts or power outages) can also interfere with the system's ability to maintain QoS. The next three sections are concerned with QoS handling in abnormal or unusual circumstances.

Table 7.1 QoS targets by media

Media	Delay, Jitter and Synchronization (reference)	Loss and/or Corruption
Perceptual forgiving A in Figure 7.1 e.g. conversational voice and video	delay <250 milliseconds jitter <50 milliseconds (Weinstein and Forgie 1983) sync <100 milliseconds (Karlsson and Vetterli 1989)	<5 % depending on encoding scheme
Perceptual precise B in Figure 7.1 e.g. telnet, real-time telemetry, computer games	delay <250 milliseconds for telnet delay <50 milliseconds for gaming (Kwok 1997) jitter <50 milliseconds sync <100 milliseconds	0 %
Cognitive forgiving C in Figure 7.1 e.g. voice messaging	start-up delay <2 seconds jitter <50 milliseconds see note below for sync	<5 % depending on encoding scheme
Cognitive precise D in Figure 7.1 e.g. E-commerce, WWW	delay <1 to 2 seconds (Scheiderman 1984) jitter does not apply see note below for sync	0 %
Social forgiving E in Figure 7.1 e.g. audio or video one-way streaming	start-up delay <10 seconds jitter <50 milliseconds sync <100 milliseconds (Karlsson and Vetterli 1989)	<5 % depending on encoding scheme
Social precise F in Figure 7.1 e.g. FTP, still image downloading	start-up delay <1 minute (Kwok 1997) jitter does not apply see note below for sync	0 %
Postal forgiving G in Figure 7.1 e.g. fax	delay about a minute jitter does not apply see note below for sync	<3 % depending on encoding scheme
Postal precise H in Figure 7.1 e.g. Email, Usenet	delay in minutes or hours (Kwok 1997) jitter does not apply see note below for sync	0 %

Note on sync: synchronization only applies when there is more than one form of media used concurrently (e.g. auditory and visual together). If a normally singular media form is matched up with another media form, a synchronization target applies. The target is usually the strictest of the targets of all the media forms involved. This is touched on further in Section 7.4.9.

7.4.7 QoS in overload

Telecommunications systems are vital tools in response to emergencies and natural disasters. Ironically, the more severe the scale of an emergency or disaster the more telecommunications networks are likely to be overloaded. Other favorite sources of overload include 'mass calling events' in telephony or 'denial of service attacks' on the Internet. These latter examples are instances of traffic being focused to parts of the network as opposed to the network as a whole. The scale of overload experienced can easily be orders of magnitude beyond the normal maximum load. This is the reason many purchasers of telephone switching equipment insist on proof that a vendor's equipment can gracefully handle ten times its rated capacity. Aside from the big events just described, capacity shortfalls occur for a variety of subtler reasons. These reasons include demand growing more quickly than the network, technology not keeping pace and the service provider's business constraints not allowing for timely purchase of upgrades.

When running below capacity it is desirable that all customers for any given service type receive the same QoS. Egalitarian handling of the traffic on a service type basis is best. All customers of a service get the same QoS.

Where relative grades of service are offered, such as premium 'gold' class to some customers relative to less-premium 'silver', the egalitarian handling applies within the class. Having relative classes generally has the effect of ensuring the lesser classes experience overload before the premium ones do. Suffice to say having multiple relative classes of service considerably complicates meeting QoS targets for the lower priority classes unless something like fair share allocation of network resources can be applied. The intent of such relative classes is to offer a better grade of service to some by limiting the impact of traffic in lesser classes.

In the remaining description of QoS in overload on the system, one can substitute the term 'class' for 'system'.

If the offered traffic is in excess of the capacity of the system then there are two fundamental approaches to protect QoS: increasing the capacity of the system and/or providing some form of load control (e.g. load shedding, load rebalancing, or admission controls).

Adding capacity to the system is a sound and simple strategy so long as is it technically feasible. This can be done by providing additional bandwidth, processing power, or through the provisioning of whatever resource is the limiting constraint. Adding capacity can ensure QoS targets are met and can preserve the egalitarian handling of the traffic. For service providers it can mean more potential revenue by carrying more traffic. For vendors of telecommunications equipment this strategy usually results in more sales. If it is possible to have a network system that will always have more capacity than the services will demand of it, then no other approach is required. However, in practice, service providers will only purchase enough equipment to handle their busiest 'normal' load. It is very rare (in many cases impossible) to provision sufficient capacity for abnormal events.

In the absence of some form of load control, QoS can collapse when offered traffic goes beyond the capacity of the system. This collapse can render 100 % of the traffic handled by the system ineffective, if the system persists in trying to handle all the offered traffic in an egalitarian manner. Good load controls segregate the traffic as close to the source as possible to allow only as much traffic into the network as the system can carry. The traffic that is carried receives egalitarian handling and meets QoS targets. The traffic that is shed is ideally given some treatment, which provides feedback to the users, that the service is not currently available or delayed beyond the norm.

It may be tempting to try to handle say 110 % of the capacity of the system in an egalitarian manner. For most services (particularly the high-revenue ones) degradation of QoS outside the target results in the service being unusable. The risk in trying to go beyond 100 % capacity is that QoS degradation will make 100 % of the carried services unusable, which can lead to 0 % revenue or worse (e.g. lawsuits or penalty clauses in SLAs).

The following principles summarize how to protect QoS:

- where possible, ensure carrying capacity of the system is greater than the offered traffic;
- carried traffic must always meet QoS targets;
- carried traffic should be handled in an egalitarian manner;
- offered traffic should be handled in an egalitarian manner so long as it is within the capacity of the system;
- offered traffic beyond the capacity of the system needs to be segregated into carried and non-carried traffic;
- segregation of traffic (e.g. load shedding or admission controls) should be done as close to the source as possible.

The objective of all overload handling is to maintain carried traffic as close to 100 % capacity as possible while maintaining QoS targets for the carried load.

7.4.8 QoS in failure conditions

Telecommunications networks and equipment are designed to minimize any outages due to hardware or software failures and to be self-healing as much as possible. Some faults are non-recoverable and clearly QoS is not an issue when a service is unavailable. When a service is recoverable the question becomes how long does the recovery take to come into effect (e.g. the re-routing of traffic or the swapping in of standby components)? This recovery time introduces a delay into the service.

For any service involving streaming of audio or video or for media in the Perceptual category (Zones A, B, C, and E in Figure 7.1) this recovery delay will usually be an interruption and will be noticed by the end-user. Any methods for mitigating delay (described on a per service type basis in section 7.4.5) will not prevent the interruption from being noticed. In most cases those mitigating methods only apply to a start-up delay.

For services not involving media in the Perceptual category and not involving streaming (Zones D, F, G, and H in Figure 7.1), the recovery delay can be mitigated as described in section 7.4.5 so long as it remains within an order of magnitude of the target.

7.4.9 QoS exceptions

The QoS targets described in this chapter are the targets that need to be satisfied in order to ensure end-users find services usable. In general, targets have been described on a per medium type basis. If a service involves more than one media type used concurrently, the strictest target in terms of delay of the media types involved should be applied to all of them. For instance: a remote collaboration service where audio, video, and a shared workspace or presentation screen are transmitted to multiple locations, the need to keep all media synchronized means that even the shared workspace or presentation screen must respond as if they are Perceptual in nature.

There is nothing preventing an SLA from specifying a QoS target which is stricter than the minimum required for human factors reasons (e.g. in order to be able to market a guaranteed premium grade service to certain customers). This means there may be instances where targets may be stricter than those described in this chapter. A good example of this would be telemedical telecommunication services. For instance, standards for image handling (e.g. X-rays) in the medical environment are much more demanding in terms of fidelity and reliability.

7.4.10 QoS trends

Whenever a new communications technology is introduced, there is a temptation to re-examine QoS targets. Users will forgive poorer quality and reliability if they feel there is a unique value in the new technology. This effect has been observed with the introduction of mobile systems. If we are dealing with the same fundamental media types, QoS targets based on human factors will eventually prevail. Over time, usability concerns and user satisfaction will drive service providers towards the same QoS targets. As soon as a service provider can consistently meet the QoS targets as described in this chapter, the forgiveness for poorer service of novel technologies is reduced.

People working on new technologies such as next generation wireless and Internet-based telephony have re-examined QoS targets due to this forgiveness of new products. Over the years this had led to the QoS targets described in this chapter being re-discovered for new technologies. This is why many of the references cited in Table 7.1 are not that recent. As it has become clear that QoS targets eventually gravitate to the same values, the ITU-T Recommendation G.1010 (ITU 2001), based on end-user multimedia needs, was put forth in late 2001.

The implication is that in the long term QoS targets are invariant and independent of technology but that there is a short-term effect for new technology that allows for poorer quality as a new technology is in its infancy. This effect maintains churn in the domain of SLAs (Service Level Agreements) because it means that as new technologies are introduced there will be give and take until the technology matures.

7.5 Further Reading

Those seeking more detailed knowledge on QoS targets should read Ang and Nadarajan (1997), Kitawaki and Itoh (1991), Kwoc (1995), Nielson (1994) and Vogel *et al.* (1995). More details on the ITU-T Recommendation G.1010 referenced in this chapter can be found at the ITU website, and for more information on SNMP refer to http://www.simpleweb.org/. Also, for further background on the cognitive categories introduced in this section, see Nielson (1997). This presents more arguments relating to appropriate responses for cognitive type services described in Section 7.4.5.

Time Affordances by Alex Paul Conn describes interface design methods for mitigating delay. This is a comprehensive look at how to make response appropriately self-evident in interface design, and relates to methods for mitigating delay described in Section 7.4.5.

8

Securing Communication Systems

Erich S. Morisse
Consultant, USA

8.1 Introduction

In the broadest digital sense, security is the practice of providing proper access to objects. These objects may be data, communication channels, program executables, or any of many other types. Security also means ensuring availability of those objects enabling those with authorization to access them, and necessitates restrictions on access from those not authorized. We then clarify that definition by noting that although access is necessarily definition-dependent upon the object, we are still able to generalize methods of control. And, while security is often pursued as its own academic discipline, it is also integral to solid communications design. By ensuring good programming practices, architectural decisions, and well-defined interaction with outside components, good security creates a robust environment for the operation and execution of digital services providing customers with a consistent and stable interaction. A well-known stable system also makes support for the services easier, faster, and more cost effective.

Herein we explore methods of determining and verifying identification of user-objects and access control on objects. We start with an overview of the prerequisite encryption technologies, and conclude with a discussion of a class of application utilizing and requiring the theory we will examine – digital cash.

8.2 Cryptosystems

The general class of cryptosystems is defined by the goal of providing a mathematically secure method of communicating over an insecure channel. An *insecure channel* is any means of communication over which an adversary is able to read, change, re-order, insert, or delete a portion of the message transmitted, and an *adversary* is anyone outside the set of intended recipients of the communication. Examples of insecure channels include speaking on the telephone, sending email, and shouting across a crowded room.

Service Provision – Technologies for Next Generation Communications. Edited by Kenneth J. Turner, Evan H. Magill and David J. Marples
 ISBN: 0-470-85066-3

The precise definition of mathematically *secure* is often argued, but is used here to follow Kerckhoffs' principle of practical, if not theoretical, unbreakability. A message is a communication of finite arbitrary size passed through a channel and can be any block of digital information or subset thereof (e.g. text file, packet radio transmission, database primitive, etc.).

The three types of cryptosystem used to secure messages that we are going to examine are the quick *symmetric key ciphers*, the broadcast-secure *public-key encryption ciphers*, and the *hash functions* used for message *signatures* and comparison. This section is intended to be a brief overview of the necessary topics we will be utilizing later in the chapter; for more extensive information, see Menezes *et al.* (1997) and Stinson (1995) among others.

You will find that throughout this chapter we utilize the *de facto* standard in encryption discussions of labeling communication partners with mnemonic and alphabetic identifiers, e.g. A is for Alice, B is for Bob, etc.

8.2.1 Symmetric key ciphers

Symmetric key ciphers (see Figure 8.1) are a general class of methods named for their utilization of the same key for both encryption and decryption.

Stream ciphers, symmetric key ciphers so named for their use in protecting data streams, are especially useful in the case where the size of the data set is not known at run-time as they usually process the data stream bit by bit. Because of the tiny size of the data blocks these ciphers operate over, they are easily implemented in hardware. Having minimal error propagation makes them a great choice for transmission methods with a high probability of loss or corruption, such as radio, and the minimal data block size also makes this style highly useful in systems with little to no buffer space.

The second principal type of symmetric key cipher is *block ciphers*. More common in open implementation than stream ciphers, these operate over larger blocks of usually 64 bits at a time. As these perform over larger data samples, these cipher algorithms can utilize the data itself as a feedback into the cipher, and the identical function can be used to operate on successive blocks. Not keeping state from block to block, block ciphers are known as memoryless. Stream ciphers, whose encryption/decryption functions may vary over time, are said to keep state. By adding memory to the block ciphers, they can use previous blocks as input and produce *streaming block ciphers*. The additional feedback provided in streaming block ciphers, of being able to more easily detect a modification of the stream by an adversary can be worth the complexity. This tamper resistance can otherwise be achieved through digital signatures discussed below.

One of the strong drawbacks of this class is that the key has to be known to both ends of the communication. And unless the recipient has enough storage for the message to be stored and processed at a later time, the key has to be known from the initialization of the

$$D(E(m,k),k) = m$$

m is a message, k is the key, D and E are de- and encryption methods respectively, and no information about m or k can be derived from $E(m,k)$.

Figure 8.1 Symmetric key cipher definition.

communication. This requires the complexities of exchanging keys, key synchronization, and general key management to be an external process as discussed below.

8.2.2 Public key ciphers

Also known as *asynchronous ciphers* this class of functions uses different keys for encryption and decryption (Figure 8.2).

Because the keys cannot be mathematically computed from one another, the system is provided with the unique flexibility of transmitting its encryption key, also known as its public key, to a recipient over an unsecured channel (Figure 8.3), without compromising the integrity of future messages. Any recipient that receives the public key is then able to create an encrypted message that no one other than the object with the matching private key can decipher, without concern for whether other objects, including adversaries, possess their public key.

It is important to note that this type of cipher does not provide authentication of the source of the key and this must be done utilizing an alternate method. This external authentication must be done, as one successful attack against this method is intercepting traffic from both communicants, and masquerading as each to the other. This *man-in-the-middle attack* (Figure 8.4) allows an adversary to listen to and modify the traffic stream.

If there is a reliable method of authenticating the public key out of band, this class of cipher is able to help provide message signatures when combined with cryptographic hashes.

These ciphers often rely on modulo operators, polynomial, or matrix mathematics, which are expensive to compute. As a result, these ciphers are much slower than symmetric key algorithms and are consequently used in more limited spaces, e.g. initial or subsequent key exchanges for symmetric key communications, and less time-sensitive operations such as email.

$$D(E(m,k_E), k_D) = m$$

m is a message, k_D and k_E are the keys, D and E are de- and encryption methods respectively, and no information about m or k_D can be derived from $E(m, k_E)$ and k_E.

Figure 8.2 Public key cipher definition.

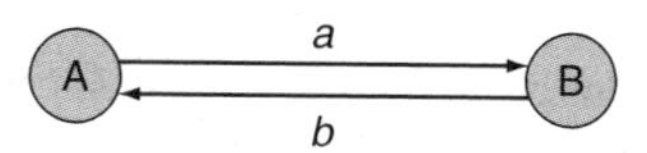

Figure 8.3 Regular communication channel.

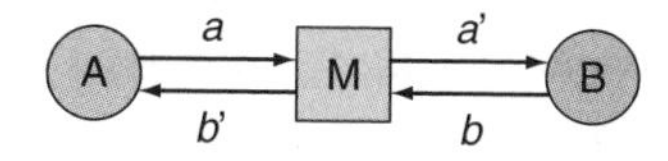

Figure 8.4 Man-in-the-middle attack.

8.2.3 Key management

KM (Key Management) is the administrative process of coordinating information regarding the keys for a given system comprising the generation, storing, distribution, and changing of keys. The process is a complex system combining all the intricacies of authentication, data warehousing, and data segregation with those of the encryption methods supported by a given implementation.

The beginning of a key, or set of keys, controlled by a KM infrastructure is created and often signed using the digital signature of the KM. A KM system that signs and distributes (usually public) keys is called a Certificate Authority (CA), as by signing the key the CA is vouching for its authenticity. Lest the keys the CA uses to sign client keys be compromised, the signing process and access to those keys must be kept separate from any public interface.

After the creation of the keys, they are handed off to another process or system which stores them for later access by the client, or others searching for the client's public key. Not all KM systems store the keys for public consumption, as many systems used for securing email do. Some, like those used for supplying the X.509 certificates used in authenticating websites with SSL/TLS, negotiate the key exchange as part of the session creation between the communicating user-objects.

In the case of public keys, distribution takes one of the two scenarios above; either the key is exchanged at the time of the communication or beforehand. In order to verify the legitimacy of the received key, both scenarios necessitate the foreknowledge of the CA's public signing key. The most common solution to this problem, identical in difficulty to determining any client public key, is through providing the CA's public key in the software that will be utilizing it.

Keys change for any number of reasons. Changing keys puts a discrete boundary on exposure for a compromised key, and it also lowers the chance of a key being compromised itself. Every message encrypted with the same key provides more data for an adversary to attempt to guess. Often, if an adversary has some knowledge of the information encrypted, she has an advantage in attempting to decipher the rest of the message. For this reason, keys are often provided with valid lifetimes, which can be any of: a beginning and end validity date, use for a specified time increment, number of messages, validity for a single connection, or number of bytes encrypted. Changes to short lifetime keys, usually used in synchronous key ciphers, are transparent to the end-user.

For software using long-life keys, usually in use with such public key systems as PGP (Pretty Good Privacy), the KM infrastructure should permit a user to explicitly expire, or revoke, their key. Role-specific keys, including job-related identities, should also be able to be revoked by authoritative parties such as an employer if employment is terminated. As a user-object may have many public–private key pairs, in combination with the possibility of early revocation of a key, a method is needed to not only determine its authenticity, but also its current validity. CRLs (Certificate Revocation Lists) were created for just this reason, and are a store of certificates revoked at the time of the creation of the list. While some implementations use real-time generated CRLs, most publish periodically, with the option of publishing a new one before the expiration of the lifetime of the current CRL.

This is only the barest introduction to KM. For a more detailed discussion please see any of the references and websites on IKE (Internet Key Exchange), PKIX (Public Key Infrastructure), ISAKMP (Internet Security Association and Key Management Protocol), OAKLEY key exchange protocol, SKEME, X.509, XKMS (XML Key Management System), or your favorite search engine.

8.2.4 *Hashes*

The cryptographic versions of hashes play a critical role in securing communication systems. Hashes are mathematical fingerprints for messages of finite arbitrary length, and cryptographically secure hashes are those from which no part of the message may be determined from the hash itself. As the fingerprint is of a known and usually smaller size than the message itself, this allows the comparison of messages in much shorter time. They make excellent checksums, detecting malicious and unintended transmission error data changes equally well.

As the hash size is shorter in length than most messages, there exists a many-to-one relationship between messages and hashes, and the possibility of collision is non-zero. A hash algorithm is considered *compromised* if one can identify any two discrete messages that hash to the same fingerprint.

8.3 Authentication

As we are operating in a digital environment, the closest we can come to authenticating an actual person is through the verification of a data-model object of collected information representing that person. Because of this gap, there is a critical need for strong methods to determine whether or not the person claiming to be represented by any particular user-object actually is the person in question. To do this, we have two non-exclusive paradigms from which to draw tests: what the person knows, and what the person possesses.

8.3.1 *Passwords*

The knowledge-based system is based upon the principle that there exists some *token* of information that the person associated with the user-object, and only that person, knows. The prime benefit to this model is its simplicity. It is easy both to understand and to implement, and a common application of this type is a password authentication system. In theory, only the person responsible for an account has this password token to access it. In application there are a lot of shortcomings to this model.

The level of protection provided by a password system is determined in large part by the quality of the passwords chosen. Since customers who outwardly value ease of use over security often determine passwords, poor quality passwords are often chosen, to the detriment of access security to their account, and to the system as a whole. As passwords easy to remember are similarly easy to guess, they are thus susceptible to *dictionary attacks*. As it is usually much easier to attack a system from within, poorly chosen passwords can severely undermine the security of a system.

After the strength of the passwords themselves, the other most significant portion of the strength of a password scheme is how the passwords themselves are stored. Because passwords have to be accessed by some system processes, they have to be available either as a local or a remote process. As a local process, usually a file, one has to be careful that read access to that file does not forfeit the security of the system. To prevent that, passwords should be stored encrypted, with modern systems storing a hash of the password, plus a *nonce*, rather than the password itself. A nonce is a small collection of random information used to make a message unique. The practice of using nonces is also called salting, and the nonce associated with password schemes is usually referred to as the *salt*.

This combination works by accepting a password from an object, combining that password with the publicly available nonce for that object, and then performing the hash algorithm over the concatenation of both. If the result of the function and the data stored are identical, the passwords are identical. The nonce is used to mitigate exposure in the case where two or more accounts may have the same password, and in this way making the attempt to guess passwords that much harder. Utilizing the known-size *fingerprint* function of a hash in a password system has the added benefit in that by setting up the system to support storage of strings equal in length to the cryptographic hash, systems can then allow passwords without an artificial upper bound on length.

8.3.2 One-time passwords

There can be legal or technical reasons that can preclude encryption from being used, in such cases channels remain unsecured. In such cases, passwords are especially vulnerable to compromise through eavesdropping. In order to provide as high a degree of security as possible, a system can utilize single-use, or one-time passwords for authentication (Figure 8.5). These disposable strings can be created using a hash that operates over the results from successive iterations, starting with the password, or passphrase (m).

Since the hash functions are one-way functions, the system preserves security by checking passwords in the reverse order of their creation (Figure 8.6). Step C_1 is initialization of communication between the user-object requesting access (client) and the object trying to do the authentication (server). The server in our example has stored password iteration 100, and so requests passphrase 99 from the client. The client, in C_2, then runs 99 iterations of its hash function over the passphrase provided by the user-object and sends the result to the server. Finally, in S_2, the server runs a single iteration of the hash function on the string from the client, and if it matches iteration 100 that it has stored, its records are updated with the string from the client, and the authentication is acknowledged as successful.

$$h_i(h_{i-1}(\ldots h_1(m)\ldots)) = h_i \text{ (stored system string)}$$

where m is a passphrase, h is a one-way hash function, and h_i is identical for all values of i with the same input m.

Figure 8.5 One-time password definition.

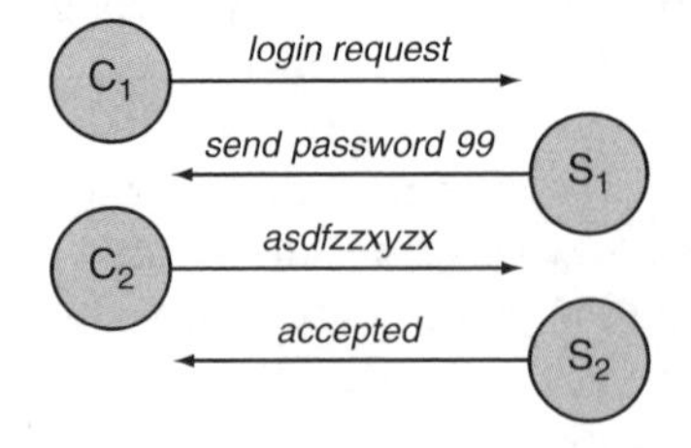

Figure 8.6 OTP exchange.

This single use of passwords prevents an adversary from listening to the session and issuing a *replay attack*. A replay attack is the retransmit of data seen to successfully elicit a desirable response, such as authenticating with a password. Consequently, one-time passwords can be used as an additional level of security in almost any system of communication, encrypted or otherwise. Note the heavy reliance on the avoidance of plain text to hash collisions, and the necessity that the hash function is one-way. Should either of these conditions be proven false, the security of the design is lost.

8.3.3 Encryption key pairs

As our first example of 'what you have' as an authorization mechanism, we use the mathematics of public key encryption discussed above, key pairs are often used for authentication in communications where the authenticatee already has access to one system to run the public key encryption software. The object that is trying to authenticate the user-object will send the user-object a nonce encrypted with the user-object's public key. The user-object must then reply to the system with the correct reply, such as the nonce deciphered and re-encrypted with the authenticating system's public key. Systems such as MIT's Kerberos also include a timestamp in the message to the user-object and place a valid lifetime on the message to prevent replay attacks.

8.3.4 Biometrics

Biometrics are the latest step to reduce the gap between authenticating a user and authenticating a user-object. By taking the authentication paradigm of using what one has to its limits, biometric methods of authentication are based on recognition of physical characteristics of the person attempting to authenticate. By digitally recording characteristics about an individual like iris patterns; fingerprint patterns; and position, size, and relative distance between key facial features; these features can be later re-measured to authenticate the user. These unique characteristics allow a user to be authenticated without having to remember passwords. Modern biometrics also include various forms of life monitoring, such as density, heartbeat, and temperature to dissuade the gruesome possibility of the metric characteristics still working after removal from the person.

8.3.5 Digital signatures

Signing a digital message requires two sides: a private method of signature by sender, and the public method of verification by the recipient. As this idea fits so neatly with the public key encryption model, digital signatures are often created using well-known algorithms on both ends, and keeping a key, instead of an algorithm, private. It is this method of signing messages that we will discuss below.

One of the key infrastructure elements that adds great strength to the systems, often handled but not specified in the protocol definitions, is a trusted third party arbiter. In implementation, this third party is often a key library for looking up public keys, and/or another signing service.

Signing services help to bridge the user-object/person gap by doing out of band identity checks. Also serving as an unbiased third party, they are able to help resolve extra-protocol disputes in authentication and non-repudiation.

Authentication

Signing a document that has been encrypted using public key encryption is implicit. If Bob can decrypt a message using Alice's public key, then it was encrypted with Alice's private key, and only Alice's private key. The act of decryption is simultaneously an act of verifying the signer. As forging such a document signature is mathematically hard, any message with such a signature shows that the user-object is the owner of the key, or more precisely, had access to the key.

An added attraction is the implicit signature change with every different message. A replay attack can be avoided even on multiple transmissions of the same message by including something as simple as a timestamp or counter in the message itself as a nonce, as long as an adversary cannot force a given, or previously seen, nonce. Thus, if a timestamp is used, it does require a stable and secure time or counter source for the originator. As one possible method of maintaining the integrity of the timestamp or counter, each only has to be unique as they are used instead of a nonce. This is often accomplished by mandating that the source can only increment, and does so at least with each message.

If an adversary can force a known nonce, on top of the negation of the added security of the nonce, it also opens the message to plain text attacks. A plain text attack is a form of information analysis on an encrypted message wherein the attempt to gain more information about the message is facilitated through partial knowledge of that message. In this case the facilitation is through the knowledge of the nonce. However, this is a wider problem and not local to only digital signatures.

Non-repudiation

Looking at authentication from the inverse angle, we have non-repudiation. In perfect non-repudiation scenarios, as with authentication too, one, and only one key can produce a signature. In such a case, you have perfect authentication. There may be cases when the system is involved with soliciting or accepting financial or other transactions that are too costly to allow a client to back out of once agreed to. In scenarios like these, the system will want to guarantee that the customer is authenticated and can deny neither their explicit, nor their implicit, acknowledgement of the transaction. Explicit in that the user-object signed the message directly, and implicit meaning that access to the signing key was granted for use to another user-object or process.

Here, a trusted third party with knowledge of the customer's public keys is particularly useful, in that the third party is able to perform the authentication actions over the same message, independently determine the validity of the signature, and bear accurate, unbiased testimony to the legitimacy.

8.3.6 Anonymity

On the flip side from authentication is a desire to keep communication partners anonymous. Often used in a series of communications that require a user-object to be the same originator as that of a previous message, public key encryption as discussed above needs some enhancements to support this idea.

Public keys usually contain an identifier tied to an individual, such as an email address. For use in a single communication, the removal of that identity string may be sufficient to provide the level of anonymity desired. However, an adversary may be able to determine information leading to the identity of the originator by tracking the use of the single key to communicate with multiple partners. A new key pair could be created for each communication, but this becomes operationally expensive. It also has the drawback of the user-object itself not being able to prove to external objects that transactions to different partners all originated from it without revealing their actual identity. Variations on two ideas have been developed to help deal with this class of scenario: *blind signatures*, and *nyms*.

Blind signatures are a class of method for signing data without ever seeing it, and in such a way that a third party will be able to verify that signature. Introduced by Chaum (1983), this variation on public key encryption utilizes commutative functions to accomplish the task. Start with a private signing function s' and a public inverse function s such that $s(s'(x))=x$ and s reveals nothing about s'. Take a commuting private function c and public inverse c' such that $c'(s'(c(x)))=s'(x)$, and $c(x)$ and s' reveal nothing about x. This set of functions allows a third party to verifiably sign data, or vouch for the originator of the data, even if the recipient receives it from an anonymous source. Consider the example where A wants to send data to C anonymously. C needs to determine the legitimacy of the data from A, and B is a trustworthy authority that C trusts to vouch for the legitimacy of A, but does not trust with the data.

- A: has data x to send to C, and also possesses functions c, c'.
- A: computes $c(x)$, sends $c(x) \rightarrow B$
- B: has functions s, s'
- B: signs $c(x)$ with s', sends $s'(c(x)) \rightarrow A$
- A: computes $s'(x)=c'(s'(c(x)))$, sends $s'(x) \rightarrow C$
- C: has B's public function s
- C: computes validity of $s'(x)$ with $s(s'(x))$

Notice this last step contains no information about A. As there is no information about A, and in particular it lacks a non-reputable signature, this is an anonymous transaction.

Nyms, introduced by Chaum (1985), are another direction in anonymous transactions, wherein a user-object maintains a suite of key pairs, and uses one pair for each organization or object communicated with, e.g. A uses key pair A_b when communicating with B, and A_c with C. The goal of nyms is to have these key pairs in no way relatable to each other, except with explicit action by A. Combining this with the commutative encryption algorithms described above, each user-object is capable of communicating with any object anonymously and is able to transfer information between objects via the blind signatures.

8.4 Access Control

Before determining the identity of a user-object, and giving any sort of access to the system, the system needs to classify what objects each user-object can access. According to Lampson (1971) and Graham and Denning (1972), the structure of the access control state-machine is defined by the 3-tuple (S, O, M), with S and O the sets of subjects and user-objects (a subset of S) respectively, and M the access matrix with one row for each subject and one column for each object. This tuple provides us with a set of cells such that element $M[s, o]$ is the set of access rights on subject s by object o (McLean 1994). This 3-tuple for a given system is often referred to as an *access control list* or *ACL*.

ACLs vary depending on the needs and function of the system. The most common rights available to be assigned include read, create, modify, delete, execute, and assign. These access types are often grouped and named for convenience (e.g. full control, read + execute, etc.), and in some systems there is also an explicit no-access right. Many systems also have mappings of groups of user-objects, one or more of which a user-object can belong to. Usually, user-object rights override group rights providing ease of configuration, without losing granularity of control.

In our generic system with ACL (S, O, M), the ACL itself is often a member of S. This being the case, it is critical to safeguard access to the ACL. Often ACLs are constructed in such a way that a user-object $o \in O$ has ownership, or responsibility delegated by the system for assigning control, over subject $s \in S$, i.e. has change, add, modify, and delete rights for M_o, the subset of M for which the object o is the owner.

8.4.1 IP network permissions

ACLs are fairly straightforward when modeling access to files on a computer. A user-object either has the right to create a new file or does not, a user-object either has the right to write to an existing file or does not, etc. Extending this model of ACL to communication networks helps us to use pre-existing models, ideas, and principles of data security and access control taken to the new medium.

One of the great difficulties in extending this model is the added layer of indirection of access to the user-object caused by communication with, and reliance upon, resources on foreign systems. We now have a user-object on a foreign machine attempting to access a local resource over a possibly insecure transmission medium. Despite this difficulty, identical principles are utilized to dictate access. This leaves us with a number of methods to filter network traffic on its journey to other systems.

The first method is not to control access at all. This is generally frowned upon in secure system design.

The second method is to do no access control at the protocol level, and to push the authentication and access control up the stack to the application. Public Web servers are a good example of this method as most do not restrict the hosts that can request messages from the server, but may have partitions of data accessible to clients, employees, volunteers, etc., which require a password or other sort of authenticator for access to the particular subset of the resource.

The third method is a decision to trust certain computers by their identifying address. In the case of IP networks, the IP address or other information available in the IP and higher-level protocol headers, e.g. UDP (User Datagram Protocol) and TCP (Transmission Control Protocol) ports, ICMP (Internet Control Message Protocol) type, etc. This is one primitive commonly used by all firewalls.

The goal of a firewall is to act as a chokepoint at the edge of a network to restrict access to resources available across that network. They are primarily used as a convenience against the high complexity of configuring security for each network system within the protected area, and are thus an example of the style of protection known as perimeter security.

These identifying data can unfortunately be fabricated easily by an adversary, which is known as *spoofing*. As a consequence, this type of access control is only useful as one step in protecting publicly available resources, where authentication is done at a different protocol level, or in conjunction with strong application level security methods.

A fourth method, often used in conjunction with one or more of the above, is to use cryptographically safe communication channels at either higher levels than TCP, such as SSL (Secure Socket Level) or TLS (Transport Layer Security, Dierks and Allen 1999), or lower with IP Security (IPSec, Thayer *et al.* 1998). There is a great trade-off in security with each protocol.

When coupled with a firewall, IPSec is used when authentication of the source is more important than the final destination application of the traffic. IPSec has two basic modes of operation, AH (Authentication Header) which signs but does not encrypt the IP packet data, and ESP (Encapsulated Security Payload) that similarly handles authentication as well as preventing the data from being viewed by an adversary with encryption. As IPSec changes the IP traffic, the traffic no longer conforms to the destination address-port tuple filters (and proxies) used by the firewall to make a determination about the traffic's access. Unless the firewall is the endpoint of the IPSec traffic, it will not be able to utilize those filters, and IPSec is thus generally not compatible with many of the commercially available firewalls, except to the extent of allowing or denying all IPSec traffic from certain addresses. For this reason, when IPSec is used in VPNs (Virtual Private Networks) and remote access, network architectures often place IPSec concentrators outside of the firewalls, or in protected networks, where the traffic will be unwrapped from the IPSec headers and then passed on to the firewall for analysis.

Operating at a higher level, SSL/TLS is able to work both with port-based filters and proxies found in firewalls, as well as incorporating encryption to secure the data transmitted. While incoming filtering application proxies are not nearly as effective on the encrypted data, the fact that they work at all is a plus. SSL/TLS is also not able to sign the IP header information that IPSec is capable of. This means that SSL/TLS is susceptible to source spoofing attacks, even though the application level data, contained in the IP payload, is protected from modification, or at least more easily detected through the use of checksums in the encryption. For this reason, SSL/TLS is much more susceptible to replay attacks. SSL/TLS seeks to mitigate this exposure to spoofing and data injection attacks through the utilization of client and server certificates, which assist in the authentication to each other of both ends of the communication traffic.

8.4.2 Symmetric key encryption

Because most security in systems is governed solely by ACLs, encryption is one of the only available methods of controlling read access to a data source on a system with a method of control external to the ACL. This protection from all user-objects within a system is most easily accomplished with static, or slowly changing resources such as files. Also existing for other resources like databases and hard drives, encryption can be used as a replacement for, or an additional level on top of, other methods of access control.

Of all methods available for access control, symmetric keys offer the smallest breadth of access control, in which a user-object either possesses the correct key or does not. And on its own, exclusive of exterior methods of control such as alternate ACL levels, external digital signatures, and revision control, possession of the key allows the user-object complete control over the document without recording the user-object agent of change or creation.

8.4.3 Public key encryption

On its own, public key encryption, like symmetric key encryption, does not provide much granularity in access control. Encrypting the file with the public key of each user-object with access to the object allows for easier revocation of access for certain user-objects through removal of their keys from the object. This may require re-encryption, but does not suffer the operational problems of securely retransmitting a new symmetric key to all of the authorized user-objects. It also provides the means of specifying authorship through the identity of the owner of the signing key, and thus if another member of the group with access to the file changes it, a change is immediately recognizable as the signing key has changed.

8.5 Security in Practice – Digital Cash

The sampling of current implementations of electronic currency is wide and varied, each pursuing different goals. As more people become active online, as more secure designs in communications systems are created, and diverse commerce applications are implemented across them, it is a simple step to see that more commerce will be transacted electronically. This section will discuss different metaphors, design principles, and environments in which current and evolving digital payment methods have been employed and are operating.

8.5.1 Design requirements

The goal of digital cash is to duplicate a subset of the functionality of the simplicity and powerful exchange facilitator we call cash. In real-world practice, cash has a number of broad functionalities that we try to extend to the digital realm. Digital cash should be:

- verifiable;
- anonymous;
- transferable;
- divisible.

By transferring this idea into the online (and offline) digital medium, we are presented with a number of challenges. Paper cash is anonymous in that it has no identity tied to it. If one makes a cash purchase in a store, the store employees may be able to provide your physical description, or show recordings from a surveillance video recorder, but your identity is not tied to your payment. As many digital cash implementations are anonymous, this presents the challenge of ensuring that the packet of bits received is indeed legitimate.

Paper cash is verifiable through its commonplace handling. We handle it regularly, and the central sources of production make it uniform and easier to recognize. Most people do not keep track of the serial numbers of all the notes they have ever possessed, and coins do not have them. This is where digital cash diverges.

As packets of bits are far easier to duplicate than the singular presentation of the special blend of papers, inks, colors, and security devices used in paper monies, digital cash has to rely on other methods for identifying uniqueness, primarily through a type of serial number. This unique identifier, in combination with digital signatures, allows participants in a transaction to ask a bank if a particular digital-cash, or *coin*, is both legitimate, and not already spent and accounted for.

Another functionality of cash oft sought to be replicated in digital cash is transferability. If Alice gives a coin to Bob, Bob should be able to give that coin to Charlie without any action or knowledge on the part of Alice. Charlie should also have no way of knowing that the coin was previously in the possession of Alice. Transferability is useful in both reducing the complexity and scaling issues of having to communicate with a bank for every interaction, as well as permitting cases where communication with a bank is not regularly available, e.g. offline.

Exact change is probably no more common in the digital world than in the real one. Many implementations use coins of pre-set denomination, and transactions may require change. In such cases, the transaction can either return one or more coins equaling the correct amount, or in some cases, split the coin into one or more pieces, and return the extra portion(s) to the purchaser. This is referred to as divisibility.

8.5.2 Purchasing protocol

The general abstraction of a purchasing scenario is as follows:

- Alice, *A*, finds a widget to purchase from Bob, *B*, of value one coin.
- *A* sends her bank *C*, a request for one coin.
- *C* creates one coin, signs it, and returns it to *A* after deducting it from *A*'s account.
- *A* performs a function *f* on the coin, and then sends the *f(coin)* to *B*.
- *B* verifies the validity of the coin using *C*'s public function.
- *B* checks the coin against the universal list to ensure the coin has not been previously spent, and conducts the sale.
- *B* sends the coin to his bank *D*, for deposit.
- *D* verifies the validity of the coin.
- *D* checks the coin against the universal list, appends it to the universal list of coins redeemed, and finally credits *B*'s account.

The precise action of these steps varies by implementation, but this is an accurate overview. Some implementations utilize digital *wallets*, which hold previously created, but as of yet unused coins from a user's *bank* for later use – a useful step in conducting transactions offline. In some implementations, *A* is anonymous to the seller *B*, and the seller's bank, *D*. In some cases *A*'s bank *C* is also anonymous to the other participants in the chain of purchase. In others, *A* is only pseudo-anonymous in that there is a trusted third party that can identify *A*, in case of intervention by authorities, or for auditing purposes. In still other implementations, *A* is not anonymous at all, and the cash scheme only verifies the coin. Among all of the implementations, the universal theme is ensuring the digital money equivalent, or *coin*, is valid. Validity, in this case, refers to being issued by a credible source, or bank, which is entitled to issue such credit on behalf of their customer.

8.5.3 Coin validity

In order to validate the legitimacy of the coin, preventing fraud and counterfeit, the participant must establish a credential with the bank through having the bank sign the coin. In the case of the schemes using blind signature schemes (Chaum 1983), this poses a problem, as the bank by definition knows nothing about the information it is signing. To get around this incongruity, banks use different signatures for different amounts, e.g. to sign a $1 coin a different signature key pair is used than for signing a $5 coin, and the signature designates the value.

Proof of purchaser

In most cases, the validation that the coin is from the account of the same user-object as the one requesting a transaction is not important. The seller receiving the coin is not concerned with the originator of the coin, only its validity. If the seller does require the purchaser to be the same as the participant paying for the object, there can be an additional initial transaction between the seller, *B*, and the purchaser, *A*, in which *B* gives *A* data to be signed by the bank. This compromises the complete anonymity of *A*, as *B* and the bank can collude to determine the identity of the purchaser, but if they do not, the information can be constructed in such a way that the bank does not learn any more information about the purchase. This is also no guarantee that the purchaser is actually the one paying for the object, but a best practice in ensuring the purchaser is at least in communication with the payer.

Proof of issuer

Utilizing the blind signature scheme discussed above, the purchaser is able to have their bank sign the coin without exposing information about the purchase. Using a group blind signature scheme (Ramzan 1999) we can create an architecture that can scale to support multiple banks, wherein only the purchaser–bank and seller–bank relationships need to be known, and only to the participants directly involved. This architecture requires new overarching entities to be introduced with responsibility to verify the banks and manage the signature keys, but is otherwise very similar. This also fits very well into the political environment of non-digital banking, where banks are overseen by governmental regulators.

Single-use coins

Currently, most coin designs are for single use. A coin exchanges hands, is redeemed, and is added to a universal list of coins redeemed. Because copying bits is trivial, this universal list is used to prevent re-use of coins. As coins are user-object independent by the blind signature scheme, the first redemption of the coin triggers the update, and unless added to that list, nothing prevents either the initial purchaser, or the seller, or anyone else in the chain from re-using it. New coins are created for new exchanges, and this overhead is linear in scaling. Divisibility and transferability, discussed below, greatly complicate this method.

8.5.4 Anonymity

Cash-in-hand transactions provide a sense of anonymity. Often this anonymity is false due to other conditions of the environment such as witnesses and close circuit cameras, but there is no personal information in the medium of the transaction itself. This ideal of removing any information about the purchaser from the actual purchase process, and only verifying the coin itself is pursued to many levels in digital cash.

Purchaser and seller

In any non-brokered transaction, at least one of the two principals has to be well known so that the other can contact them, thus limiting the possibility for all around anonymity. The principal form of anonymity in digital cash is that of the potential purchaser from the perspective of the seller, and it is this scenario Chaum had in mind when he came up with the blind signature protocol. The reverse, anonymity of the seller, is simpler technically, but harder to ensure from an information analysis perspective if portions of the chain can be observed.

For a seller to remain anonymous the actual transaction is simple – there is no information about the identity of the seller exchanged in the transaction; the object sold is picked up at an anonymous drop-off. The difficulty in maintaining complete anonymity arises when the seller communicates the anonymous coin to the bank, as they have to provide an account in which to deposit.

Issuer and seller

In some cases, it is desirable for the bank issuing the coin to the purchaser to also remain anonymous to the seller. Lysyanskaya and Ramzan (1998) and Ramzan (1999) explore this idea by extending the blind signature principle to multiple governed coin issuers: group blind signatures.

Most systems rely on a single authority to issue and verify coins. Ramzan's thesis promotes the unique authority from the level of the issuers to a central authority for verifying the issuers themselves. This extended abstraction to Group Digital Signatures (Ateniese and Tsudik 1999) permits much greater extensibility in the scale of support for service customers wherein the need for a single central bank is eliminated.

8.5.5 Transferability

At each node the coin passes through, there is the possibility that a node will keep a copy of the coin after the transaction is complete. As all bits are identical, and there is no way to tell which is the 'real' coin, most systems implement a universal list of coins that have been used. This eliminates the possibility of double spending, even with existing copies of coins.

To both incorporate this prevention of double spending and allowing coins to be traded between user-objects while maintaining the anonymity of the coin is highly complex and causes the coin to grow in size as unique information is attached to each transaction (Chaum and Pedersen 1992). This adds the artificial limitation of coin size, and violates the absolute anonymity of a coin, as, should a coin return to a user-object through circulation, the user-object would be able to recognize that coin through its chain of signatures. Because of these limitations and added complexity, transferability is often avoided, and only dealt with extensively in systems that are used in implementations necessitating offline (no immediate access to the universal list) functionality.

8.5.6 Divisibility

In the world of digital cash, divisibility is receiving change from a transaction; it is making sure that coins of value X and Y can be either procured anew from the bank in redemption of, or through splitting, a coin of value $X+Y$. In a completely online scenario where coins are single user and not transferable, divisibility matters little; the bank can provide change. Offline scenarios are more complex.

In the case where the sales agent needs to issue change to a purchaser, and cannot contact the bank to do so immediately, as in the metaphorical case of from a cash drawer, the agent has to possess sufficient, and exact, change (from new coins) to provide to the purchaser. Unfortunately, this does violate the anonymity so cautiously devised previously by identifying the purchaser as the customer in a specific transaction. The customers can be tracked through records of the coins issued as change. Jakobssen and Yung (1996) present an alternative wherein the coin itself can be subdivided, albeit in a fashion partially limiting anonymity in that the portions of the original coin are linkable to each other.

8.5.7 Revocability

As the evolution of digital cash systems continues, researchers explore technical limitations on previous systems imposed by legal and social concerns. Jakobssen and Yung (1996) take on the legal necessity of revocable anonymity, both of coins and of the identity of the spender. They introduce a third party of unassailable integrity, their ombudsman, who signs withdrawals of coins, but not the coins themselves. The bank records the signature, and the session or withdrawal ID associated with it. Upon collaboration between the two, such as in a case of a court order for a case prosecuting blackmail, money laundering, or theft; the two can determine a list of all coins from the same withdrawal as the one presented, as well as the identity of the originator of the coin.

8.6 Future

In just over the past decade since the NSF lifted the commercial ban on the Internet, we have seen an almost urgent growth in the field of communications. Hardware is becoming a commodity, and broadband users routinely have routers, switches, firewalls, and LANs at home – elements not long ago relegated to only companies with a full-time IT department dedicated to supporting those elements. With this commoditization comes strong customer demand for ease of use and breadth of performance – people who cannot program the clock on their VCR now use computers daily. The IT departments will continue to be responsible for more services all the time, and depend on consistent ease of use to support their business roles.

Much of this ease of use is a necessity for systems and their support mechanisms to 'just work.' This means more robust designs that have to be simple to set up. As networks grow larger, more connected, and more systemically and geographically diverse, secure communications become a necessity just to support them. Internal systems are now, and will continue to grow, more connected to the Internet, potentially exposing critical business systems and information to everyone with a computer and an Internet connection. And the use of biometrics will increase in installation footprints due to their simple use for authentication.

Because of the relative anonymity of the Internet and speed at which computer transactions take place, auditing will no longer have the luxury of taking place after the fact. Auditing will be integral to the access control mechanisms in place and will interact and respond to the services parceling the information. The access control mechanisms will grow from resource to content- and context-sensitive as information itself starts to have control mechanisms distinct from its containers, especially as the boundaries between computing spaces blur.

As the customer demand for consistent and highly reliable systems rises, security will become even more important in the design requirements. Providing access control is providing robust availability to those with authorized access and keeping those without out of the system altogether.

8.7 Further Reading

Much of system and network security relies on cryptography, and we recommend Menezes *et al.* (1997), Schneier (1994), Stallings (1995) and Stinson (1995) as background for a solid understanding, while Diffie and Hellman (1976), Feller (1968), Graham and Denning (1972), Kerckhoffs (1883a, 1883b), Lampson (1971), Rivest *et al.* (1978) and RSA Laboratories (1993) provide details for those interested in the historical context of the development of the field. Digital cash principles can be pursued in more depth in Chaum (1983), Chaum (1985) and Jakobssen and Yung (1996).

9

Service Creation

Munir Cochinwala, Chris Lott, Hyong Sop Shim and John R. Wullert II

Telcordia Technologies, USA

9.1 Overview

Communication service providers are continually looking to develop new services. This desire for innovation creates a need for tools to make the process of service creation faster, easier, and more reliable. The need has been addressed in the telephone network with graphical Service Creation Environments (SCEs) that support the development of Intelligent Network services. The Internet also supports communications services, where service creation is supported with scripting tools, protocol stacks, and defined interfaces. New tools are needed to support services delivered over integrated voice and data networks. In addition, the tools must be extended to support more of the steps involved with provisioning and managing emerging services.

9.2 Introduction

9.2.1 What is service creation?

Communication service providers are always looking for new ways to generate revenue from their network infrastructure. Services are a fundamental driver of revenue, and these services evolve continuously. Each service provider must keep up with services offered by competitors in order to retain existing customers. To attract new customers, service providers strive to invent new ways for subscribers to communicate with each other. In both cases, service evolution is heavily affected by network capabilities – new features in the network infrastructure may be prerequisites or enablers for innovative services.

With all these reasons for building new services, providers need flexible, powerful, and efficient methods for delivering them. These methods are collectively referred to as *service creation*. While each service provider has a unique process in place for creating services, there are several steps that all these service creation processes are likely to include.

Service Provision – Technologies for Next Generation Communications. Edited by Kenneth J. Turner, Evan H. Magill and David J. Marples
 ISBN: 0-470-85066-3

The initial step is the definition of the service concept. This process could go from high-level storyboards and mocked-up user interfaces to the specification of detailed requirements.

Based on the defined concept, the next step is to construct the service logic. This involves the design and development of the service functionality, the specification of the required data structures, the definition of the necessary messaging between components, and possibly a mock up of the user interface. Systems referred to as Service Creation Environments tend to focus on this portion of the process.

The next step is closely related to the creation of the service logic: defining the processes for provisioning, activating, managing, and billing for the service. Provisioning and activation involve populating the defined data structures with system-related and subscriber-related data, respectively. Managing a service includes monitoring a service for faults and ensuring that it is performing adequately. Billing implies definition of the items that need to be tracked and recorded in order to bill subscribers properly for usage.

The final step in service creation is testing – ensuring that the system as built performs in a manner consistent with the defined concept. This step may have multiple substeps, depending on the level of integration necessary to support the service.

There are other steps that must be taken before a service can begin to produce revenue for a provider. These steps, which are part of the process referred to as *service introduction*, are described in detail in Section 9.6.

One factor that complicates the concept of a uniform service creation process is that the enhanced services that are the result of service creation can be at many layers (Önder 2001). Services can range from connectivity-related services at the network layer, such as VPNs (Virtual Private Networks) and bandwidth on demand, to end-user applications such as Caller ID and instant messaging. Creation of services at these different levels can require very different techniques and expertise, so the image of a single service creation process is somewhat idealized.

9.2.2 Historical perspective

Modern service creation began with the introduction of SPC (Stored Program Control) switching systems in the mid-1960s. Stored program control was the first step in transitioning service logic from being hardwired to being programmable (Telcordia 2003b). The distinguishing feature that stored program control brought was the incorporation of memory into the switch, enabling the construction of services that depended on management of state (Farley 2001).

Service creation became more clearly recognizable as a separate operation with the introduction of the IN (Intelligent Network) within the global telecommunications network in the 1980s, as described in Chapter 4. The advent of the Intelligent Network separated the logic used to implement services from the switching systems used to connect calls. The Intelligent Network thus created the necessary framework to consider services separately from switching. In addition, the ITU Recommendations for the Intelligent Network specifically describe the functionality of an SCE (Service Creation Environment, ITU 1993j). A Service Creation Environment provides a graphical user interface for developing and testing new services. The concept supports the rapid construction and functional validation of new services before deployment.

9.2.3 Simplifying service creation

The appearance of service creation in the Intelligent Network recommendations reflected the need of service providers to roll out new services quickly. This need has driven vendors of service execution platforms to develop and enhance these Service Creation Environments in an effort to simplify the process of service creation.

Graphical Service Creation Environments simplify the service creation process by allowing service designers – instead of programmers – to implement services. Rather than writing software to implement each new service, designers use an SCE to arrange building blocks that represent service-independent logic. The service-independent nature of the building blocks is critical to the success of SCEs. The blocks must be usable by multiple services, ensuring that the software code that they represent can be written once and used many times. As a specific example, when the Intelligent Network concept was introduced, the triggers (events to which a service could react) were service specific. For instance, Intelligent Network switches generated events that were customized for a particular service, such as making a query when seeing one specific area code in a dialed number. These service-specific events eliminate any chance of software re-use. One of the key features in AIN (Advanced Intelligent Network) was the standardization of these building blocks, ensuring that multiple services could use the same program code.

The use of graphical interfaces and re-usable building blocks is not limited to services in public telecommunications networks. Internet communications services have also benefited from the abstraction of specific functionality. The use of object-oriented methodologies, in languages such as C++ and Java, and standardized object interfaces, such as Java Beans and COM objects, has enabled something approaching a building-block approach to development of Internet communications. The emergence of Web services promises to extend this model.

9.2.4 Service creation in multiple networks

The following sections discuss how the service creation process is implemented for several different types of network. These discussions cover telephone services in the Public Switched Telephone Network (Section 9.3), data communications services using the Internet (Section 9.4), and service creation in the emerging integrated networks (Section 9.5). These discussions focus on the logic definition step and tools that support it. This is where the unique aspects of these different network technologies make the largest difference. Section 9.6 presents a discussion of the more complete service introduction process, and Section 9.7 concludes the chapter.

9.3 Services in the Public Switched Telephone Network

The PSTN (Public Switched Telephone Network) offers many services to wired and wireless telephone subscribers. This section discusses the network architecture that makes these services possible, and describes the Service Creation Environments used to create these services.

9.3.1 Example services

The word ‘service’ can mean different things to different people. One way to illustrate the range of concepts that represent services in the PSTN is by example. Examples of services offered by modern intelligent telephony networks include:

- Basic translation services such as 800, 900, and Call Forwarding. These services work by translating the number dialed by the caller into a different number, possibly with special billing arrangements.
- Routing services (a variation on basic translation services) such as Advanced 800 and Personal Number. These services direct calls to different destinations based on the time of day, day of week, holiday, or caller's geographic location. In wireless networks, the HLR (Home Location Register) service transparently routes incoming calls to the wireless base station closest to the subscriber.
- Charging services such as Reverse Charging, and Automatic Credit/Calling Card. These services allow people to make calls from any telephone without relying on coins, etc.
- Privacy and security features such as Caller ID, Call Intercept, Call Barring, or Closed User Group. These services help subscribers receive desired calls and avoid unwanted calls.
- Multiple call and queuing services such as Call Waiting, Automatic Callback (outgoing call busy), Automatic Ringback (incoming call missed), and Multiparty. These services make it easier for subscribers to communicate with each other.

These telephony services are built on top of the basic capabilities provided by end office switches/exchanges such as providing dial tone, gathering dialed digits, and establishing circuits to carry voice traffic.

9.3.2 Intelligent Network architecture

In order to understand the concept of service creation in the PSTN, it is necessary to understand how and where PSTN services are executed. As described in the AIN overview (Chapter 4), the architecture of modern Intelligent Networks involves a switching system, a signal transfer point, and a Service Control Point (SCP). The SCP is a network element that supports or controls call processing in end office (exchange) switches and thereby centralizes the 'intelligence' of an Intelligent Network. A single SCP can be queried by many end office switches/exchanges. An SCP has data and possibly logic for all of the services offered to subscribers via the switches. A Service Control Point implements all of the services discussed above, either partially (providing only data to a querying switch) or completely (directing all call processing at the switch).

Basic SCP functionality

Switches query SCPs for assistance with processing calls. Early SCPs, such as those that implemented 800 number translation services in the U.S., stored only subscriber data. Each switch implemented the translation service logic separately, querying the SCP for required data. Because the SCP controlled the data but not the execution of the service, the SCP could not instruct the switches that a prefix such as '866' should also be translated and treated as a toll-free call. That change in the service had to be implemented at the switch, possibly requiring assistance from the switch vendor.

As network operators upgraded their switches to support protocols such as AIN 0.1, it became possible to store both service logic and subscriber data on SCPs. In other words, the network operator could provision data and service logic directly on the SCP; as a result, execution of services moved out of the switches and into the SCP. With service logic available on the SCP, it became possible for a network operator to implement identical services in many switches without relying on the switch vendors. For example, a network operator could implement a feature across the network to forward calls if the called party's line was busy. By removing the dependency on switch vendors, network operators could introduce new services very rapidly when compared to previous roll out rates. This capability drove the need for environments to ease the creation of these services.

9.3.3 Telephony creation environments

Modern Service Creation Environments enable telephone company personnel to perform the following activities rapidly and easily:

- design new telephone services (service creation);
- test a service design's functionality (service validation/testing);
- enable and configure designed services for customer use (service provisioning and activation);
- verify that live services are functioning as expected in the network (service assurance).

SCP vendors offer graphical Service Creation Environments to support all of these activities. A Service Creation Environment is a computer application that allows a network operator to build service logic. Service logic is the term used for the special-purpose applications that control the functionality of the telephone network. Instead of requiring users to write code in a low-level programming language, Service Creation Environments provide high-level means to access generic network functions. This high-level programming model provides operators a high degree of flexibility and makes it relatively straightforward for a network expert to create new services.

Many companies include Service Creation Environments with their service platforms. For example, Lucent offers eSAE (Enhanced Services Authoring Environment), Telsis offers the Ocean fastSCE and both Hughes Software Systems and Ericsson offer products called the Service Creation Environment.

Another example of such a Service Creation Environment is Telcordia SPACE ® (Service Provisioning and Creation System, Telcordia 2003a). The SPACE system is a tool for designing, creating, testing, validating, and activating services for customers. We will use the SPACE system to provide concrete details about the features available in modern Service Creation Environments.

Service creation

In the SPACE system, instead of writing statements in a textual programming language or entering data in tables to create services, users build graphs that represent the call processing logic. The graph is a flow chart and each node in the graph can perform a generic network action (e.g. send a signal) or affect the flow of control in the graph (e.g. an if–then statement). A set of call-processing graphs in the Telcordia SPACE system is called a CPR (Call Processing Record).

Figure 9.1 shows an example CPR. The user creates the graph in this CPR by inserting nodes and connecting them to each other. One of the nodes, an example of a SIB (Service Independent Building Block), is labeled *AnalyzeRoute*. A SIB is an elemental service component that serves to isolate a service capability from the underlying technology. When the

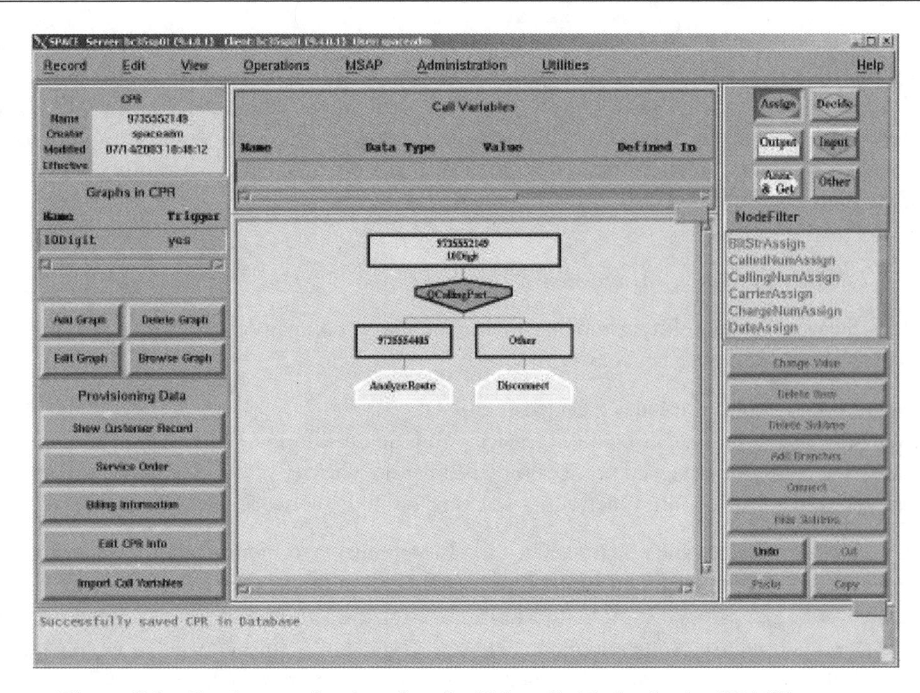

Figure 9.1 Service creation interface for Telcordia Technologies SPACE system.

service is executed, the *AnalyzeRoute* node causes an AIN output message to be sent to a switch, instructing the switch to process the call normally with no further need to contact the SCP. The *AnalyzeRoute* node is part of a simple graph that directs processing of calls that terminate at a particular phone number. The graph that connects the SIBs, representing the interactions between the SIBs, is often referred to as the Global Service Logic. The Global Service Logic specifies the service flow by defining the entry and exit points, the order of operations and the related service and instance data. Although a simple service can be implemented with a single graph like the one shown here, a production-grade service will generally involve multiple graphs.

Modern SCEs provide a clear separation between service logic and service data. Such separation, a basic feature of modern software design methodologies, allows multiple subscribers to share the same service logic. The Telcordia SPACE environment separates logic and data through a mechanism called *template CPRs*. These records contain subscriber information, or the specific data needed for service implementation. An example of such subscriber data might be the telephone number to which all calls should be forwarded if the phone rings during the night. The execution of the service logic for a particular service begins with the retrieval of the template CPR to gather the subscriber data. Execution then proceeds to generic service logic, for example the call-forward-at-night service, which uses the subscriber data appropriately. While service logic and subscriber data are logically separated, both are stored in a single database to ease provisioning, administration, and service assurance.

Service validation

Service validation refers to capabilities that allow users to test their services prior to deploying them. Often this involves simulation or emulation without a complete set of network equipment. In the SPACE system, users can initiate a simulated call by choosing a specific service definition, represented by a call processing record, and entering all the parameters (data) that arrive with the triggering call event processed by the graph. The GUI then steps through all of the logic building blocks in the service, giving a visual display of the path taken through those blocks. If additional input messages are required during call processing, the user enters data associated with each as the processing flow reaches the appropriate graph node. Such a graphical validation feature is essentially a visual debugger for the service creation system.

Service provisioning and activation

As mentioned above, template CPRs in the SPACE system contain subscriber-specific information. Activating a service for a subscriber involves associating a template CPR with some call model trigger for that subscriber's telephone number. For example, a call blocking service may use template CPRs associated with the AIN '10-digit' trigger; data from the template-based CPR will describe the hours when calls should be blocked, etc. In normal operation, flow-through activation is supported by Service Management systems. Generally, an order processing system will direct a Service Management system to activate a service for a particular subscriber. The Service Management system checks with data repositories to verify that the step is allowed (e.g. prevent against changes not ordered by the customer) and eventually sends an order to an SCP to create records as needed.

Service assurance

To verify that live services are performing as expected, operators must watch the traffic that they process. The SPACE system allows users to insert data-collection nodes in call processing graphs. These nodes instruct service assurance systems to collect data from a live service. Analyses and reports based on those data, some of which can be performed from the service creation system, form the basis for service assurance.

9.3.4 Integrated services

While the telephone network existed as a separate entity for many years, more recently we see the beginnings of integration between the telephone network and the Internet. While truly integrated services, as described in Section 9.5, are still to come, service providers are already offering services that cross this former boundary. The SMS (Short Message Service) that is popular with many cellular telephone users routes data traffic using the PSTN signaling capabilities. The tools for service creation are evolving to support this integration and to reduce further the amount of independent software that must be developed to create new services.

The emergence of these integrated services is a hint that the PSTN will not live forever as an isolated network. The challenge going forward is to provide adequate service integration with this ubiquitous network in the near term and the extension of its reliability and ease of use in the longer term.

9.4 Internet-Based Data and Communication Services

While Intelligent Network capabilities added flexibility to the PSTN, the Internet was designed from the beginning to support a wide variety of services. The Internet achieves this goal by deliberately separating the service intelligence from the data transport role of the network. The Internet has existed for decades since the 1960s. However, the idea of using the Internet as a general-purpose service platform for the 'mass market' became viable only in the mid-1990s, when a special overlay network, called the World Wide Web or simply the Web (Berners-Lee 1994, Fielding *et al.* 1999), and its client application, called the Web browser, hit the scene. In this section, we describe existing and emerging technologies for creating and providing services on the Web and Internet.

9.4.1 *Service architecture*

Many Internet-based services employ a *client-server architecture* (Orfali and Harkey 1998). A client is a software/hardware entity that provides the means (often graphical) for end-users to subscribe to and access their services. A server is a software/hardware entity that provides a set of (pre-defined) functions to connected clients (thus to end-users). In a typical usage scenario, a client sends a *request* to a server, which performs a set of operations and returns results to the requesting client in a *response*. Typically, a large-scale service may operate with a large number of servers and support different types of client application, each of which is designed to run on a particular end-user device, e.g. desktop PCs, Personal Digital Assistants (PDAs), and mobile phones.

Note that this distinction between clients and servers only applies to services, not to the Internet. On the Internet network, both servers and clients are networked hosts (devices or endpoints) with specific addresses. These addresses are used in transporting data packets from a source host to a destination host, in a process sometimes called *routing*. For all practical purposes, routing can be viewed as the only service provided by the Internet, with individual service providers using this service in the creation and operation of their own, value-added services.

Thus there exists an inherent separation between the routing capabilities of the Internet and the services that operate using the Internet. One indication of this separation is that the Internet does not look at the service-related contents of data packets being routed, while servers and clients do not need to know how their requests and responses (in the form of data packets) are transported. Such a decoupling does not exist in the PSTN, where network elements themselves, such as SSPs (Service Switching Points), not only route telephone calls but play the role of servers.

9.4.2 *The World Wide Web as a service platform*

Prior to the development and proliferation of the Web and Web browsers, service providers had to develop and maintain resources for their services *end-to-end*. While this process is still in use today, many applications have been able to leverage the Web infrastructure to simplify the service creation process. The elements of end-to-end service creation include:

- *Custom client applications*, which run on end-user hosts and enable end-users to submit service requests and receive and view responses.

- *Service protocols*, which specify how service functionalities are delivered to end-users by describing in detail the names, formats, contents, and exchange sequences of request and response messages between clients and servers.
- *Service logic*, which specifies how servers should process service requests from clients, which may, in turn, involve additional service protocols needed for sending requests and receiving responses from other services (potentially of different providers).

Such end-to-end service development is a result of the limited (albeit general) capabilities of the Internet and often results in end-users having to learn how to use new applications when subscribing to new services. Furthermore, it is often complicated, if not impossible, to re-use the client applications and service protocols of one service for other services. Thus, developing new services often required developing new client applications and service protocols, which was wasteful of development resources and presented difficult management issues for service providers. Ideally, for a new service, a service provider should be able to focus only on development of its functionalities and service logic, while knowing that the service would work with existing client applications and service protocols.

Since its introduction in the mid-1990s, the Web has become the standard service platform on the Internet. The Web provides a universal service protocol, in the form of HTTP (Hypertext Transfer Protocol, Fielding *et al.* 1999), and client application, in the form of the Web browser. The Web facilitates fast time-to-market for new services by enabling service providers to focus on developing new service logic and dramatically reduces the 'learning-curve' for end-users by having services present a consistent user interface.

9.4.3 Service integration

Providing services on a large scale often involves a number of constituent services interacting with each other. For example, an e-commerce service may use a security service in order to authenticate user identities and authorize user access to its resources. Furthermore, interacting services are likely to be from different providers. The process of enabling interoperation between different services is called *integration*. Creating integrated services can be hampered by miscommunications and incompatibilities among systems. Among the key technologies developed to address integration problems is the concept of *interfaces*.

An interface refers to a set of function calls or method calls exposed by a particular service component, e.g. a Web server, user directory, or database. The definition of a method call consists of its name, an ordered list of input and output parameters and their respective data types, and the data type of any return value. Thus, an interface can be viewed as a set of method signatures but not method definitions. The *server* of an interface refers to a set of objects that implement the interface methods, while the *client* of an interface refers to those objects that invoke interface methods. As long as the interface remains unchanged, its server and client can always interact with each other. Thus it is often said that an interface is a contract that strongly binds the server with the client. As such, the server should not (but unfortunately can) 'break' the contract by modifying or removing the method signatures of the interface without properly notifying the client.

Associated with interfaces is the concept of an ORB (Object Request Broker). An ORB handles message passing between the server and client of an interface, so that invoking methods on the server is syntactically similar to invoking methods on the local object. Typically, ORBs are required to run on both the client and server, and in theory, may be from different vendors, as long as they implement the same interface technology. However, mainly due to implementation incompatibilities, that promise has not been fully realized with most existing interface technologies, often forcing service providers to use the ORB of a single vendor in order to integrate their services. Fortunately, Web services provide a promising solution to this long-standing issue.

9.4.4 Web services

Web services (Champion *et al.* 2002) have recently emerged as a promising technology for providing interfaces[1]. Merging the ability of XML (Extensible Markup Language) to encode a wide variety of information in a structured manner with the ubiquity of the Web, Web services promise to provide interfaces that are loosely coupled with user components. That is, not all changes to a Web services interface would require re-compilation of user components, to the extent that the 'old' methods of the interface that are invoked by existing clients remain intact. Furthermore, clients and servers of a Web service may be built and run on different platforms, e.g. a client on a Java Virtual Machine and a server on a Windows host. Both of these features are possible in theory but have been difficult to realize in practice with many existing interface technologies, e.g. CORBA (Common Object Request Broker Architecture, OMG 2002). CORBA specifies a standardized architecture for networked application interoperability.

In this section, we give an overview of the process of creating Web services and a few critical issues to consider when creating and/or using Web services. Note that we do not give a ubiquitous 'Hello World' tutorial on Web services, of which several are available online, e.g. Armstrong *et al.* (2002). We assume that the reader is familiar with the basic concepts of distributed computing architecture and technologies and XML.

Process of creating and using Web services

The technologies most often cited when discussing Web services are UDDI (Universal Description, Discovery and Integration, UDDI 2002), SOAP (Simple Object Access Protocol, Gudgin *et al.* 2002), and WSDL (Web Services Description Language, Chinnici *et al.* 2003). UDDI is used to advertise and search for Web services of interest in public service repositories, SOAP specifies formats of messages to be exchanged in using Web services, and WSDL describes the request messages a given Web service accepts, the response messages it generates, and its location on the Internet.

The process of creating and deploying a Web service is as follows:

1. Create WSDL interfaces.
2. Publish WSDL interfaces using UDDI or other means, including manual, point-to-point notification of their availability.
3. Install on a server and wait for client requests.

[1] Web services should *not* be confused with services provided over the Web, which are often referred to as Web-based services.

Given the current state of support technologies and development tools for Web services, it is not an exaggeration to say that the process of creating WSDL interfaces is much like specifying interfaces in any other distributed computing technologies, such as CORBA and Java RMI (Remote Method Invocation, RMI 2002). For example, in Java, the same process is used to create both WSDL interfaces and RMI server interfaces. What is different is the tool (or the parameters given to the same tool) used to compile and generate end products. In the case of a WSDL interface, what is produced is an XML file, called a WSDL file, which conforms to the WSDL specification (Chinnici *et al.* 2003).

In theory, WSDL files can be published and searched for in public databases, using the UDDI specification. There are websites that are dedicated to publishing and discovering Web services, e.g. (XML-RPC 2003). However, as this part of Web services technologies is not yet as mature as the WSDL and SOAP. Use of manual or other external means of publishing and discovering Web services, e.g. the Web and email, is still prevalent in practice.

The process of using a Web service consists of the following steps:

1. Locate its WSDL file.
2. Generate client-side stubs from the WSDL file.
3. Write client application logic using the client-side stubs.
4. Have the client invoke the Web service on the server.

Again, this process is not dissimilar to writing client applications to CORBA or Java RMI server objects. If one substitutes 'CORBA interface' for 'WSDL file' in the above steps, we would have a general description of how to write CORBA clients. The same holds true for creating RMI applications.

As described earlier, the true strength of Web services is the loose coupling they allow between servers and clients. Servers and clients can be implemented on diverse platforms using different tools and products from different vendors, and a WSDL file can change without inadvertently affecting existing clients (as long as the signatures and functionalities of the methods that existing clients use do not change). Both of these 'features' of Web services are also possible, in theory, using other distributed computing technologies, but are not readily realized in practice. This is largely due to the fact that their message transport mechanisms are proprietary and that request and response messages are binary-encoded, all of which make them favor homogeneous operating environments, in terms of hardware platforms, operating systems, and middleware. In contrast, the *de facto* message transport mechanism in Web services is HTTP (Fielding *et al.* 1999). Furthermore, by design, plain SOAP request and response messages, which form the message bodies of an HTTP GET or POST request and response messages respectively, are text-encoded. This means that a given HTTP transport cannot distinguish between Web service request (response) messages and plain HTTP request (response) messages. In turn, this means that Web services can (and do) leverage the proven ability of HTTP to work in heterogeneous environments.

Interoperability – sample issues

The preceding discussion paints the promising picture of the capabilities that Web services will deliver. The lesson in this section is simple: Web services are not yet as interoperable as promised. This is not surprising given that many Web services technologies and support tools are in their infancy. As basic technologies and specifications continue to evolve, the industry is working hard to ensure interoperability of Web services tools and services of different vendors, e.g. WS-I (WSI 2003).

Compatibility between implementations of Web services is still an issue. For example, Figure 9.2 shows two different ways of defining a complex type in a WSDL file. In (a), the type *MyHeaderType* is separately defined from its use in the element definition of *MyHeader*. In (b), the type definition is embedded in the definition of the same element.

Both (a) and (b) are syntactically correct and have the same semantics. However, client-side stubs generated with some compilers cannot recognize references to child elements of *MyHeader* when format (b) is used. These same compilers generate proper stubs for format (a). Other compilers interpret both correctly. Even without knowing the reason for this discrepancy, this anecdote illustrates the level of immaturity of Web services technologies with respect to interoperability.

Another interoperability issue deals with XML, the fundamental technology of Web services. UDDI, SOAP, WSDL, and other related technologies all use XML as the basic grammar for definition of message/content structures and specification of requirements.

There are many design choices in implementing Web services, particularly in the definition of the WSDL. One factor that can be overlooked in this design process is the size of the XML document for data items to be sent between the client and the server. XML by its nature has not been optimized for size. For example, in one implementation of a Web service, 6 KB of data produced a 500 KB XML document.

The size of data exchange between clients and servers can be a critical issue in the design of a Web service. For example, large documents are not suited for use on resource-limited platforms (e.g. handheld devices) or in bandwidth-constrained environments. In addition, larger documents place a greater processing and storage load on servers. Thus, the design of the WSDL can have an impact on the platforms that can run the resulting service.

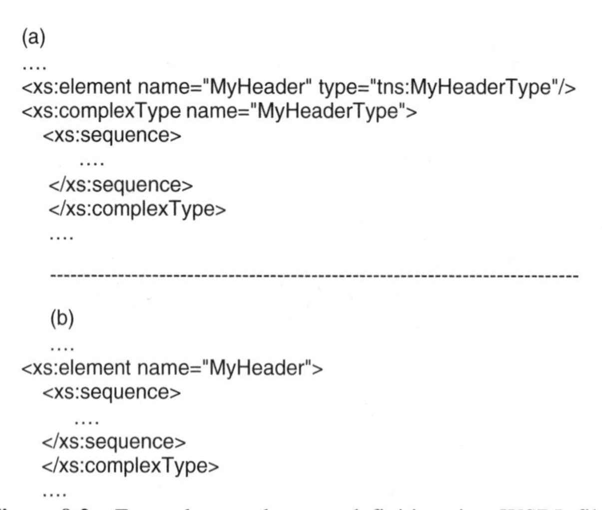

Figure 9.2 Example complex type definitions in a WSDL file.

Web services offer a promising future for Internet based services. However, they do not address the issues of interoperability that will arise with integrated networks that cross IP and PSTN boundaries.

9.5 Integrated Services

The integration and inter working of the telephone network and the Internet enables convergence of voice and data services. Moreover, the explosion of information appliances, such as PDAs and smart telephones, also provides new service opportunities. New appliances and the proliferation of mobility-enabled networks promise ubiquitous access to services independent of location. The underlying technological trend is the very rapid growth of data networks. The percentage of network bandwidth occupied by voice traffic is predicted to decline precipitously: from 40 % today to less than 1 % by 2010 (ETSI 2003d). These trends make it clear that telecommunication services must encompass all types of media: voice, video, and data should be handled in an integrated manner and, most importantly, in a way that is transparent to the end-users of services.

9.5.1 Challenges for an integrated network

The challenge for communication service providers is to deploy new, innovative services that integrate the traditional ease of use and reliability of the phone network with the vast new potential of the data network and the Web. Using the flexibility of the NGN (Next Generation Network), a new integrated multi-service communications framework can be created. Such a framework would enable service providers to bundle new and traditional services. To deliver such bundles, providers need to offer services that work seamlessly across diverse networks, devices, and providers. Providing services across a variety of environments and available resources in converged PSTN and Internet networks poses great challenges due to the different philosophies and architecture of the two networks.

Services

In the PSTN architecture, services can only execute in centralized servers within the network. This is in contrast to the open, flexible architecture of the IP world; services can reside and execute at the edge of the network. This aspect has two distinct dimensions: service execution location and service control.

- *Service execution location*: In the PSTN world, a service is a monolithic entity that resides at centralized servers and is operated only by trusted providers. In the IP world, any third party service provider can supply a service, and a service can execute anywhere in the network, based on availability of resources and user preferences. A service may also be a distributed entity that concurrently executes in multiple environments. The goal is to move towards the IP model while retaining the reliability and trust of the PSTN. Furthermore, service definition and creation should be independent of execution architectures.
- *Service control*: A telephone call in the traditional architecture has a notion of caller and called party, and a server controls the execution. In the IP world, communication is client-server or peer-to-peer, and the need for a central controlling agency is obviated. As

mentioned in the previous section, communication in the IP world can be done in a variety of customized ways, as with instant messaging, or using a set of standard protocols. Having the protocol and/or call as an integral part of the service is not conducive to service creation. In an integrated world, the network and protocol layer should be separated from the service. The goal is to create a service and have it execute on a wide variety of networks.

Resources

Another major feature of an integrated framework that distinguishes it from the traditional architecture is the mechanism for allocating resources. In the traditional architecture, resources are allocated at the beginning of a computation (call set-up) and are kept until the end of the computation. In the IP world, resources are allocated in a more dynamic fashion, in that the resources can be allocated and de-allocated while the service is active. Dynamic allocation of resources and the ability to exchange for equivalent resources should be a characteristic of an integrated environment. Dynamic allocation requires that a service specify resources by their capabilities rather than by their identities. This creates some specific requirements for the integrated environment:

- A language for resource specification is needed. The language needs to have the flexibility to define equivalent resources. For example, a phone could be a PSTN phone or an IP phone. Furthermore, the language should allow for specification of resource degradation. Resource degradation is a way to substitute a lower quality of resource than that requested. For example, a high bandwidth channel is requested for video but a lower bandwidth is deemed acceptable.
- Resource pre-emption and scheduling are needed to support substitution of resources, leasing of resources for a limited duration, or relinquishing resources to higher priority tasks, such as a 911 (emergency) service.

Security

Security in the traditional architecture is purely physical: restriction of access. In traditional computer systems, security is maintained by associating actions with principals and annotating resources with lists of principals (called access control lists). When an application tries to access a resource, the system checks that the principal responsible for the action is on the list of allowed accessors; if the principal is not on the list, the access fails. However, this makes it difficult for one party to hand access off to another or across networks, which is an essential paradigm of an integrated system. Access and/or authorization capabilities must be an integral part of service definition.

Adaptability

Implementing multiple versions of the same service to support different environments is unproductive. Ideally, services should be built once and then dynamically adapt themselves to the available resources of specific environments.

- A new application should be able to readily integrate the capabilities of other services, whether an emerging service that was just created or a telephone call.

- Communications applications should adjust dynamically to changing environments, networks, and availability of resources. For instance, while roaming, a service should always appear to be the same from the user's perspective, regardless of the current access point and the physical realization of the service, which will vary depending on the availability of resources (devices, bandwidth, access rights) in different provider environments.

One approach for creating these new applications is object-oriented techniques, which show great promise for building new classes of service that dynamically adapt to changing networks and devices.

9.5.2 Research trends

There are no tools or products that allow for creation of new services in an integrated environment. Most of the current work is in the research arena, and only addresses a subset of the above problems.

Industrial bodies and researchers are actively exploring integrated service creation and management. The TMF (TeleManagement Forum) has developed an object-based information model (SID 2002) for definition, creation, and deployment of services. A related TMF catalyst project, called PROFITS (PROFITS 2003), demonstrated rapid service introduction, including aspects of the service development required for service fulfillment, assurance, and billing. This was accomplished by use of an enhanced service model that integrated high-level services with the underlying network and related business process items.

Research communities have been developing new approaches and mechanisms that apply concepts and methodologies from diverse fields of study to service creation. For example, Luck *et al.* (2002) present a model-based service creation environment in which components are layered based on a semantic level of abstraction. Services are created as using components of the highest level of abstraction, each of which is then mapped to components of the next level of abstraction. Shen and Clemm (2002) describe a profile-based approach to service creation. Each service profile is associated with a provision profile, which specifies service and network devices to be configured and an ordered sequence of operations to be performed on these devices. Information in provision profiles is specific to access network and device technologies, whereas information in service profiles is technology-neutral.

Cochinwala *et al.* (2003) describe a framework and platform, called ACAP, which applies object-oriented principles to developing and delivering adaptive services. ACAP is based on objects and object migration. ACAP meets the goal of seamless integration of the phone network and data network, while incorporating the best that each network has to offer. The framework also meets the goal of ubiquitous access by end-users, independent of providers, networks and devices. However, the new framework requires a significant change in the infrastructure used for call (communication) set-up.

Creation and deployment of services in integrated networks are still research issues. The challenge is to define a general service platform that provides systematic support for abstracting the underlying network technologies and complexities. As services will be deployed on diverse networks, support for mapping to diverse networks and resources is also needed. The problem can generally be solved for a specific technology, service, or network; the challenge is to provide a generalized solution.

9.6 Service Introduction

Today, there is no systematic and consistent way of introducing a broad range of new higher-level services into telephone and data networks. The term 'introduction' is used to be distinct from 'creation' to describe the process here. Service creation has tended to focus on the development of the functionality or logic of services. This is one important portion of the more complete process required to actually introduce a service to the market. In addition to defining the functionality of the service, it is necessary to define other service aspects, including:

- the data needed for the service and the manner in which this data is provisioned;
- the dependencies the service has on other services and elements;
- the procedures for responding to failures and the performance requirements of the service.

These other aspects are frequently supported by operations support systems.

OSSs (Operations Support Systems) are software products designed to perform many of the back-end functions necessary for effective service delivery. Many operations support products are able to handle a pre-defined collection of services, but lack the capability to be extended to the new services produced using service creation techniques. For example, a network configuration tool might have built-in support for the configuration of virtual private networks, but be incapable of simply extending that support to bandwidth-on-demand.

An effective service creation process integrates the existing network capabilities (connectivity, routing, addressing, and security services) with the higher-level services. This service creation process extends the traditional AIN and Internet service creation processes described above. With integrated networks, a richer set of services can be created. These networks provide dynamic allocation of resources, and new services are supported on a wide variety of network servers and devices. These networks, however, also require coordination across multiple OSSs and coherent support for service activation, assurance, and billing. An integrated service introduction architecture is necessary to support effective service introduction in these networks.

9.6.1 Service support infrastructure

Several elements are required in such an integrated service introduction architecture for high-level services. An instance of such an architecture, shown in Figure 9.3, includes the following components:

- *Service APIs*: Service creation uses a set of APIs to configure both execution environments and management environments to support the service. The service execution environment implements the service logic. The management environment, including both network management and service management systems, handles the activation, assurance, and billing flows needed for each new service. The purpose of these APIs is to enable execution platform and OSS support for new services, without requiring service-specific development.
- *Service model*: The service creation process uses an object-oriented service information model to define components that can be composed to create a new service. These components provide an abstract view of the networks, resources, and services that can be synthesized.

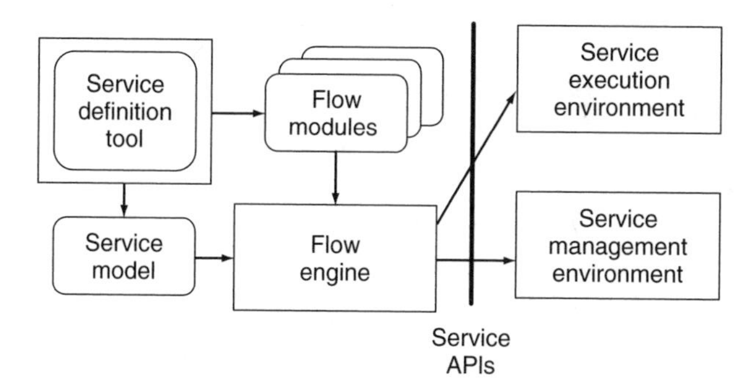

Figure 9.3 High-level architecture for service introduction.

- *Service definition tool*: A network provider uses a service definition GUI to tie together components provided by the service model to create a new service. This service definition is stored in a service catalog, and is used by a flow engine to configure OSSs and to orchestrate service fulfillment, assurance, and billing.
- *Service management*: Fulfillment, assurance, and billing capabilities integrate existing OSS functionality, and are configured by the flow engine, driven by the service definition.

As this discussion makes clear, the process of introducing services into use goes well beyond creating the service logic. Current research aimed at managing this full lifecycle of services will hopefully result in greater integration and automation to simplify the processes for creating, deploying, and managing new services.

9.7 Conclusion

The search for new communications services is a continuous quest. The need to design, develop, and deploy such services will not go away, and there will be continued pressure to make these service creation processes easier, faster, and more reliable. The integration of network technologies will increase the set of capabilities that can be applied to new services, but at the same time will increase the complexity of creating them. Computer tools and techniques currently available for building PSTN and Internet services, such as Service Creation Environments, scripting tools, and defined interfaces, will need to be enhanced to support service development in these more complicated environments.

9.8 Further Reading

This short chapter can only begin to discuss the topic of service creation.

Much can be learned about the current state of service creation in the switched telephone network by reading about the history, which is well presented by Farley (2001). For those looking for more tutorial material, the IEC Web ProForum tutorial (Telcordia 2003b) is a good starting point.

Those wanting to delve into the full details of Web services are referred to the W3C working drafts and candidate recommendations, such as Champion *et al.* (2002), Chinnici *et al.* (2003) and Gudgin *et al.* (2002). A collection of more tutorial papers on Web services for various purposes can be found in Telektronikk (2002).

10

Service Architectures

Gordon S. Blair and Geoff Coulson

Lancaster University, UK

10.1 Introduction and Motivation

The telecommunications industry is undergoing a period of rapid change with deregulation, globalization and, crucially, diversification – particularly in the area of service provision. In particular, telecommunications operators are increasingly offering a wide range of services, including multimedia services, over a diverse communications infrastructure, e.g. incorporating wireless networks. To date, such services have often been developed in a rather ad hoc manner, with little attention paid to service engineering or indeed to underlying service architectures. This chapter focuses on the latter topic, i.e. the emergence of appropriate service architectures to support the creation of next generation telecommunications services. Chapters 7, 9, and 11 discuss complementary issues of service engineering. Particular attention is given to the emergence of *middleware* platforms and technologies that underpin such service architectures.

The goals for service architectures (and associated middleware platforms) are as follows:

- To support the *rapid creation* and subsequent *rapid deployment* of new services, for example by providing a *higher level programming model* for the creation of services and also by encouraging *re-use*.
- To overcome problems of *heterogeneity* in the underlying infrastructure in terms of network and device types, system platforms and programming languages, etc., and hence to offer greater *portability* and *interoperability* of services.
- To provide more *openness* in service construction, for example to promote *maintenance* and *evolution* of often highly complex software.

The chapter is structured as follows. Section 10.2 discusses early developments in the area of service architecture, with Section 10.3 then examining existing and emerging architectures, focusing mainly on distributed object and component technologies. Emphasis is placed on the rationale for such approaches. Following this, Section 10.4 investigates the application of such technologies in the telecommunications industry. Section 10.5 discusses some challenges

Service Provision – Technologies for Next Generation Communications. Edited by Kenneth J. Turner, Evan H. Magill and David J. Marples
 ISBN: 0-470-85066-3

facing existing service architectures, including the need to support user and device mobility and also the increasing prevalence of multimedia services. A number of emerging solutions are also considered, including reflective middleware technologies and the Object Management Group's Model Driven Architecture. Finally, Section 10.6 contains some concluding remarks.

10.2 Early Developments

Given the nature of the area, it is essential that service architectures are supported by appropriate standardization, whether in terms of *de jure* or *de facto* standardization processes. This is essential to ensure portability, interoperability and openness as discussed above. The first relevant step in standardization was the creation of the Reference Model for OSI (Open System Interconnection) by ISO (International Organization for Standardization). This standard defines the now famous seven-layer model for communicating processes as illustrated in Figure 10.1.

This layered structure should not be viewed as an implementation architecture; rather it is a reference model providing concepts and terminology for reasoning about communications protocols, and also an abstract framework (cf. meta-architecture) which can be populated by a selection of appropriate technologies. For example, the TCP/IP protocol stack can be viewed as an instantiation of the lower layers of this architecture. (Note that similar comments apply to the ISO/ITU-T RM-ODP as discussed in Section 10.3.1 below.)

While OSI offers an open approach to service construction, it does not in itself constitute a service architecture for next generation telecommunication services. For example, OSI is fundamentally based on abstractions over message passing, and does not attempt to hide this from the user. In other words, the approach does not provide the higher level programming abstractions we seek. In addition, OSI is limited in scope, for example when dealing with broader service creation issues such as security and fault tolerance.

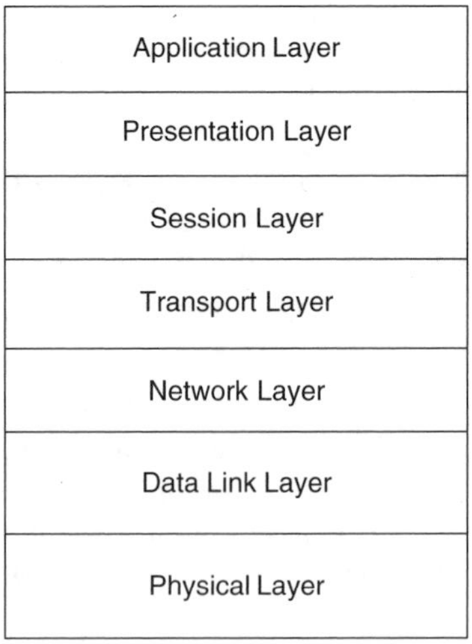

Application Layer
Presentation Layer
Session Layer
Transport Layer
Network Layer
Data Link Layer
Physical Layer

Figure 10.1 The ISO OSI seven-layer model.

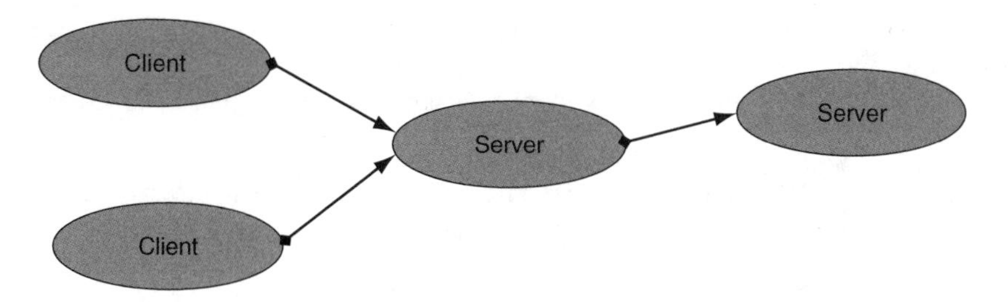

Figure 10.2 The client-server model.

To overcome such limitations, subsequent research has focused on the development of distributed systems technologies that have the goal of offering a higher level of *transparency* (i.e. hiding aspects of distribution), including support for non-functional properties like security and fault tolerance. The first approaches in this area were based on the classic client-server model as depicted in Figure 10.2.

In this model, processes can take the role of either clients (requesting a service) or servers (providing a service). Such roles can also change over time; for example, server can become a client of another server. The model is supported by an appropriate protocol supporting the associated request/reply style of interaction, most typically a *remote procedure call* protocol that makes interactions appear to the programmer to be similar to a standard procedure call in the host language.

The client-server approach has been highly influential and has led to the emergence of *middleware* technologies like the Open Group's DCE (Distributed Computing Environment, see Figure 10.3). As can be seen, this is based on an RPC (Remote Procedure Call) protocol together with a platform-independent thread service, enabling the creation of multi-threaded servers (and indeed clients). Crucially, the architecture also comprises a

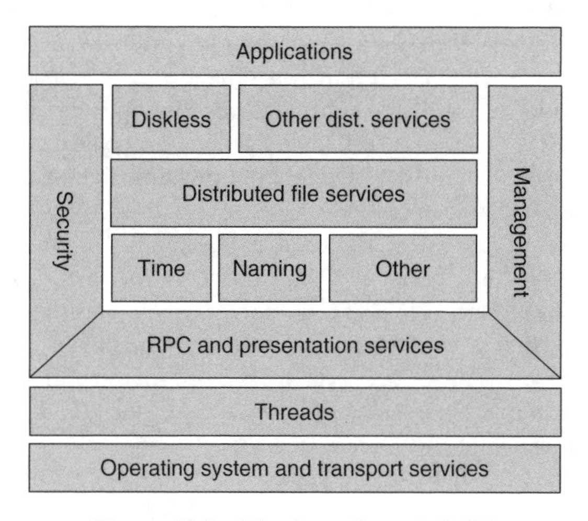

Figure 10.3 The Open Group's DCE.

series of value-added servers that can be accessed by applications offering security, persistency of data, etc.

DCE has been successfully deployed in the telecommunications industry. The approach is still however rather limited:

- it is still quite low level in terms of the programming model offered to service developers;
- there is little in the way of software engineering methodologies and tools to support the client-server approach;
- non-functional properties must be programmed explicitly, e.g. by making calls to appropriate system servers (security servers, etc.).

In addition, DCE is inflexible to deploy due to the fact that, although the various modules shown in the figure are separable *in theory*, they are tightly coupled *in practice*. Thus, for example, the RPC service cannot be isolated, because this requires the security service, which in turn requires the time service, etc.

As a result of these limitations, attention has more recently focused on more supportive service architectures based on *distributed object technologies* or *component-based software development*. We discuss these developments in some depth in Section 10.3 below.

10.3 Current Architectures

10.3.1 RM-ODP – a meta-framework for distributed computing

Introduction

RM-ODP (Reference Model for Open Distributed Processing) is a joint standardization activity by both ISO and the ITU-T. It is therefore a *de jure* standard produced following a period of international consultation to reach a technical consensus.

The aim of RM-ODP is to enable 'the development of standards that allow the benefits of distribution of information processing services to be realized in an environment of heterogeneous IT resources and multiple organizational domains' (ISO/ITU-T 1995a). It is important to stress however that RM-ODP does not itself prescribe particular standards for open distributed processing. Rather, as with OSI, it provides a framework to enable specific standards to emerge and hence should be considered a meta-standard for open distributed processing. It is therefore essential for RM-ODP to be sufficiently generic to enable a range of standards to be accommodated. The main value of RM-ODP is in providing a common set of concepts for the field of open distributed processing.

The standard consists of four parts. The heart of the standard is Foundations (Part 2, ISO/ITU-T 1995b), which defines basic modeling concepts for distributed systems. This is followed by Architecture (Part 3, ISO/ITU-T 1995c) which constrains the basic model by introducing concepts which a conformant RM-ODP system should embody. The standard also includes an Overview (Part 1, ISO/ITU-T 1995a), providing motivation and a tutorial introduction to the main concepts, and Architectural Semantics (Part 4, ISO/ITU-T 1995d) that provides a formalization of RM-ODP concepts.

Note that many of the RM-ODP ideas to be described below were initially developed in the ANSA/ISA project (van der Linden 1993). This work also produced a prototype platform, ANSAware (APM Ltd 1993), as a partial implementation of the ANSA architecture.

Major concepts in RM-ODP

An object-oriented approach. The most important contribution of RM-ODP is to develop an *object-oriented approach* to distributed computing, hence tackling the major limitations of the client-server approach outlined above. In particular, the motivations for adopting such an approach were as follows:

- To provide a higher level of abstraction for the programming of distributed systems (incorporating features like encapsulation, data abstraction, support for evolution and extensibility).
- To enable the use of object-oriented analysis and design methodologies in a distributed setting.

One major problem in object-oriented computing is the lack of an agreed terminology for the subject; for example, there are several interpretations of the term *class*, e.g. as a template, as a factory, as a type, etc. One major contribution of the RM-ODP standard is therefore to provide a precise and unambiguous *object model* tailored to the needs of open distributed processing. This model features for example definitions of terms like object, interface, role, template, factory, class, type, subclass, and subtype. A complete presentation of the object model is however beyond the scope of this chapter. The interested reader is referred to Blair and Stefani (1998) for a comprehensive treatment.

One interesting feature of the RM-ODP object model is its support for *multimedia programming*. In particular, multimedia is supported by three complementary features:

1. As well as supporting conventional operational interfaces (i.e. interfaces containing operations that can be invoked in a request/reply fashion), RM-ODP also offers *stream* and *signal* styles of interface. Stream interfaces support the production or consumption of continuous media data such as audio or video, whereas signal interfaces serve as emitters or receivers of asynchronous events.
2. RM-ODP enables the creation of *explicit bindings* between interfaces, the end result being the creation of an object that represents the communications path between the interfaces. There are three styles of such binding, i.e. operational, stream, or signal, corresponding to the styles of interface discussed above.
3. The explicit binding mechanism described above provides intrinsic support for *QoS (Quality of Service) management*. First, the QoS of a binding can be specified at creation time and appropriate steps taken to ensure that desired levels of guarantee are met (e.g. resource reservation or admission control). Second, as the binding is an object, appropriate steps can be taken to manage the binding at run-time (e.g. through monitoring and adaptation of QoS).

Again, further discussion of this aspect of the object model can be found in Blair and Stefani (1998).

The concept of viewpoints. The second important contribution of RM-ODP is its development of a *viewpoint-based methodology* for the design of distributed systems. Viewpoints are intended to deal with the inherent complexity of distributed systems by partitioning a system specification into a number of partial specifications, with each partial specification (or viewpoint) offering a complete and self-contained description of the required distributed system targeted towards a particular audience. The terminology (language) used for each description is therefore tailored towards its target audience. Viewpoints are as central to RM-ODP as the seven-layer

Table 10.1 RM-ODP viewpoints and languages

Viewpoint	What it addresses	Example modeling concepts
Enterprise	Business concerns	Contracts, agents, artifacts, roles
Information	Information, information flows and associated processes	Objects, composite objects, schemata (static, dynamic, invariant)
Computational	Logical partitioning of distributed applications	Object, interface, environmental contract
Engineering	Distributed infrastructure to support applications	Stubs, binders, protocol objects, nodes, capsules, clusters
Technology	Technology procurement and installation	Implementation, conformance points

model is to OSI. It should be stressed however that viewpoints are not layers. Rather, they are more correctly viewed as projections of the underlying system.

RM-ODP defines five viewpoints, namely the Enterprise, Information, Computational, Engineering, and Technology viewpoints. The different viewpoints have corresponding viewpoint languages, each of which shares the general object modeling concepts described above; each language can be thought of as a specialization of this general model for the particular target domain. A summary of the various viewpoints and models is given in Table 10.1.

The different viewpoints outlined above effectively create a *separation of concerns* in the specification of a distributed system. The five viewpoints collectively provide a complete specification of the system which would enable an RM-ODP compliant implementation to be developed.

The one added difficulty of introducing such viewpoints is that of ensuring consistency between the different specifications. This is a difficult problem and comprehensive solutions are currently beyond the state of the art. One approach is to consider the use of appropriate formal specification techniques for each viewpoint and then employ mathematical analysis to ascertain consistency (Bowman *et al.* 1996).

Distribution transparency and associated functions. To support a desired level of abstraction in RM-ODP, it is necessary to offer a high level of *distribution transparency*, i.e. to hide the complexities of distribution and heterogeneity from the user. The RM-ODP approach is to support *selective transparency* whereby the programmer can elect for a given level of transparency. A number of different levels of transparency are possible; for illustration, a selection of the main distribution transparencies is given in Table 10.2.

RM-ODP also defines a number of so-called *functions* that enable desired levels of transparency. Essentially, the requirements for distribution transparency are expressed in the Computational viewpoint, and the Engineering viewpoint is then responsible for meeting the desired level of transparency by employing the appropriate transparency functions. The standard also defines other supportive functions, most notably the *trading* function which acts as a broker for service offers in the distributed environment. This service is central to RM-ODP (and to the Computational viewpoint in particular) and hence the trading function merits a separate document in the standardization of RM-ODP (ISO/ITU-T 1997).

Table 10.2 Distribution transparencies in RM-ODP

Transparency	Concern	Effect
Access	Means of access to objects	To mask differences in data representation or the invocation mechanism employed
Location	The physical location of objects	To enable objects to be accessed by a logical name
Failure	The failure of an object	To mask invocation failure from the user through an appropriate recovery scheme
Migration	The movement of objects	To mask the fact that an object has moved
Relocation	The movement of objects involved in existing interactions	To mask the fact that an object has moved from current users of the object
Replication	Maintaining replicas of objects	To hide the mechanisms required to maintain consistency of replicas
Persistence	Maintaining persistency of data across interactions	To hide the mechanisms required to maintain the persistency
Transaction	Maintaining consistency of configurations of objects	To hide the mechanisms required to maintain consistency of configurations over multiple invocations

RM-ODP and TINA

As mentioned above, RM-ODP is a meta-standard that can be specialized for a number of domains. Crucially, given the goals of this chapter, there has been a significant effort in specializing the concepts of RM-ODP for the telecommunications industry, resulting in the TINA architecture (Telecommunication Information Networking Architecture, Dupuy *et al.* 1995, Chapter 11). This architecture was created by the TINA consortium (TINA-C) which featured many of the world's leading telecommunications companies, centered on a core team located at Bellcore in New Jersey. TINA is intended to provide a framework for the development of future telecommunications networks. The aim is to define an architecture to support improved interoperability, to be able to re-use both software and technical specifications and to have greater flexibility in the design and deployment of the distributed applications that comprise a telecommunications network. A further aim is to provide a path of evolution and integration from existing telecommunications architectures such as the IN (Intelligent Network, Abernethy and Munday 1995) and the TMN (Telecommunication Management Network, ITU 1991).

The overall TINA architecture is illustrated in Figure 10.4.

TINA places a DPE (Distributed Processing Environment) at the heart of its architecture. This DPE adopts the RM-ODP object model and is also organized along the lines of the RM-ODP Engineering Model. The TINA DPE provides a uniform platform for the execution and deployment of the different distributed applications that can be found in a telecommunications network. TINA has identified three broad classes of application, with associated architectural models as defined below:

1. The *network control architecture* defines a generic set of concepts for describing and controlling transport networks in terms of two frameworks. The *network resource information*

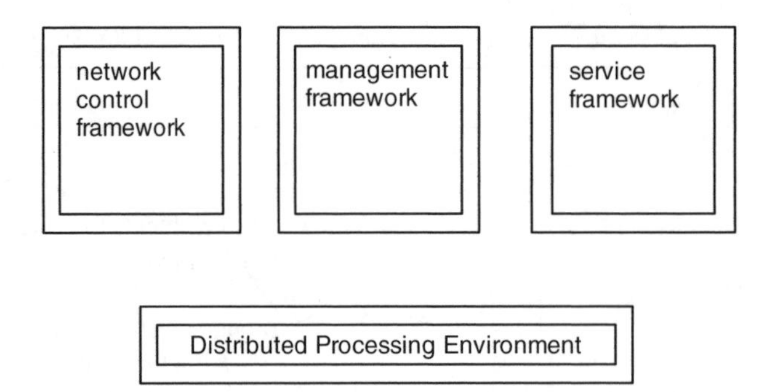

Figure 10.4 The TINA architecture.

framework is inspired by ITU-T Recommendation G.803 and defines a generic network architecture, organized around transmission independent and switching technology independent notions of network elements, how they are related (aggregations) and topologically interconnected, and how they can be configured to provide an end-to-end path. The *connection management framework* then describes generic components for the control of end-to-end communication paths in a network. This framework is organized around notions of connection graphs that describe logical and physical topologies of network connections.

2. The *service architecture* defines the principles, concepts, and basic classes necessary for the design, deployment, operation, and management of information services in a TINA environment. It includes notions such as user agents, service and communication sessions, access management, and mediation services.
3. The *management architecture* defines a set of principles for the management of the components in the distributed computing infrastructure, and in the network control and service architectures described above. The architecture covers two main areas. The first, referred to as *computing management*, is responsible for the management of computers, the DPE and related software. The second, referred to as *telecommunications management* is concerned with the management of information network services and software related to the underlying network. The latter is further divided into service, network, and element management in a similar manner to the above-mentioned TMN. Finally, the architecture partitions management into a number of *functions* including the OSI functions of fault management, configuration, accounting, performance, and security.

The breadth of the architecture should be apparent from the above description of the main components in TINA. This illustrates the level of complexity of dealing with heterogeneity in modern telecommunications environments.

A final noteworthy contribution of TINA is the definition of a role-based business model for telecommunications. This model identifies the following roles: *service provider* (provides and manages the service); *consumer* (the customer of a service); *retailer* (sells the service to consumers on behalf of service providers); *broker* (puts consumers in touch with retailers); and *communication provider* (provides the network connectivity to let it all happen). This simple model gains its power and flexibility from the fact that roles are distinguished from actual stakeholders; for example, the service provider and communication provider roles may both be played by a single telecommunications company or, equally well, by separate telecommunications content provider companies. This property makes the business model applicable in a very wide range of situations as further discussed in Section 10.4.2.

10.3.2 Specific distributed object technologies

Introduction

The work on RM-ODP has encouraged the emergence of a range of specific middleware technologies adopting the object-oriented paradigm. The most significant players in this area are listed below:

The Object Management Group's CORBA: The Common Object Request Broker Architecture is a standardization activity promoted by OMG (Object Management Group). This organization is sponsored by a number of IT and telecommunications companies with the aim of promoting an open object-oriented framework for distributed computing. The particular goal of CORBA is to provide the mechanisms by which objects can transparently interact in a distributed environment.

Microsoft's COM and DCOM: COM (Component Object Model) is functionally quite similar to CORBA but is generally restricted to Microsoft platforms. It does however have some significant differences, including support for multiple interfaces (as in RM-ODP) and also interoperability at the binary level. Distribution is supported by DCOM (Distributed Component Object Model), which in turn is based heavily on the Open Group's DCE RPC technology.

Java RMI: RMI (Remote Method Invocation) is tightly bound into the Java language and supports the transparent remote invocation of methods defined on Java objects, whether in a separate address space or on a separate machine. RMI is now supported by other facilities such as Jini and Enterprise Java Beans offering a comprehensive platform for distributed computing (albeit restricted to a single language environment).

Further details of these technologies can be found in Emmerich (2000).

In this chapter, we focus on CORBA because of its unique role in providing a language and platform independent environment for distributed objects. It is also interesting to note that for similar reasons, the TINA consortium has identified CORBA as the technology of choice to serve as a basis for the TINA DPE (see below).

Focus on CORBA

OMG and the Object Management Architecture. CORBA is supported by the OMG as part of an initiative to develop a comprehensive OMA (Object Management Architecture) for distributed object-oriented computing. The OMG is a non-profit organization sponsored by over 800 organizations including computer manufacturers, software companies, telecommunications companies, and end-users. The OMG's overall objective is to 'promote the theory and practice of object technology for the development of distributed computing systems'. The approach adopted by OMG to achieve this goal is 'to provide a common architectural

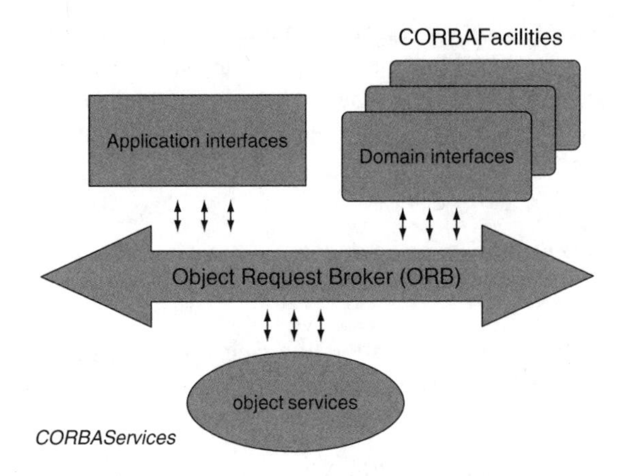

Figure 10.5 The Object Management Architecture.

framework for object-oriented applications based on widely available interface specifications... conformance to these specifications will then make it possible to develop a heterogeneous application environment across all major hardware platforms and operating systems'.

The Object Management Architecture is illustrated in Figure 10.5.

As can be seen, the ORB (Object Request Broker) has the central role in this architecture, providing the mechanisms whereby objects can transparently make requests and receive responses (we examine this in more detail below). The ORB is then supplemented by the following:

- The CORBA developer can provide an arbitrary number of application objects offering specific services through *application interfaces*.
- In doing so, they can make use of a range of object services (*CORBAServices*), including support for functions like naming, trading, transactions, and security.
- There are also a range of specific interfaces for specific application domains (*CORBAFacilities*), including areas such as financial services, healthcare and real-time systems. These interfaces are specified by associated *Domain Task Forces*. Crucially, there is a very active task force for the telecommunications domain as discussed in Section 10.4.1.

Interestingly, the original scope of the OMG has been extended to encompass design notations for distributed objects. In particular, the OMG produced a standard for the UML (Unified Modeling Language) in 1997 and continues to support extensions and refinements to this standard.

The CORBA object model and IDL. The CORBA object model defines an *object* as 'an identifiable, encapsulated entity that provides one or more services that can be requested by a client'. To access an object, clients issue a *request* for a service. Requests consist of the target object, the required operation name, zero or more actual parameters, and an optional *request context*. The request context is used to specify additional information about the interaction. This can convey information about either the client or the environment; for example, the context can be used to define a user's preferred priority or transactional information. Should a request fail, an *exception* is raised in the client object.

Objects in CORBA support a single interface described in terms of the OMG's IDL (Interface Definition Language). IDL is a language independent, declarative definition of an object's interface. It is not expected that the object implementation shares the typing model of IDL. Rather, a mapping must be provided between IDL and each host language. IDL closely resembles C++ but with added features to support distribution.

The main features of IDL are best illustrated by example. The following IDL specification defines a simple stock control system:

```
// OMG IDL definition
interface inventory
{
// attributes and type definitions
const long MAX_STRING=30;
typedef long part_num;
typedef long part_price;
typedef long part_quantity;
typedef string part_name<MAX_STRING+1>;
struct part_stock {
part_quantity max_threshold;
part_quantity min_threshold;
part_quantity actual;
};
// operations
boolean is_part_available (in part_num number);
void get_name(in part_num number, out part_name name);
void get_price(in part_num number, out part_price price);
void get_stock(in part_num number, out part_quantity quantity);
long order_part(in part_num number, inout part_quantity quantity, in
account_num account);
};
```

The `in`, `out` and `inout` annotations define whether the associated parameter carries data into the object, returns results from the object, or a combination of both. The remainder of the IDL is relatively straightforward and is not described further.

Note also that IDL supports interface inheritance whereby a derived interface can be created by inheriting the specification of one or more base interfaces. This is not shown in the above example.

The Object Request Broker. The role of the ORB (Object Request Broker) is to enable requests to be carried out in a heterogeneous distributed environment. A client can issue a request on an *object implementation* and the ORB deals with finding the object, sending the request to the appropriate object and preparing the object to receive and process the request and return the result back to the client. The ORB therefore implements a level of distribution transparency (actually location and access transparency in terms of Table 10.2).

The overall architecture of the ORB is captured in Figure 10.6.

The *ORB Core* is the lowest level in the ORB architecture. This is generally implemented as a library that is linked into both the client and the server. This library for example offers

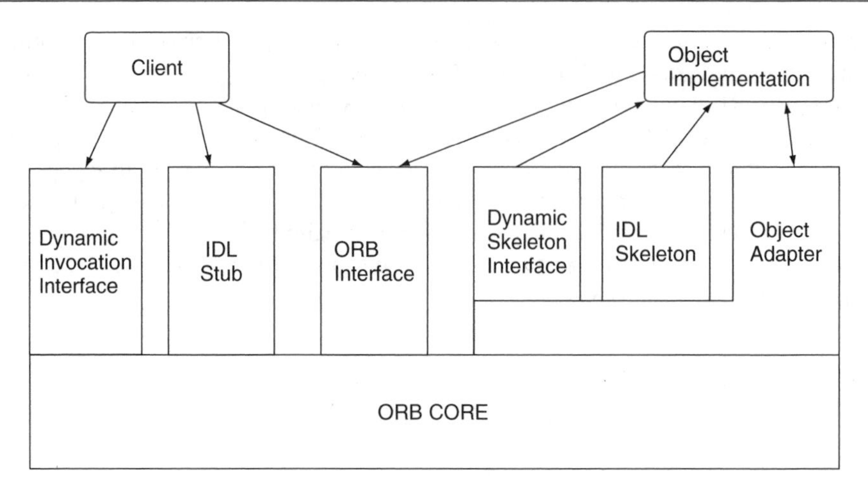

Figure 10.6 The Object Request Broker.

support for sending requests using CORBA's GIOP (General Inter-ORB Protocol).[1] An application programmer will typically not access the ORB core directly but will access the higher level interfaces indicated in the diagram.

IDL Stubs provide the mechanism for clients to be able to issue requests transparently, dealing with the *marshaling* of parameters and the *unmarshaling* of results (marshaling is the mapping of typed parameters on to a flat, programming language independent, message format; unmarshaling is the inverse process). The stubs are normally pre-compiled from the IDL definitions of the required object implementations and linked into the client's address space. Hence, the set of stubs is *static*. The *IDL Skeleton* interface offers the same service at the receiving end of the request. Skeletons identify the required procedure, unmarshal the parameters, up-call the object implementation to request that the operation be carried out and then deal with marshaling the results and returning them to the client. Again, this mechanism is static in nature.

To overcome the static nature of stubs, the client can use the *Dynamic Invocation Interface*. Using this interface, the client must manually construct a request indicating the required operation, the parameters, etc. This is obviously more difficult but the flexibility is sometimes required in applications that must operate on objects not known at compile time (e.g. debuggers and browsers). CORBA also supports an analogous Dynamic Skeleton Interface on the implementation side.

Finally, objects have access to an *Object Adapter* interface. This interface effectively provides a 'life-support environment' for CORBA objects. In earlier versions of the architecture, this was under-specified leading to a number of disparate implementations of the interface. To overcome this, the OMG have now specified a POA (Portable Object Adapter), offering fully specified support for functions like object activation and passivation, and the management of object references. To encourage flexibility, the POA is now policy driven in areas such as persistency and thread management.

[1] OMG also defines the IIOP (Internet Inter-ORB Protocol) which is an instantiation of GIOP that runs over the Internet's TCP/IP protocol stack.

10.3.3 From objects to components

Motivation

Distributed object technologies have proved to be highly successful in encouraging the uptake of distributed systems technology in a wide range of application domains (including telecommunications). However, a number of significant problems have been identified with the distributed object approach:

- little support has been provided for the third party development of objects and for their subsequent integration into applications;
- it is difficult to deploy and subsequently manage large-scale configurations of objects.

In addition, significant difficulties have been reported in dealing with the extensive range of interfaces and services offered by current platforms. For example, when dealing with CORBA, the programmer must deal with a number of complex interfaces including the POA (see Section 10.3.2 above), as well as explicitly inserting calls to the various CORBA services to obtain desired nonfunctional properties like transactions and persistence.

Recently, a number of *component technologies* have emerged in response to these difficulties. Szyperski (1998) defines a *component* as 'a unit of composition with contractually specified interfaces and explicit context dependencies only'. In addition, he states that a component 'can be deployed independently and is subject to third party composition' (Szyperski 1998). A key part of this definition is the emphasis on *composition*; component technologies rely heavily on composition rather than inheritance for the construction of applications, thus avoiding the so-called *fragile base class problem*, and the subsequent difficulties in terms of system evolution (Szyperski 1998). To support third party composition, components support explicit *contracts* in terms of *provided* and *required* interfaces. The overall aim is to reduce time to market for new services through an emphasis on programming by assembly rather than software development (cf. manufacturing vs engineering).

In terms of middleware, most emphasis has been given to *enterprise* (or *server-side*) *component technologies*, including EJB (Enterprise Java Beans) and CCM (CORBA Component Model). In such enterprise technologies, components typically execute within a *container*, which provides implicit support for distribution transparencies in terms of transactions, security, persistence, location transparency, events, and resource management. This offers an important separation of concerns in the development of server-side applications, i.e. the application programmer can focus on the development and potential re-use of components to provide the necessary application logic, and a more 'distribution-aware' developer can then provide a container with the necessary non-functional properties. As well as support for distribution, containers also provide additional functionality, including lifecycle management and component discovery.

In the following section, we describe the CORBA Component Model in more detail. Note that this is conceptually similar to the Enterprise Java Beans model, but generalized to offer full language independence. Other important component technologies include Sun's Java Beans technology, a component technology restricted to a single address space, and Microsoft's .NET, which, like Java, relies on the existence of an underlying virtual machine.

Focus on the CORBA Component Model

Overview. CCM (CORBA Component Model) is an integral part of the CORBA v3 specification. Other important extensions to the architecture include improved *Internet integration* through, for example, firewall support, a URL-based naming scheme, links to appropriate Java technologies, and also *Quality of Service support* in the form of an asynchronous messaging service and minimum, fault-tolerant and real-time specifications. CCM provides a language-independent, server-side component technology supporting the implementation, management, configuration, and deployment of future CORBA applications (Wang *et al.* 2001). There are essentially three main parts to the technology:

- an underlying component model;
- a container framework offering implicit security, transactions, persistency, and event management;
- a packaging technology for deploying binary, multilingual executables.

We look at each in turn below.

The component model. A component in CCM is defined in terms of a number of interfaces of different styles (cf. RM-ODP). In particular, the following styles are supported:

- *Facets* are interfaces that the component provides to the outside world, corresponding to 'provided' interfaces as discussed above. A given facet looks very much like a traditional CORBA interface; it can be invoked synchronously via a CORBA request or asynchronously via the asynchronous messaging that is part of CORBA v3.
- *Receptacles* are equivalent to 'required' interfaces; before a component can operate, its receptacles must be connected to facets offered by other components. This embodies an important part of the contract in terms of the usage of this component.
- The CCM also supports a publish-subscribe style of interaction through *event sources* and *sinks*. Event sources are emitters of information, which are then loosely connected to zero or more event sinks through an appropriate event channel, cf. the Observer pattern (Gamma *et al.* 1994).
- Finally, a CORBA component exposes a set of *attributes* that can be used by various configuration tools to enable declarative configuration of the component.

The above features are supported by an extended IDL that has new keywords representing the new concepts. For example, facets and receptacles are declared using the keywords `provides` and `uses` respectively. The interested reader is referred to Seigel (2001) for a treatment of this extended IDL.

Internally, a CORBA component consists of implementations of each of the facets and event sinks, together with the set of attributes. A component also exposes additional interfaces, including the *home interface*, which supports lifecycle operations such as the creation or deletion of particular instances, and the *equivalent interface*, which allows legacy software (including, interestingly, EJB software) to be presented as CORBA components.

The container framework. A CORBA component can be placed in an appropriate container that provides implicit management of that service. In particular, the container performs the following tasks on behalf of components:

- the management of resources, e.g. in terms of activation or passivation of instances;
- the setting of appropriate POA policies;
- the calling of CORBA services at appropriate points to achieve the desired non-functional properties of the component, e.g. in terms of security;
- the container also undertakes to inform the component of important events; the component can provide handlers to respond to such events.

The component/container approach represents an extremely high level of transparency in terms of distributed systems development. With this approach, the programmer can focus on the development or re-use of components, and the associated interconnection logic, and can then rely on the infrastructure to manage the complexities of distribution. This overall approach is summarized in Figure 10.7, see Wang *et al.* (2001) for further discussion.

A given container can also be configured according to key *policies*. For example, the *servant lifetime policy* can be used to establish a policy for activation or passivation of components (i.e. activate on first invocation/passivate on request completion; activate on first invocation/ passivate on transaction completion; activate on first invocation/destroy explicitly; activate on first invocation/de-activate when container needs to reclaim resources). Policies can also be established in other areas including security, transactions, and persistency.

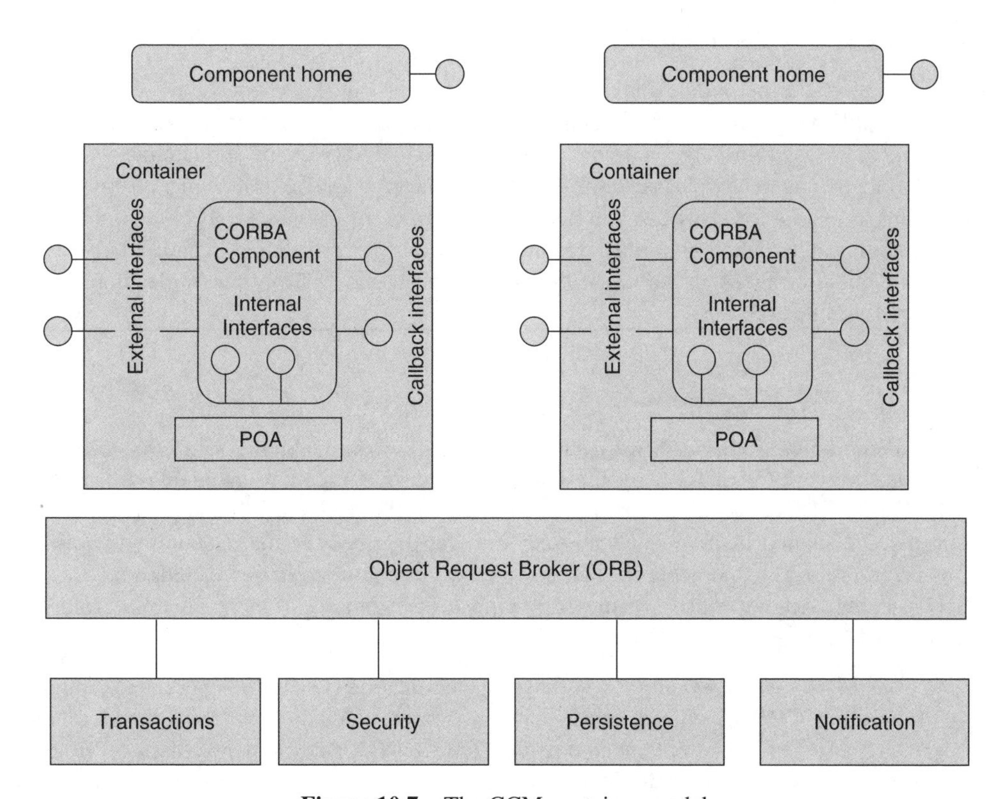

Figure 10.7 The CCM container model.

The packaging technology. Once completed, components can be packaged together with appropriate meta-data in an assembly file (CAR file). The meta-data comprises information on the component, its dependencies, and its assembly instructions. In all cases, this meta-data is captured using XML templates.

More specifically, the meta-data consists of the following key elements:

- a *software package descriptor* containing a description of the overall package in terms of, for example, the author, the implementation tools, the IDL, etc.;
- a *component descriptor* describing, for example, the required container policies;
- a *component assembly descriptor* describing a set of components and how they should be interconnected;
- a *property file descriptor* describing initial attribute values for a given component.

Some of this information is collated from a programmer supplied *CIDL* description. CIDL (Component Implementation Definition Language) is a declarative language that complements the (extended) CORBA IDL, and which describes additional properties of the eventual implementation.

This information is then sufficient for deployment tools to unwrap the component and install it in the distributed environment, thus considerably easing the task of installing distributed systems software.

10.4 Applying the Technologies

It is fair to say that the middleware technologies described in this chapter have so far been primarily applied to business-oriented computing environments. For example, they are often used to facilitate interworking in company networks, to integrate legacy applications into distributed systems, and to provide back-end Web services in areas such as accounting, e-commerce, and database integration. However, in addition to this, there has already been a significant use of middleware technology in the telecommunications industry. This was initially seen in the development of TINA in the mid-1990s as described above. More recent developments are described in this section.

10.4.1 The OMG's Telecommunications Domain Task Force

A clear example of the trend toward the use of middleware technology in telecommunications is the work carried out by the OMG's Telecommunications Domain Task Force (OMG 2001a). This task force, which dates back to 1997, seeks to influence the development of CORBA and its associated services to meet the needs of the telecommunications industry. It has been responsible for the introduction of a number of new standards into the OMG's stable. Among these are the following core standards (others are discussed in Section 10.4.2):

- The *Control and Management of A/V Streams* specification (OMG 1997) defines a framework in which CORBA objects can manage audio and video streams; the streams themselves are assumed to be implemented outside the CORBA environment. 'Manage' in this context refers to functions such as establishing connections, starting and stopping the flow of data, and adding new parties to multi-peer sessions.

- The *Notification service* (OMG 1998a) and the associated Management of the Event Domain specification (OMG 2000a) standardize publish-subscribe oriented communication. The Notification service lets consumer objects 'subscribe' to a 'topic' and thereby obtain information made available by producer objects that have previously 'published' the information under that topic. The service holds published information persistently so that it is not necessary for the publishers and subscribers to interact synchronously. This communication style is particularly useful for loosely coupled systems for which scalability is a prime concern (e.g. control and management in telecommunications networks).
- The *CORBA/TCAP* (OMG 1998b), *CORBA/TMN* (OMG 1998c) and *CORBA-FTAM/FTP Interworking* (OMG 2000b) specifications define ways in which CORBA can interoperate with, respectively, the TCAP, Telecommunications Management Network, and File Transfer Access and Management/File Transfer Protocol standards to facilitate the use of CORBA in network management. Also in this area, the Telecom Log Service (OMG 1999) enables the logging of management events and the querying of log records.

Most recently, the Telecommunications task force has finalized a standard for Wireless Access and Terminal Mobility (OMG 2000c). This defines an architecture and interfaces to support wireless access and end system mobility in CORBA. Mobility of both clients and server objects is supported though an architecture that supports handoffs between base stations in a similar way to Mobile IP (Perkins 1996). GIOP tunneling[2] is employed between the base stations involved in a CORBA invocation to hide the fact of mobility from both client and server.

10.4.2 Middleware technology in open service provision

Recently, a number of initiatives have emerged that are aimed at using middleware technology in *open service environments* in telecommunications networks. An open service environment is one in which anyone, not only the telecommunications network provider, is permitted to create and manage services (e.g. a multimedia content archive or a private virtual network).

TINA has been fundamentally influential in this area through its role-based business model that was briefly discussed in Section 10.3.1 (see also Chapter 11). Given this business model, it is natural to identify a number of standardizable interfaces in an open service environment. For example, it is necessary to define interfaces between the consumer and the retailer, and between the service provider and the communication provider. In this environment, Parlay (Parlay Group 2001) and JAIN (Java API for Integrated Networks, Sun Microsystems 2003) are standardization initiatives defining interfaces that allow controlled access by third parties (e.g. service providers and retailers) to the network functionality owned by telecommunications companies. Parlay defines its interfaces in CORBA IDL, while JAIN is a Java-based technology and defines its interfaces in Java (interestingly using the Java Beans component model). But both initiatives have broadly similar aims and there is current work on convergence between the two. Both Parlay and JAIN are covered in more detail in Chapter 11.

---

[2] Generally, the term *tunneling* refers to an encapsulation of one protocol within another. A tunnel is used to ship data between two administrative domains that use a protocol that is not supported by the Internet that connects them.

In a similar area, the OMG's Telecom Service Access and Subscription (OMG 2000d) standard has recently been adopted. This standard includes a set of interfaces whereby communications providers can offer third party enterprises secure access to telecommunications network functions like call control and user location. The standard also supports the straightforward integration of billing and payment for such services.

Work is also beginning on populating the other parts of the TINA business model (i.e. not just interfaces to the communication provider role). For example, the OMG has recently issued a white paper on an 'Open Services Marketplace' that aims to develop CORBA specifications in this area. This work, however, is still at a relatively early stage and specifications are yet to emerge.

10.4.3 Middleware technology in programmable networking environments

A final area of current research interest in applying middleware technology to telecommunications is the use of these technologies in *programmable networking environments*. These provide network programmability at a lower level than the Parlay and JAIN initiatives mentioned above. The primary aim here is to render networks more programmable in terms of their fundamental packet forwarding behavior. The basic approach is to open up the network at the level of routers and switches so that new functionality can be introduced onto routers or switches. For example, packet schedulers, traffic shapers, routing algorithms, and firewalls can be dynamically loaded and configured using this approach.

A good example of recent work in this area is Columbia University's Genesis project (Campbell *et al.* 1999). This introduces the abstraction of *network spawning* in which new 'child' virtual networks can be created dynamically on top of an existing 'parent' network (the ultimate parent is the physical network). The degree of resource sharing between parent and child can be closely controlled so that child networks are capable of providing predictable Quality of Service, e.g. in terms of throughput and delay. Thus, spawned networks can add QoS support and dynamicity to the traditional private virtual network abstraction that typically is concerned only with security. Genesis uses CORBA to control the process of network spawning and each newly spawned network has its own CORBA-based control infrastructure.

More information on the area of programmable networking can be found in the proceedings of the Open Signaling (OPENARCH 2001) or Active Networking (IWAN 2000) conference series.

10.5 Meeting Future Challenges

Although there is a steadily increasing uptake of middleware technology in telecommunications, the telecommunications domain is evolving so fast that keeping up is a major challenge. In this section, we examine a number of specific areas in which this evolution is particularly challenging. Then we briefly discuss approaches that are being adopted in the middleware research community to address these challenges.

10.5.1 Emerging challenges

The first major problem area is *mobility*. Although the above-mentioned OMG standard provides basic support for mobility in terms of supporting handoffs and GIOP tunneling, much remains to be done. One key characteristic of mobile applications is that their connectivity can vary widely over the lifetime of an application session. When a user's machine (e.g. a mobile PDA) is connected to a fixed, wired network, connectivity may be good, with high throughput and reliability, and low latency and jitter. At other times, however (e.g. when the user is traveling), the PDA may experience either total disconnection, or intermittent low-grade connection (low throughput and reliability, and high latency and jitter) through a wireless LAN or mobile phone link. Traditional request/reply communications protocols like CORBA's GIOP are ill-suited to such an environment. They do not work at all during periods of disconnection and work inefficiently over low-grade and intermittent connections. What is required is 'built-in' tolerance to temporary disconnection and the ability to communicate asynchronously during periods of disconnection.

Another mobility related challenge is to provide distributed systems services that are sensitive to the user's current *location*. For example, if a user accesses a roadmap service, the underlying service discovery infrastructure should implicitly return a map that relates to the user's current location. Similarly, if a user prints a document, the request should be transparently routed to a local printer in the same room or building. Current service discovery mechanisms like the CORBA naming or trading services do not meet these requirements as they can only deal with relatively static services and have no concept of location. Although work in this area is proceeding (e.g. see Arnold *et al.* 1999), there is still much to be done before location-based services can be widely deployed.

A further requirement arising from mobility is the desire to establish *ad hoc networks* of mobile users who need to come together for some transient purpose – e.g. a business meeting or a team of mountain rescue personnel who need to collaborate on a rescue mission. In such ad hoc networks, each user should have transparent access to services provided by other users as well as to other services that happen to be locally situated.

A second major challenge arises from the emergence of *ubiquitous computing*. This involves large and highly heterogeneous networks of primitive devices like domestic appliances, devices attached to in-car local area networks, and wearable computers. Such an environment imposes particularly demanding requirements for de-centralized management, which current middleware technologies do not deal with well. Current systems are typically managed manually and usually break when key services become unavailable or networks partition. Furthermore, the ubiquitous computing environment demands that the middleware runs on very primitive devices with few resources (especially memory and CPU capacity). This is difficult or impossible with current technologies that employ client and server side libraries occupying many megabytes of RAM.

A further challenge arises from the increasingly central role of *multimedia* – particularly continuous media like audio and video. Although the above-mentioned CORBA standard for the control and management of A/V streams enables the *control* of media streams, it does not provide a framework for in-band processing of such streams. For example, one cannot implement sources, sinks, or filters for streams in the CORBA middleware environment. This can easily lead to fragmented system design and competing resource management strategies in end systems.

A final challenge arises from the need to integrate middleware technologies into the broader application-level software environment – especially the Internet and the World Wide Web. This area has received attention in CORBA v3 as discussed in Section 10.3.2 above. However, there is little experience with these extensions and fully seamless integration is still to be achieved.

10.5.2 Emerging solutions

The problem areas identified above are now widely acknowledged by the middleware research community and work is underway to address them. In this section, we focus on two complementary research approaches that promise to address many or most of the above problems, these are: *reflective middleware*, and the quest for *model driven architectures*.

The basic approach of *reflective middleware* is to 'open up' the middleware platform implementation at run-time so as to be able to inspect, configure and reconfigure the platform internals. In other words, the approach is to abandon the previously dominant 'black box' paradigm in favor of a 'white box' paradigm in which the platform implementation is made selectively visible to applications. The essence of the approach is to provide a 'causally connected self-representation' (CCSR or 'meta-model') of the middleware platform. 'Causal connection' means that changes made to the model are 'reflected' in the represented platform, and vice versa. As an example, given a meta-model that represents the current structure of the middleware in terms of a topology graph of constituent components, we could inspect the structure of the system simply by observing the structure of the meta-model, and we could implicitly remove, replace, or add components simply by manipulating generic graph operations defined on that meta-model.

Reflective middleware is typically structured in terms of re-usable components that can be appropriately composed to yield a desired middleware profile. Note that in this approach the middleware platform *itself* is built using components; this differs from the existing commercial component models discussed above which provide application level component functionality *on top of* a traditional black box middleware platform. As an example of the use of componentized reflective middleware, a cut-down profile can be specified for a PDA or sparsely resourced ubiquitous computing node (e.g. client-only libraries can be specified, or multithreading or media -stream capability can be left out). To assist the composition process, the component models, although lightweight in comparison to the commercial application level models, typically support the provided/required style of component composition discussed in Section 10.3.3. This explicit specification of requirements makes it easier to predict accurately the effects of composing sets of components in a given manner.

The fact that reflective middleware facilitates the *dynamic* manipulation of components is particularly useful in a mobile computing environment in which it is desirable to maintain application continuity – albeit at reduced quality – when passing from one connectivity domain to another. For example, as one moves from a good to a poor quality network connection, one might alter the compression scheme used by a video stream by replacing one compression component with another. This, however, is only one example; given appropriate system support there are numerous situations in which it is useful to reconfigure running middleware platforms. In particular, the reflective middleware approach seems well placed to address the dynamicity requirements arising from ubiquitous computing and continuous media support. It also appears promising in managing the evolution of software over longer time periods. Further details of research on reflective middleware can be found in the literature (e.g. Coulson 2000, RM 2000, Kon *et al.* 2000, Blair *et al.* 2001).

The concept of *model driven architecture* (OMG 2001b) was first proposed by the OMG in late 2000. Essentially, the MDA concept is to raise the level of programming abstraction so that an abstract service specification (expressing ad hoc application logic) called a PIM (Platform Independent Model) can be automatically mapped, through a sophisticated tool chain, to a suitable middleware based implementation – which is called a PSM (Platform Specific Model). The motivation for the MDA concept is the proliferation of middleware level technologies that are already available (e.g. CORBA, Microsoft's DCOM, Java RMI, or Web-based platforms) or becoming available (e.g. XML/SOAP, Microsoft's .NET or Sun's ONE). It was felt that a new approach was needed to avoid the need to re-tool every time a new technology comes along, and to facilitate interworking between these technologies.

In more detail, PIM-level specifications are expressed in the OMG's UML (Universal Modeling Language) and then 'compiled' to generate an implementation in a pre-selected set of (PSM) middleware technologies. Currently, CORBA is the only PSM technology supported but the list is expected to grow as the model driven architecture paradigm gains acceptance. As well as mapping to one or more selected PSMs, the tool chain also generates a set of 'bridges' that enable transparent interworking between parts of the system that might be implemented in different middleware technologies. A standard set of distributed systems services is implicitly available, via bridging, from any middleware implementation environment. At present, these are the standard CORBAServices which, however, have been re-named Persistent Services in the MDA. Note that the 'compilation' process may not be entirely automatic in that some 'glue' code may need to be provided by the programmer. However, the amount of work required would be expected to be minimal compared to the task of achieving the same result entirely manually.

Currently, the MDA is in an early state of development and the first MDA tools produced by OMG members are not expected until early 2002. However, several key parts of the architecture are already standardized. These are UML, a standard called XMI that specifies how to interchange XML meta-data, a meta-modeling repository called the MOF (Meta-Object Facility), and the associated CWM (Common Warehouse Model). Overall, MDA promises to be an important step forward in the automation of middleware-based system development, particularly in combination with the reflective component-based technologies discussed above. This combination promises not only a significantly improved development environment but also an environment in which existing distributed software can be adapted, reconfigured, and evolved over time.

10.6 Conclusion

This chapter has examined the role of service architectures in supporting the creation and management of next generation telecommunications services. Particular attention has been given to the potential support offered by emerging middleware platforms and technologies.

A range of middleware technologies has been considered:

- client-server technologies such as DCE;
- distributed object technologies such as CORBA, Java RMI, and COM/DCOM;
- component-based technologies (particularly enterprise technologies) such as the CORBA Component Model, Enterprise Java Beans, and .NET.

These represent a gradual evolution of middleware technologies towards the needs of modern service creation and deployment. In particular, the authors believe that component technologies are particularly well placed to meet the needs of the telecommunications industry in terms of offering a high level programming model, supporting a third party style of development, encouraging re-use and the rapid creation and evolution of services, ensuring that services are portable and interoperable, and also supporting their automatic deployment.

The chapter has also examined existing practice in applying such middleware technologies in the telecommunications industry. While the use of such technologies is not yet widespread, there are a number of interesting developments including the work of the OMG Telecommunications Domain Task Force, the use of middleware technologies in opening up networks (e.g. through JAIN and Parlay) and in managing programmable networking environments. There are, however, a number of research challenges that have not been met. Most importantly, existing architectures tend to be too complex and heavyweight for the telecommunications domain. This has, for example, been a major inhibitor in the uptake of technologies such as TINA. In addition, there is little support in the key areas of mobile and ubiquitous computing, or multimedia services. The chapter concludes that such challenges demand a new approach to service architecture and points to interesting work in the field of reflective component-based middleware that offers one promising way forward to meeting the middleware needs of the future telecommunications industry.

11

Service Capability APIs

John-Luc Bakker and Farooq Anjum

Telcordia Technologies, USA

11.1 Introduction

Communications networks are being radically transformed in many ways. Advances in technology have made it possible for these systems to change from being closed systems which are owned, operated, and programmed by a few companies or individuals towards open systems that allow, in principle, anyone to offer new services to users. This transformation, which represents a major development in the evolution of networks, will be key to realizing the tremendous potential of telecommunications and the Internet today. The reference points that provide access to the telecommunications capabilities and enable this vision are the context and motivation for this chapter.

A major capability needed to make this transformation possible is a secure and assured means for users to control the network functionality that is offered by a variety of network elements such as switches and databases. This can be made possible by making use of the concept of APIs (Application Programming Interfaces). Service capability APIs can be considered to be the set of functions offered by a network to allow a programmer to write programs that provide new, useful, innovative, and (hopefully) lucrative services to users. This chapter is about *open standard* APIs for converged networks that are accessible to network operators. Converged networks consist of, and can take advantage of, multiple underlying technologies, including packet networks like ATM and IP networks, the PSTN (Public Switched Telephone Network), and wireless networks. The promise of converged networks lies not in their ability to interconnect different networking technologies per se, but with the help of open standard APIs, enable a wide range of innovative and advanced services to be developed and offered to end-users.

We would like to remark that the APIs addressed here are not intended to open up the signaling infrastructure of telecommunications networks to public usage. Rather, network capabilities are intended to encapsulate and be visible using object technology, in a manner that is secure, assured, and billable. This approach allows independent service developers to develop applications supported by the network without compromising the network and its services.

Service Provision – Technologies for Next Generation Communications. Edited by Kenneth J. Turner, Evan H. Magill and David J. Marples
 ISBN: 0-470-85066-3

For most of its existence, the PSTN has been a monolithic entity that offered essentially only one service; a service that provides the ability to make two-party voice calls, i.e. POTS (Plain Old Telephone Service). However, numerous services were added to satisfy user demand. In practice the provision of these rich and flexible features or services is hard to do, especially in a system as large and complex as the telephone system. Introducing a new service, even one that is not very different from existing services, often takes a significant investment of money, time, and effort. Implementing a new telephone service can take at least eighteen months, and often much longer. The reason is not the complexity of call control. The key reason is that the PSTN as a whole is inherently inflexible, and is not easily programmable. This is the case even with innovations like the AIN (Advanced Intelligent Network) architecture – see Chapter 4.

The key to improving the introduction of new services in a network lies in whether or not its capabilities are easily programmable. The ability to program a new service is set by the API which defines the capabilities that are programmable. The API's capabilities, in turn, are limited by the capabilities of the underlying network elements. If the network has functionality that is not made available to programmers via the API, then such functionality cannot be used by the services to be developed in the API's Service Creation Environment. It might be possible that the vendors of the network elements (e.g. switches, network databases) have access to that functionality and are capable of using that functionality. Such vendor-provided services are often specific to the vendor's network element, the operator's configuration, and perhaps even tailored to regulatory demand. Clearly applications depending on such services are not portable.

To reiterate, we consider APIs for converged networks. Hence, in this case, service capability APIs have to marry two service creation models: the Internet service creation model, and the telecommunications service creation model. The former is based on open standards and common protocols. It has a wide base of developers and low development costs resulting in a large set of new, popular, innovative services. Internet services have a short peak in popularity, and are subsequently replaced by improved versions of the capability or sometimes even lose their appeal altogether. On the other hand, the telecommunication service creation model is often based on proprietary or regulatory extensions to standards and protocols. New services are expensive, time consuming to create, and can only be authored by a small set of experts. The telecommunications services are based on a vast collection of resources and capabilities primerly used to create efficient point-to-point communications between fixed and/or mobile network subscribers according to regulatory demands.

This chapter provides an overview of the standardization activities ongoing in the area of service capability APIs for converged networks. First, JAIN (Java API for the Integrated Network) is a set of activities specific to the Java Programming Environment. Secondly, there is joint work being carried out by a number of bodies on a set of comprehensive and powerful APIs which are both programming language and service architecture independent. The joint activities are performed by working groups of the Parlay Group, 3GPP, 3GPP2, and ETSI. Finally, more recently, work has started on Web services as a specific service architecture for embedding abstract and easy-to-use service capability APIs. All these efforts share some aspects of the TINA-C (Telecommunications Information Network Architecture Consortium) activity. The next section (Section 11.2) introduces the TINA-C. The TINA Consortium created several deliverables, amongst them the TINA Business Model. The TINA Business Model is a timely and valuable document that identifies roles, stakeholders, and their relations. We describe TINA briefly from the objective of pointing out the common theme with the environment that is being specified by JAIN (Section 11.3) and the activities from and around The Parlay Group (Sections 11.4 and 11.5).

11.2 Telecommunications Information Network Architecture

The first TINA Workshop was held in 1990. At this workshop, the telecommunications community came together to assess the common need for improving the way services are designed, deployed, and offered. The focus was on defining a common software architecture for rapid and efficient service creation (Berndt *et al.* 2000).

In 1993, the TINA Consortium was created and chartered to develop a common architecture. Some 40 telecommunications operators, telecommunications equipment and computer manufacturers joined the consortium. Initially, a core team, consisting of engineers from member companies, was created at BellCore. Core team members cooperated on the development of the common architecture. The TINA Consortium was disbanded in 2000 and the effort ended completely in 2002.

TINA was guided by four principles:

- *Object-oriented analysis and design*: object-oriented modeling and design were applied throughout the definition of both the architecture and its reference points. At the time, 'new' software paradigms, such as object orientation, were met with considerable skepticism within the telecommunication industry. The TINA core team recognized that its complex systems would have to integrate many existing subsystems with systems that were not designed using object-oriented analysis and design. When this integration effort involved crossing administrative or technology domains, object-oriented technologies would be used to specify this reference point.
- *Distribution*: in order to open up the products of the traditional vendors, it was envisaged that implementing a 'distribution transparency' would be key. The various components would have to be distributable such that different applications could be accommodated based on traffic characteristics, network load or survivability, and specific customer demand. To support this distribution, the DPE (Distributed Processing Environment) was introduced.
- *Decoupling of software components*: decoupling enhances re-usability of software. A proper design of the software components would make them less dependent on the underlying network technologies. Hence, the components could be re-used when telecommunications networks would partly, or entirely, migrate from circuit switched networks to packet networks.
- *Separation of concern*: TINA recognized three levels of abstraction. First, the DPE, secondly a resource layer and services layer, and thirdly a service generic component and a service-specific component. The DPE level provides common middleware functions for seamlessly interconnecting the various components. Indeed the applications do not have to implement common functions but can rely on the DPE to offer them. For the second level, TINA recognizes a resources layer and a services layer (much like AIN). However, the services layer can control the resources layer both through the DPE, and through the use of high level APIs as reference points (rather than through low-level application protocols). The third level is to encourage an explosion of rapidly created new services. Common functions such as session control can be defined as re-usable, generic components.

11.2.1 The TINA architecture

The architecture also defines a number of clear separation points between the various stakeholders in the telecommunications industry. These clear separation points are called reference points. They provide a clear separation of the roles of each player in the industry and hence encourage players to enter the marketplace and prosper. One of TINA-C's core results, the TINA Business Model (TINA Consortium 1997) with its reference points, is depicted in Figure 11.1. The TINA Business Model is a high-level enterprise model which suits a variety of multimedia information/communication services and businesses. This model provides a set of business roles and a set of reference points. The reference points are specified as OMG IDL (ISO/IEC 1999) interfaces between the interacting roles.

Five business roles are identified: consumer, broker, retailer, communication provider, and third party service provider (the named boxes in Figure 11.1 represent business roles and the named lines between them represent reference points). A number of reference points were identified, the most important and well thought through reference point is Ret, which is short for the Retailer's reference point. The other reference point names are not expanded upon further here, suffice to say that reference points exist between a number of the business roles. The interested reader is referenced to (TINA 1997). Note that a stakeholder in this business may assume more than one role, for example the role of the retailer and communication providers are often assumed to be one entity. (TINA 1997) stipulates:

- A stakeholder in the consumer business role takes advantage of the services provided in a TINA system by paying for the usage of the services. The consumer business role has no interest in making revenues by being engaged in the development of a TINA system (as a specific technology). This type of stakeholder will be the economical base for a TINA system.
- A broker provides and manages references to the various parties in a TINA system, which is analogous to providing both White Pages and Yellow Pages information.
- A stakeholder in the business role of communication provider owns (manages) a network (switches, cross-connects, routes, and trunks). Finally, for its consumers' devices, a communication provider can offer location information, e.g. through position approximation techniques, and status, i.e. on/off/engaged.

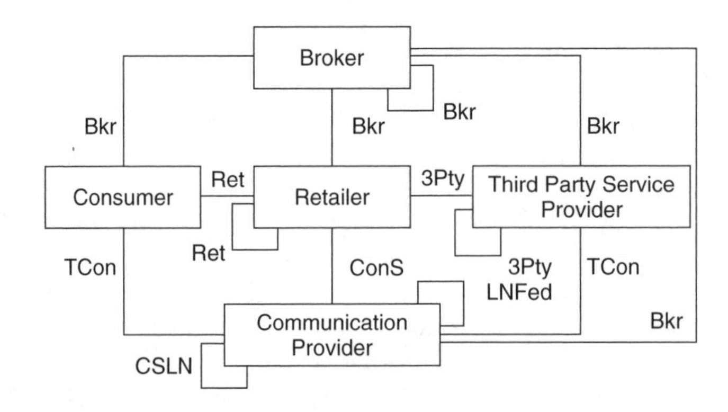

Figure 11.1 TINA business roles and reference points.

- A stakeholder in the retailer business role serves stakeholders in the consumer business role and third party business role. A retailer typically maintains customer relationships, re-sells third party services to consumers, and provides billing and charging to consumers and third parties.
- The aim of a stakeholder in the business role of a third party service provider is to support retailers or other third party providers with services. These services can be regarded as 'wholesale' services. The third party service provider can be a service logic provider or a content provider or both.

11.2.2 TINA and the service capability APIs

In the previous section we have seen the TINA Business Model with the five business roles and their reference points. The work of the Parlay and JAIN groups is commonly seen as an offspring of the TINA architecture. However, the goals of Parlay and JAIN were more modest when compared against TINA. The Parlay and JAIN APIs are focused on making the network capabilities programmable for stakeholders in other business roles, not on defining common, re-usable components or mandating a DPE throughout the system. Additionally, and more importantly, the Parlay and JAIN APIs did not focus on programmability towards the consumer, but only towards (third party) application providers. As a result, the existing *User Network Interface (UNI)* was maintained.

Some of Parlay or JAIN's services can only be supported by specialized telecommunications resources and capabilities, while others consist of more general servers running, for example, databases. An example of the first category is the call control capabilities which make it possible to monitor, re-route, and abort calls, and the user interaction capabilities which allow the presentation of and retrieval of information from the customer. Other capabilities like content-based charging, which facilitates the interaction with a billing system are more generic (for an overview on JAIN call control APIs, see Section 11.3.1, and for a list of all Parlay services see Section 11.4.5). As we have seen in the previous section, the content-based charging capability is a typical TINA retailer responsibility, while the call control and user interaction capabilities are typical TINA communication provider responsibilities. Hence, an operator hosting the Parlay and JAIN APIs includes both the retailer and communicating provider business roles in its business administrative domain.

In contrast, the Parlay and JAIN third party's business administrative domain contains the retailer and third party service provider business roles. The third party service provider enhances the operator's network with content and/or service logic. Typically, a third party service provider provides wholesale services. As the Parlay and the JAIN third party may have business relations with other operators, and so address markets beyond mass marketing and carrier grade services (such as providing niche services or non-operator branded services), it must implement the retailer business role. Hence, the reference points making up the business relation consist of Ret, 3Pty, and TCon (compare Figure 11.1 and Figure 11.2). Exactly these reference points are addressed by the Parlay Group and JAIN.

Figure 11.2 also shows the consumer. In the traditional Parlay and JAIN APIs, the consumer interaction with the operator and/or the application service provider are not in scope beyond that of user interaction through SMS, WAP Push, or even voice response. We re-emphasize that the Parlay and JAIN APIs do not require modification of the UNI.

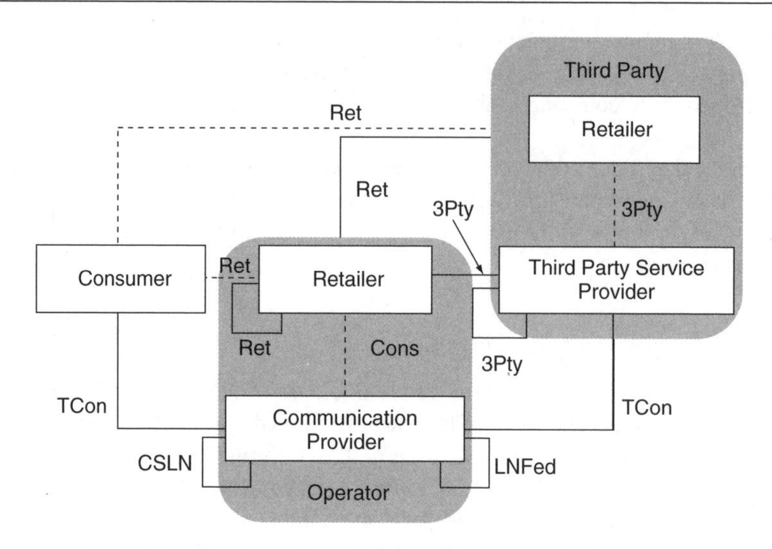

Figure 11.2 A service capability API's mapping to TINA.

Figure 11.2 does not explicitly show the broker. Initially, it is assumed that business relations between third parties and operators are established through more traditional means instead of through electronic, Internet-based brokers.

So what capabilities are made programmable through the Parlay and JAIN APIs? The remainder of this chapter will describe the Parlay and JAIN APIs in more detail with limited reference to the TINA Business Model. More emphasis will be given to the TINA Business Model when introducing Parlay X later in this chapter. Parlay X is seen as the next wave in programmable APIs. For now, it is sufficient to understand that the TINA Consortium depicted the future organization of the communications business, but that it was probably ahead of its time. Meanwhile, other groups in the industry have adapted the TINA concepts into near-term profitable and sustainable architectures, while creating industry acceptance for the resulting standardized APIs and products.

11.3 Java APIs for The Integrated Network

In this section we discuss the work of JAIN. JAIN is an initiative led by Sun Microsystems to create an open value chain from third party service providers, facility-based service providers, telecommunications providers, and network equipment providers to manufacturers of telecommunications, consumer, and computer equipment. The JAIN effort is focused on the creation of a network-agnostic level of abstraction and associated Java interfaces for service creation across PSTN, packet (e.g. IP or ATM) and wireless networks.

The primary benefits/goals JAIN aims to bring are service portability, convergence, and secure network access:

1. *Service portability*: *write once, run anywhere*. Technology development is currently constrained by proprietary interfaces. This increases development costs – time to market – and maintenance requirements. With JAIN, proprietary interfaces are being reshaped to uniform Java interfaces delivering portable and future proof applications.

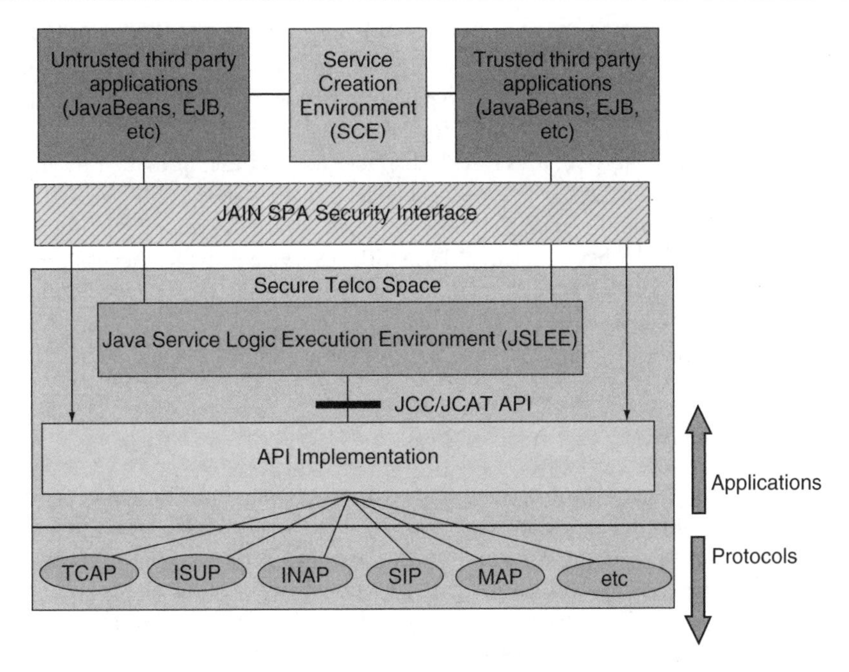

Figure 11.3 JAIN architecture.

2. *Network convergence: any network.* By delivering the facility to allow applications and services to run on PSTN, packet, and wireless networks, JAIN will speed network convergence. As demand for services over IP rises, new economies of scale are possible as well as more efficient management and greater integration with IT.
3. *Secure network access: by anyone.* By enabling service applications to directly access network resources and devices to perform specific actions or functions, opportunities will be created for developers and users. The service applications reside outside the scope of the operator. The market opportunity for new services is huge when controlled access is provided to the available functionality and intelligence inside the telecommunications networks.

The JAIN standardization effort is organized in two broad areas:

- PEG (Protocols Expert Group) standardizing interfaces to PSTN and IP signaling protocols. See, for example, JAIN SIP (JAIN 2002c) and JAIN TCAP (JAIN 2001).
- AEG (Application Expert Group) dealing broadly with the APIs required for service creation within a Java framework. Each Expert Group is organized as a collection of Edit Groups dealing with specific protocols or APIs. Examples are JCC (Java Call Control, JAIN 2002a) and JCAT (Java CAll control exTensions, JAIN 2003a). JCC and JCAT will be further discussed below.

A JAIN service provisioned network includes support for service creation, service logic execution, and connectivity management (policies). A Service Creation Environment (SCE) supports the development of new application building blocks and the assembly of applications from these building blocks (JAIN 2003c). Applications are then deployed into a Service Logic Execution Environment (SLEE JAIN 2003d). The SLEE exposes the APIs of the middle layer, e.g. with JAIN Call Control APIs, directly to the applications. Figure 11.3 shows the JAIN architecture.

The JCC and JCAT APIs define the JAIN Call Control layer. The JCC API defines the interface (e.g. call triggering) for applications to initiate and manipulate calls, while JCAT defines the facilities for applications to have extended control over end-user terminals. Informally, call control includes the facilities required for observing, initiating, answering, processing, and manipulating calls. The JCC/JCAT API abstracts the facilities provided by the lowest layer. At the lowest layer are the subsystems that handle protocols such as SIP, MGCP, or ISUP, or they handle hardware entities such as trunk interfaces or circuit switching hardware.

The figure also shows a subsystem called JAIN SPA (Service Provider Access). JAIN SPA is based on the Parlay API 1.2 specification issued by the Parlay Group (see Section 11.4), and is designed to allow non-trusted applications access to network resources in a secure and assured manner. As such, it contains strong facilities for authentication and authorization.

11.3.1 Call control in JAIN

Call control is the traditional bread and butter of a network operator. The earliest attempts to abstract service control from content are embodied by AIN. The AIN standardization paradigm was successful and was also applied in wireless networks: WIN (Wireless Intelligent Network) (Faynberg *et al.* 1997), and CAMEL (Customized Application of Mobile Enhanced Logic) specifically for GSM networks. In wireless networks more capabilities such as short messaging and tracking the position of mobile terminals were made available. Opening up the service control network to third parties is both a source of new revenue and an operation with enormous risks associated to it. (Accidental) errors in the software may cause (parts of the) service control network to become unavailable or malfunctioning, which may result in short term loss of (voice) revenue, violation of regulator enforced regulations, and (longer term) potential increased subscriber churn.

The voice service state model is complex and contains many 'programmability points'. It is a delicate balance to identify and make programmable the part of the voice service state model that results in high-revenue applications as well as sustainable load on the services control network. In general, the call control services support the following functionality:

- management function for call-related issues, e.g. enable or disable call/session-related event notifications; and
- call control, e.g. route, disconnect.

The JCC and JCAT APIs define a programming interface to next-generation converged networks in terms of an abstract object-oriented specification. As such, these are designed to hide the details of the specifics of the underlying network architecture and protocols from the application programmer. Thus, the network may consist of the PSTN, a packet (IP or ATM) network, a wireless network, or a combination of these, without affecting the development of services using the API. The API is also independent of network signaling and transport protocols. As a result, the network could be using various call control protocols and technologies, for example, SGCP, MGCP, SIP, H.323, ISUP, DSS1/Q.931 and DSS2/Q.2931, without the explicit knowledge of the application programmer. Indeed, different legs of a call may be using different signaling protocols and may be on different underlying networks.

It is assumed that the network will notify the platform implementing the API of any events that have occurred, and hence the platform will be able to process the event and so inform the application using the API. In addition, the application will be able to initiate actions using the API. In response, the platform will translate the actions into appropriate protocol signaling messages for the network. It is the job of the platform to interface to the underlying network(s) and translate API methods and events to and from underlying signaling protocols. Note that this translation is vendor-specific and is not specified by the API, thus different platform vendors may differentiate and compete based on the attributes (e.g. performance) of their translation.

Traditionally, the word 'call' in the PSTN evokes associations with a two-party, point-to-point voice call. In contrast, in this chapter and within the JAIN JCC and JCAT Expert Groups, we use the word call to refer in general to a multimedia, multi-party, multi-protocol communications *session* over the underlying integrated (IP, ATM, PSTN, wireless) network. By 'multi-protocol' we mean here that different legs of the call, representing the logical connection to individual parties of the call, may be affected by different underlying communications protocols over different types of network. Note however, that JCC/JCAT is not intended to provide full control of multimedia streams. Aspects such as synchronization facilities or control of different substreams would be provided by additional packages built on top of JCC/JCAT.

The JCC API is a Java interface for creating, monitoring, controlling, manipulating, and tearing down communications sessions in a converged PSTN, packet-switched, and wireless environment. JCC provides basic facilities for applications to be invoked and to return results before, during, or after calls. Hence, applications can process call parameters or subscriber-supplied information, and engage in further call processing and control. JCAT uses these facilities and extends them. Note that in this context, applications may be executing in a coordinated, distributed fashion across multiple general-purpose or special-purpose platforms. JCC provides facilities for first party as well as third party applications, and is applicable to network elements (such as switches or Call Agents) both at the network periphery, e.g. local exchanges or Class 5 (end-office) switches, and at the core, e.g. trunk switches or Class 4 (Tandem) switches. In addition, JCC supports the origination and termination of first party and third party calls.

JCAT is an API for call control, and builds on the existing JCC specification. JCAT has JCC as its core call control model. JCAT extends the JCC call control model with terminal modeling and it enriches JCC's state diagrams to support a diverse range of applications. JCAT is intended to support the features commonly provided currently by telecommunications carriers and regarded as switch-based, CLASS, as well as IN/AIN features. Support for different features comes at the cost of feature interaction. JCC/JCAT specification assumes that feature interaction itself is not the concern of the API but should be handled by the application developers (further details on feature interaction are given in Chapter 13). JCAT provides facilities for modeling terminal features, features presented on terminals, and modeling the relationships between addresses and terminals, e.g. allowing multiple addresses per terminal and indeed vice versa. JCAT is compatible with facilities for interacting with the user during an active call. These user interaction facilities are expected to be in the form of a separate optional package that is outside the scope of JCAT.

JCC/JCAT allows applications to be invoked or triggered during session set-up in a manner similar in spirit to the way in which AIN services can be invoked. JCC/JCAT thus allows programmers to develop applications that can execute on any platform that supports the API, increasing the market for their applications. It also allows service providers to offer services, rapidly and efficiently, to end-users by developing the services themselves, by outsourcing development, purchasing services developed by third parties, or a combination thereof.

11.3.2 Hands-on

The JSP (Java Specification Process) requires that when a Java specification of the API is made available, it must be accompanied by both an RI (Reference Implementation), and a TCK (Test Compatibility Kit). The RI speeds up the adoption process as a prospective user can easily download and experiment with the API and its associated implementation. Additionally, for vendors wishing to comply with the API a TCK is made available such that they can certificate their products (JAIN 2003b). The TCK and RI for the JCC specification are available (JAIN 2002b). At the time of writing the TCK and RI for JCAT are not yet available. Finally, see Ghosale *et al.* (2003) for implementation considerations of Java APIs and JTAPI (2003) for an implementation with access to Java source code.

11.4 The Parlay APIs

11.4.1 History

Initially, the Parlay Group consisted of five companies: British Telecommunications, Microsoft, Nortel Technologies, Siemens, and Ulticom. The Parlay Group responded to the following drivers:

- Regulatory bodies are expecting network operators to open up their networks to third party service providers;
- The rapid increase in the number of service providers;
- The model for delivery of communication services is moving towards that of a Service Provider Architecture. That is:
  - — acknowledging that the services are moving to the rim of the network. Examples are SCP-based services (AIN services), AIN services at the periphery (CTI and CSTA-based services), and IP-based call centers;
  - — a change in the edge of service delivery. Enterprise and personal functionality requirements are pressing inwards towards the core network. Examples are switch-based CTI extensions, TAPI, and PBX-based services.

The Parlay Group began work on the Parlay specification in 1998. Their intention was to create open technology independent APIs which would enable application development spanning multiple networks. Such applications would be developed outside the traditional network operator space. Phase 1 developed the following APIs: Framework, Call Control, and Messaging and User Interaction. Version 1.2 of these APIs was released in September 1999.

Six new members were added to start Phase 2 in May 1999: AT&T, Cegetel, Cisco, Ericsson, IBM, and Lucent Technologies. In January 2000, Parlay 2.1 was released. It focused on support for packet networks and wireless networks. Also in 1999, ETSI and 3GPP started joint work on OSA using the Parlay initial specifications as input documents. As the work on Parlay Phase 2 commenced, the first release of the Parlay API was taken into ETSI with the intention of standardizing the work (note that the Parlay Group is not a standardization body).

Parlay Phase 3 was announced in June 2000 and focused on the expansion of the recognition and the visibility of Parlay. As a result, the membership increased, there was a further extension of the capabilities of the APIs, and there was an alignment with 3GPP and ETSI. There was also a dialogue and input from JAIN participants. To prevent scattering and fragmentation, Parlay, 3GPP, ETSI, and JAIN member companies agreed to concentrate their efforts on a joint programmability specification. This resulted in the formation of the Joint API Group. This group is responsible for all common APIs between 3GPP, Parlay, and ETSI. Other APIs continued maturing within Parlay and these were brought into the Joint API Group when requirements for them emerged in 3GPP The Parlay 3 specification work was finished one year later in July 2001.

Parlay Phase 4 started in September 2001 and was made available in November 2002. Its main features were backward compatibility, alignment, and further maturing of the APIs. In addition, Phase 4 added the Policy Management API, the Presence and Availability Management API, and introduced support for new middleware technology known as Web services. The Web Services introduction was two-fold (Bakker *et al.* 2002). First, the regular Parlay APIs were published in WSDL (Web Service Description Language) (W3C 2001) to enable interaction between the Parlay Gateway and its application through HTTP and SOAP (Simple Object Access Protocol). Secondly, a new activity was started that resulted in a new suite of APIs called Parlay X (Parlay Group 2002, 2003). Parlay X Web services will be further discussed in Section 11.5.

11.4.2 Motivation

Today's applications and services exist within the operator's domain and are primarily built using AIN or its wireless offspring CAMEL and WIN. However, network operators are increasingly integrating more mobility aspects and data networks. The services for mobile networks and data networks have a much shorter lifecycle than typical mass-market and carrier grade AIN services, thus it is important to rapidly and cost-effectively create and deploy new and innovative applications while not compromising the services already delivered to the operators' subscribers and end-users. To entice new service providers it is important that the applications are not under the control of the operators. Business relations can be created with the new service providers, hence creating a new revenue stream for operators and sharing the commitment and risk involved when developing and deploying new applications.

In order to create a 'killer application' environment, the following points were kept in mind:

- a broad scope of network functionality must be made programmable, so in principle all relevant functionality inside a telecommunications network should be accessible through APIs;
- any API must be developed using mainstream technology to attract large numbers of developers; and
- a proper abstraction level must be found such that developers without telecommunications expertise can easily deploy the APIs and applications built for one type of network on other types of network. Additionally, the abstraction level should also hide the underlying network details leading to a reduction in interdependence. It is necessary that the operators be able to make modifications to their networks without breaking the functionality of large numbers of third party applications.

Additionally, the operators require 'openness' at the Service Management level. In order to offer Parlay or proprietary services implemented by different vendors to their service providers, they must be able to add (and remove) services easily. Finally, and most importantly, operators must be assured that their networks are not vulnerable as a result of opening up control to third party service providers and their developers.

11.4.3 Architecture

Figure 11.4 shows a generic architecture that enables programmability of abstract network capabilities. The figure shows three layers: the application layer with the Application Server, the services layer with the Programmable Gateway and the resources layer with the network protocols and their corresponding networks (Bakker *et al.* 2000). The services layer or Programmable Gateway abstracts the functions provided by the resources layer sufficiently such that non-telecommunications experts are able to develop and host services on top of the services layer. The Programmable Gateway is also in charge of secure access, as well as assurance through enforcing Service Level Agreements.

In the remainder of this section we discuss Parlay at a very high level as shown in Figure 11.5. Our goal in using this figure is to explain the steps involved in the use of Parlay and thereby compare it with the general architecture given earlier. Step 1 in Figure 11.5 indicates that third party service providers or vendors can develop a Parlay service and register it with Parlay's registering component. The registering component is called the Parlay Framework. This provides discovery, authorization, authentication, service agreement, and fault management functions for the Parlay services. Step 2 shows a Parlay service initiating a usage session with the Framework. During the usage session the service is available for a Parlay application to use. Step 3 shows a Parlay application authenticating and entering a service agreement. In the service agreement, an application indicates the constraints and assurances it expects. Subsequently, the operator will accept or reject the service agreement. If the service agreement is accepted, a reference to the service the application wishes to access is made available. Finally, step 4 shows the application accessing the service it signed the service agreement for.

It is important to realize that the capabilities provided by the services layer are constrained by the capabilities of the resources that are abstracted (shown by step 5 in Figure 11.5). For example, traditional PSTN voice networks will not support the multimedia capability in

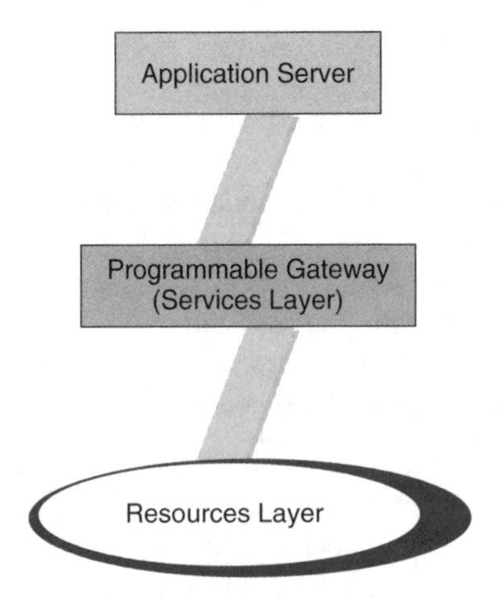

Figure 11.4 General Programmable Gateway architecture.

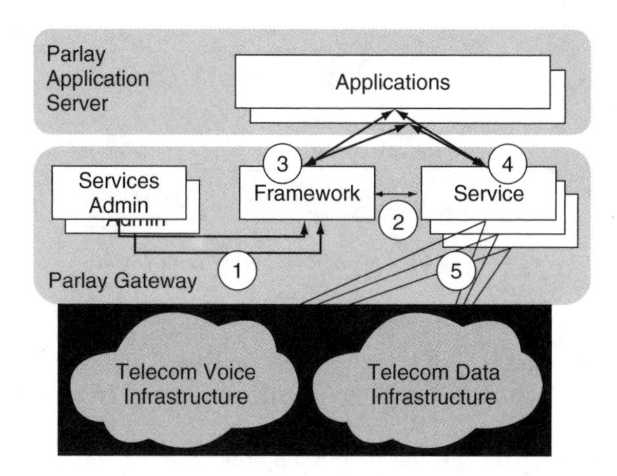

Figure 11.5 High-level overview of Parlay.

which subscribers can handle high-quality voice and video streams. Yet generic capabilities can be identified. For example, the call control capability for PSTN, wireless, and data networks is concerned with the same key functions: call forwarding, call rejecting, or call proxying. Additionally, both wireline and wireless networks support charging and billing functions. This shows that a common set with abstract capabilities exists in all communications networks. Making this abstract set with capabilities programmable will enable applications to be re-usable over multiple networks.

In subsequent sections we will look at the capabilities of each of the Parlay services in more detail. But first we will explore the procedures in place to generate these APIs and their programming language realizations.

11.4.4 Procedure

The Parlay Group is not strictly a standardization body, in fact The Parlay Group releases industry recommendations. Other bodies have picked up The Parlay Group's APIs and have attempted to standardize them based on their requirements. For example, ETSI SPAN 12, 3GPP2 TSG-N OSA, and 3GPP CN5 have adopted The Parlay APIs. ETSI, 3GPP2 and 3GPP are standardization bodies and have their own requirement processes. The ETSI requirements are driven by fixed networks, while the 3GPP and 3GPP2 requirements are regionally driven by wireless and 3G networks. Initially, The Parlay Group, ETSI, and 3GPP diverged from their common base as each was driven by unique requirements (3GPP2 joined later). Thus, there was now a risk of developing three different APIs all having the same purpose. It was seen to be appropriate to combine the groups into a single joint working group in the context of what is called OSA (Moerdijk and Klostermann 2003).

The joint working group process works as follows: member company representatives work on aligning the requirements brought in by Parlay, ETSI, 3GPP2, and 3GPP, and subsequently on realizing them. After satisfying the requirements, the member companies offer the resulting work for endorsement by the respective bodies. This arrangement guarantees that the APIs endorsed by Parlay, ETSI, 3GPP, or 3GPP2 are identical or a subset of the full API that satisfies

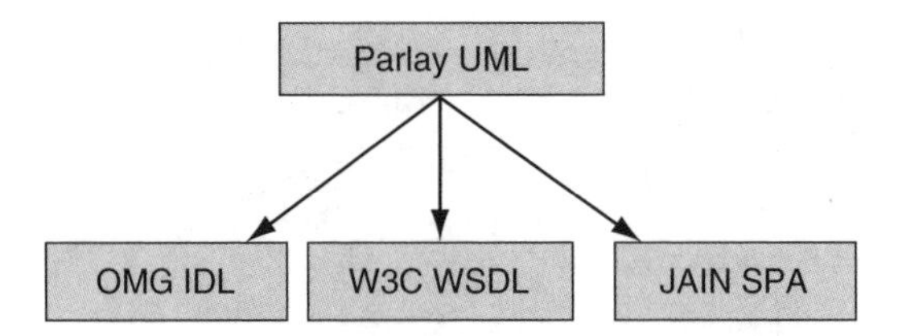

Figure 11.6 Parlay technology realizations.

all requirements. For example, 3GPP does not endorse the Conferencing Call Control API, the Connectivity Management API, and the generic Messaging API, as there was no 3GPP requirement to do so.

The joint working group models the API in the programming language neutral UML. The UML model is a single source for the specification and its programming language realizations. Three programming language realizations exist: OMG IDL, W3C WSDL, and Java (see Figure 11.6). These realizations are to be generated automatically from the UML model each time a new release of the Parlay APIs is made available.

11.4.5 Overview of Parlay APIs

In the previous sections we have introduced the history, the scope, and procedures behind the Parlay APIs. In this section we look at the various Parlay Service APIs starting with the Framework. Following the Framework discussion we will discuss the other Parlay services, most notably the hierarchy of call control services, user interaction, user location and status (or mobility), terminal capabilities, data session control, generic messaging, connectivity manager, account management, charging, policy management, and PAM (Presence and Availability Management). We will indicate the documents and their location as we describe each service. Mostly, we will refer to the 3GPP documents, for those Parlay services that do not exist in 3GPP we will cite the ETSI/Parlay documents.

Note that Parlay also defines both a set of common types and data (3GPP 2002b) to be used by all APIs, and a more specific common data set to be used by only the call control hierarchy of the APIs (3GPP 2002d). Finally, an overview document (3GPP 2002a) complements the Parlay specifications by bringing all APIs and bodies that contributed to the set together.

Framework

The Framework (3GPP 2002c) provides applications with the basic mechanisms that enable them to make use of the services offered by the network. Before an application can use the network functionality made available through Parlay APIs, authentication between the application and Framework is needed. After authorization, the discovery function enables the application to find out which network service APIs are provided by the Parlay Gateway. The network services are accessed by the methods defined in the Parlay interfaces and will be discussed in subsequent sections.

Next we provide a brief description of each of the Framework interfaces.

Trust and security management: The trust and security management interfaces provide:

- the first point of contact for a client to access a Framework provider;
- the authentication methods for the client and Framework provider to perform an authentication protocol;
- the client with the ability to select a service capability feature to make use of; and
- the client with a portal to access other Framework interfaces.

The Framework authenticates new services when they become available, as well as applications that wish to make use of the services that can be discovered through the Framework. Following successful authentication, a Framework client (be it an application or a service) can request access to any of the other Framework interfaces.

Registration: Registration of new service implementations to the Framework. Before a service can be brokered (discovered, subscribed, accessed, etc.) by an application, it has to be registered with the Framework. Services are registered against a particular service type, e.g. Multimedia Call Control. Therefore service types are created first, and then services corresponding to those types are accepted from the services for registration in the Framework. The Framework maintains a repository of service types and registered services. In order to register a new service in the Framework, the service supplier must select a service type and the property values for the service. The service discovery functionality described above enables a service to obtain a list of all the service types supported by the framework and their associated sets of service property values.

Service discovery: Discovery of APIs provided in the target network. After successful authentication, applications can obtain available Framework functions and use the discovery function to obtain information on authorized network service capability features. The discovery function can be used at any time after successful authentication.

Service lifecycle manager: Creation of a new instance of the service implementation. The `IpServiceInstanceLifecycleManager` interface allows the Framework to get access to a service manager interface of a service.

Integrity management: Load balancing, fault management, heartbeat. This interface is used to inform the application of events that affect the integrity of the Framework, Service or Client Application. It includes fault management, load management, heartbeat management, and some OA&M functionality.

Event notification: Notifications for specific events (e.g. registration of new service implementations). This interface is used by the services to inform the application of a generic service-related event.

Service agreement management: Management of service agreement between application and Framework domains. Before any application can interact with a network service capability feature, a service agreement must be established. A service agreement may consist of both an offline (e.g. by physically exchanging documents), and an online part. The application has to sign the online part of the service agreement before it is allowed to access any network service capability feature.

Service subscription: Management of subscriptions to services. In some cases, the client applications (or the enterprise operators on behalf of these applications) must explicitly subscribe to the services before the client applications can access those services. To accomplish this, they use the service subscription function of the Framework for subscribing or un-subscribing to services. Subscription represents a contractual agreement between the enterprise operator and the Framework operator. In general, an entity acting in the role of a customer/subscriber subscribes to the services provided by the Framework on behalf of the users/consumers of the service. In this model, the enterprise operators act in the role of subscriber/customer of services and the client applications act in the role of users or consumers of services. The framework itself acts in the role of retailer of services.

It is important to note that the Framework is designed such that it allows for multiple vendors, and even the inclusion of non-standardized APIs, which is crucial for innovation and differentiation. Any services that can be brokered through the Framework must implement the `IpServiceInstanceLifecycleManager` interface. This interface is part of the service lifecycle manager package. Additionally, such services must extend the `IpService` interface. The `IpService` interface is the parent interface of each Parlay service (described below) and serves as a marker interface (the `method createServiceManager()` on the `IpServiceInstanceLifecycleManager` interface returns a reference to an instance).

Call control APIs

There are four call control APIs defined by Parlay. These are:

- generic call control;
- multi-party call control;
- multimedia call control; and
- conferencing call control.

The first three have been adopted by 3GPP and are available in the OSA suite. Generic call control is focused on two-party sessions. This is the principal hallmark of early wireless and CS-1 based networks. Additionally, besides being limited to two-party sessions only, early wireless and CS-1 based networks did not support third party initiated sessions. These constraints are reflected in the Generic call control API (3GPP 2002e). Generic call control provides:

- *Simple call control*: simple call control is defined as control over two-party calls without leg manipulation. Additionally, in some wireless networks there is no support for application-initiated calls, as the network technology does not support this capability.
- *Call gapping (load control)*: Call gapping is a load control technique employed by the network elements to reduce the load presented by a number of service requests. The network element (e.g. the gateway) will not service any requests from any other network elements for specified intervals.
- *Call supervision (for pre-paid charging)*: call supervision is available in the APIs in the case of network support for a pre-paid service. The gateway will be notified when the time for the call is up, or network elements may even sound an alarm and eventually terminate the call when the supervision criteria are met. In all three cases the gateway will be notified.
- *Static triggers*: for example call origination.

- *Dynamic triggers*: for example busy, no answer.
- *Call routing/redirecting*: an application may control the routing of the call and effectively redirect it.
- *Advice of charge*: some network elements and their communication protocols support the advice of charge feature (e.g. ISDN). Typically, ISDN terminals are equipped with a small screen that can display information such as the charging policy that will be applied to the call.
- *Charge plan*: charge plans can be set by the application on a per-call basis. Hence, the application can determine charging characteristics such as premium service or reverse billing.

The Multi-party Call Control API (3GPP 2002f) extends – but does not inherit from – the capabilities of the Generic Call Control API with support for *n*-party sessions, third party initiated sessions, and a richer call model. A mapping from MPCCS to 3G networks is made available by the 3GPP (3GPP 2002g). In addition to all the capabilities of generic call control, multi-party call control introduces:

- Call leg manipulation: includes an interface called `IpCallLeg` for call leg manipulation. This interface provides programmability of call capabilities on a per-call leg basis.
- Functionality on both call and call leg levels:
  - supervision;
  - retrieval of call-associated data: e.g. reports such as duration of the call or call leg;
  - advice of charge: enables the representation of call charges per call or per call leg;
  - charge plan: enables split charging if the charge plan is different for the two call legs of the call.
- Functionality on call leg level only:
  - routing/redirecting;
  - dynamic notification reporting;
  - query last redirected address, so if a call is redirected, the original redirected address is made available as the last redirected address;
  - attach and detach legs to or from a call.

Multimedia call control (3GPP 2002g) on the other hand, extends multi-party with control over sessions that contain multiple media. A typical multimedia session may contain separate voice, white board, and video sessions. If a subscriber initiates requests for setting up sessions with multiple media, or dynamically wishes to add more media, the Multimedia Call Control API may deny or modify these requests. This is based on criteria such as network load, access network bandwidth, or subscriber's subscription status. Multimedia call control extends the multi-party call control by inheritance and adds the following capabilities:

- Volume-based supervision on call level applies in cases where call supervision (i.e. prepaid service) is not time-based but volume based.
- Allow or deny the addition of media streams to call legs so media streams with particular characteristics may be denied (or allowed) following a decision of the application.
- Request explicit subtractions of established media streams.
- Media stream notification reporting, based on establishment of media streams.
- Media stream monitoring, based on addition/subtraction of a media stream of an ongoing call.

In addition, the Conferencing Call Control API (ETSI 2003a) is intended to expose capabilities of specialized conferencing resources. It adds control for (say) a conference chair to allow and deny participation, creation, and merging of subconferences, conferencing resource reservation, floor control, and chair assignment. Conference call control extends (through inheritance) multimedia call control and adds:

- floor control, chair, and speaker capabilities;
- split or merge subconferences;
- resource management.

Finally, the document by 3GPP (2002d) contains common types for all four call control APIs.

User interaction

The user interaction service (3GPP 2002h) is specified by the Generic User Interaction API. The API is used by applications to interact with end-users. The user interaction service allows an interaction with end-users that are participating in a call controlled by the call control service. In addition, it can initiate a user interaction session with an end-user that is not participating in a call. Examples of the latter case include an automated survey application where the application interacts with the user interaction service to present questions and retrieve answers from end-users. Another example could be a wake up call. This API is not necessarily voice based, it can also be mapped to SMS (Short Message Service), USSD (Unstructured Supplementary Service Data), or WAP (Wireless Application Protocol) for user interaction. The API allows for simply sending information, or for sending information *and* collecting a response from an end-user. The Call User Interaction Service Interface provides functions to send information to, or gather information from, the user (or call party) to which a call leg is connected. This API also allows the user interaction service to record messages and to delete them, for example, after they expire.

An example of a user interaction session with an end-user that is currently participating in a call is, for example, to warn the end-user that his pre-paid account status may cause the call to be terminated. This is an example of two collaborating but distinct Parlay services. The call control service would supervise the call and request to be notified if a time limit is met. The application would then interact with the user interaction service and attach a user interaction function to the call leg that connects with the end-user whose account balance is becoming insufficient for continued participation in this call. The user interaction service could be instructed to inform the user of the account balance status and offer the possibility to add more funds to the account, e.g. through subtracting funds from a pre-provisioned bank account. After receiving and processing the user's response, the application continues to monitor the call leg until the account status reaches another threshold or until the end-user disconnects.

Mobility

The mobility service (3GPP 2002i) consists of two separate APIs: mobile terminal status, and mobile terminal location. The Mobile Terminal Location API supports registration of location events. More specifically, the application specifies an area and mobile terminal addresses, and if a specified mobile terminal enters or leaves the area, the application will be notified and can take subsequent action such as sending an SMS message using the user interaction service. Additionally, the application may request to be periodically notified of the location of a mobile terminal. Finally, the application can request the location of end-users on demand. These interfaces are set-up in an abstract, technology/application aware manner:

- The generic UL (User Location) service provides a general geographic location service.
  - UL has functionality to allow applications to obtain the geographical location and the status of fixed, mobile, and IP-based telephony users.
  - Supports direct location requests, and triggered (based on specified location boundaries) and periodic location requests.
- The ULC (User Location CAMEL) provides location information, based on network-related information, rather than the geographical coordinates that can be retrieved via the general UL.
  - Using the ULC functions, an application programmer can request the VLR Number, the location Area Identification and the Cell Global Identification and other mobile-telephony-specific location information.
  - Supports direct location requests, and triggered (based on location change) and periodic location requests.
- ULC was designed such that it can make use of the GSM-specific CAMEL (currently maintained by 3GPP). In the ANSI-41 world, 3GPP2 is looking into defining a specific variant of user location that exposes the unique capabilities of WIN-based location determination and retrieval.

In the case of an emergency call (e.g. 911, 999, or 112) the network may locate the caller automatically. The resulting location is sent directly to an application that is dedicated to handling emergency user location. If the dedicated emergency call application is using the API, the location is sent to the application using a call-back method specified in the `IpAppUserLocationEmergency` interface.

The US (User Status) API provides a general user status service. US allows applications to obtain the status of fixed, mobile, and IP-based telephony users. Examples of reported status include reachable/not-reachable and busy.

Terminal capabilities

The Terminal Capabilities API (3GPP 2002j) allows an application to retrieve the known capabilities of the specified terminal. The terminal capabilities are made available by a WAP Gateway/PushProxy in CC/PP format. Relatively new functionality, supported by the terminal capability API, is triggering for dynamic changes in terminals. Note, however, that the underlying mechanisms that support dynamic triggering are currently not fully standardized.

Data session control

The API for DSC (Data Session Control) (3GPP 2002k) controls a non-voice value added service that allows information to be sent and received across a (telephone) network. Typically, the Data Session Control API is used in mobile networks and controls an overlaid packet-based air interface on the existing circuit-switched network, i.e. not SMS. Thus, data originating from a mobile terminal is relayed in packets to a packet data network, for example in GPRS or 1×RTT networks. The term 'data session' is used in a broad sense to describe a data connection/session. For example, it comprises a PDP context within GPRS.

This API provides a means to control the establishment of new data sessions. In the GPRS context this means that the establishment of a PDP session is modeled and not the attach/detach mode. As in call control, change of terminal location is assumed to be managed by the underlying network and is therefore not part of the model. The underlying assumption is that a terminal initiates a data session and the application can reject the request for data session establishment, can continue the establishment, or can continue but change the destination as originally requested by the terminal.

Immediacy is a key feature for this type of connection. Data traffic must be offloaded quickly to packet networks such as the ubiquitous Internet. Subscribers demand access to their favorite sites and applications on the Internet without delay. In general, an operator will wish to control the access to the packet data networks and the tariffing in place. Different tariffing policies may be activated based on time of the day, load of the network, and/or subscriber's subscription. The DSC API is much less involved than the call control APIs.

Generic messaging

The GMS (Generic Messaging Service) (ETSI 2003b) essentially supports mailbox management. It is used by applications to send, store, and receive messages. GMS has voicemail and electronic mail as the messaging mechanisms, and so the messaging service interface can be used by both.

The Messaging API assumes the presence of a network element in the network with mailboxes, folders, and messages within these folders. Messages can be stored in each folder and they usually have properties associated with them. Additionally, a messaging application can request to be informed of newly arriving messages that match certain criteria. For example, the application can request to be notified of incoming voicemails with a high priority from particular sources. If voicemails matching these criteria arrive, the application is notified and it can subsequently (say) send an SMS message to a mobile terminal to inform the end-user that a voicemail has arrived. SMS messages can be sent to the end-user's mobile terminal using the user interaction service.

Connectivity manager

The Connectivity Manager API (ETSI 2003c) is expected to be used by enterprise operators that have established a service agreement with network operators. The connectivity manager includes the APIs between the enterprise operator and the provider network. This allows the two parties to establish QoS (Quality of Service) parameters for enterprise network packets traveling through the provider network. The service agreement allows the enterprise operator to setup and tear down provisioned QoS virtual private pipes. Elements that can be specified for a virtual private pipe include attributes such as packet delay and packet loss. Characteristics of traffic that enters the virtual private pipes at its access point to the provider network can also be specified with attributes such as maximum rate and burst rate.

The functions provided by this API are also made available though the Policy Management API discussed below. The policy management service work was started after completion of the connectivity service.

Account management

The account manager interface (3GPP 2002l) provides methods for monitoring accounts. It supports operations such as the retrieval of transactional history and balance queries. In addition, it supports the enabling or disabling of charging-related event notifications. This information can be made available to an application, for example, for presentation on a Web page or for conversion to voice so that voice portals can access the account details.

Charging

The Charging API (3GPP 2002m) is used by applications to charge for their use. The user being charged can be the same as the user using the application. It is also possible that another user will pay the charge. The charging API supports both scenarios in which the charging for service use is split amongst multiple accounts, and simple charging scenarios where one account is charged. The charging session represents these accounts. In the case of split charging, e.g. all participants in a conference call pay their share, the charging methods provide operator specific clues for the Charging API to decide how to distribute the charge over the accounts.

The charging session interface contains both the charging methods and the debit and credit functions to support each of the methods. There are four charging methods: currency amount charging, unit charging, immediate charging, and charging against reservations. The use case for charging against reservations is obvious, simply consider an end-user wishing to download a movie. Before engaging in the transaction the movie provider may want to ensure that the end-user has enough funds, hence the movie provider will attempt to reserve the funds necessary to download the movie and charge against the reservation in increments. Other services, such as receiving an SMS message may be charged immediately, i.e. without reservation. Finally, the application can either charge in a currency amount or in units. Units and currency amounts can both be charged immediately or through a reservation.

Policy management

The policy management service (3GPP 2002n) contains APIs to create, update, and view policies, but also supports interactions between the policy and the application. The application can specify its interest in particular policy events. The following capabilities are provided through the policy management interfaces:

- publishing of policy events supported by a service;
- subscription to policy events supported by a service;
- generation of events;
- obtaining statistics associated with the use of policies; and
- handling of service agreements that may be used to convey authorization for access or subscription to policy information or to modify or create policy information.

The centerpiece of the policy management service is the policy information model. The policy information model is based on the IETF Policy Core Information Model but extends it to include support for events.

PAM

PAM (Presence and Availability Management API, 3GPP 2002o) provides functionality for maintaining, retrieving, and publishing information about:

- digital identities;
- characteristics and presence status of agents (representing capabilities for communication, content delivery, etc.);
- capabilities and state of entities; and
- presence and availability of entities for various forms of communication and the contexts in which they are available.

11.4.6 Hands-on

Unlike the JAIN community, The Parlay Group does not mandate reference implementations or test compatibility kits. However, recently implementations of the Parlay APIs have become available free of charge. Those that wish to get their hands dirty building some Parlay applications without ordering a programmability gateway or full scale, carrier grade application server right away are referred to Ericsson's Parlay simulator (Ericsson 2002), Appium's Parlay simulator (Appium 2003), Lucent's Parlay simulator (Lucent 2003), or Open API Solution's Parlay simulator (OpenAPI 2003). Note, however, that this list is not exhaustive.

11.5 Parlay X Web Services

WWW, as one of the many applications of the Internet, has been around for a while. Web pages are often published in HTML (HyperText Markup Language) and transported over HTTP (HyperText Transport Protocol). In a way, its ease in achieving ubiquity has made many believe that XML-based techniques must be considered an integral part of the new distributed computing paradigm. Web services is an XML-based middleware technology, based on common Internet technologies such as HTTP or SMTP (Simple Mail Transfer Protocol). HTTP, SMTP, or other protocols can serve as suitable 'transport protocols'. Programming interfaces are commonly specified in WSDL. SOAP is a layer of XML structuring on top of these transport protocols, and amongst other message paradigms, SOAP provides remote procedure call semantics.

Unlike the Java and OMG IDL Parlay realizations, and indeed the JAIN APIs, the Web services service architecture does not standardize an execution platform. Rather, it specifies the interoperability requirements 'on the wire'. The particular API a developer will use in the execution platform depends on the tools that have generated it. At the moment, there is no single body that pursues standard bindings between WSDL specification and the various programming languages.

In the remainder of this section, we will look at a current Web services-based open APIs initiative in the telecommunications industry. We will present Parlay X Web services, a WSDL defined set of interfaces that model Web service enablers.

11.5.1 Motivation

Consider a Web server that allows end users to charge for services they consume to their prepaid accounts, or a customer support page that creates, by the press of a button, a voice call between an end user and a customer service representative. Developers in this environment often use Web services technology to communicate with different capability servers, i.e. they use SOAP and WSDL. Web services provide a set of capabilities and technologies that result in a very compelling foundation for supporting converged telecommunications/IT applications.

The Parlay APIs expose capabilities of the telecommunications network in a network technology and programming language neutral way. In fact, the contributors to the APIs go to great length to ensure that the common aspects of (converged) mobile, fixed, and managed packet networks are made programmable for a large audience of developers of carrier grade systems. Yet, some more specific but highly valuable capabilities remain unsupported. As an example, the support for SMS and MMS (Multimedia Messaging Service) is inadequate in the Parlay and JAIN APIs. Additionally, the developer who wishes to program simple applications, such as setting up a third party call (click-to-dial) using the Parlay APIs, needs to go through elaborate interactions with different components and interfaces. This is illustrated in Lagerberg *et al.* (2002) and Jain (2003a).

These observations motivated the need for APIs that are predominantly simple and, consequently, restricted in their capabilities. Developers that need access to an advanced means of control would not use these simple APIs, they would use the existing Parlay APIs. Additionally, the rigid dogma that favors exposure of common network capabilities over specific capabilities needed to be relaxed. Finally, the resulting APIs would predominantly be used in a multi-portal or Web environment. The Parlay community proved to be open to these views and approved the establishment of a group, the Parlay X Group (Parlay Group 2002; 2003), in late 2001, which was chartered to create APIs that incorporated these views. It was felt that new markets are made available through Web services with Parlay X application definitions. Given a set of high-level interfaces that are oriented towards the skill levels and telecommunications knowledge levels of Web developers, the Parlay X APIs offer accessibility of the network capabilities to a much wider audience.

11.5.2 Architecture

Figure 11.7 shows where the Parlay X applications are situated with respect to the Parlay X Gateway. Note that many of the Parlay X capabilities will be mediated by the Parlay Gateway. Some, however, are not supported because of Parlay's focus on common capabilities as opposed to network-specific capabilities. Such capabilities cannot be mediated through the Parlay Gateway, rather they need to be made available through the resources layer. The Parlay X Web services specification considers this an implementation issue and does not mandate how these capabilities are made available to the Parlay X Gateway.

11.5.3 Framework

One of the main attractions of the continued embracing of the Internet service creation paradigm, is the virtually unbounded variety of business relationships that can be established. In such a huge and diverse space as the Internet it is important to have efficient discovery mechanisms to find the services one is looking for. In TINA-C this role is modeled by the Broker. In Web services, the interfaces assumed by such a broker can be UDDI (Universal Description, Discovery and Integration) (UDDI 2002), or performed by xMethods (2003) or SalCentral (2001). The UDDI standard is intended to provide central directories

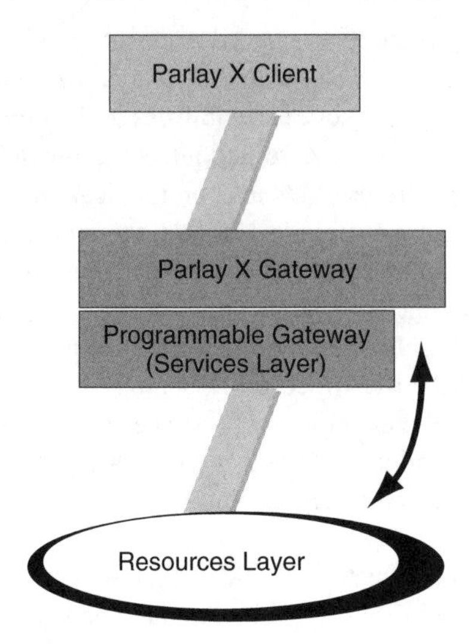

Figure 11.7 Parlay X Gateway accessing specific network capabilities directly.

where consumers of Web services can access various services, either within a company's firewall, via an extranet, or on the public Internet. Service providers can register them and make them available via UDDI. Such discovery services can be seen as the first level of discovery.

A second level of discovery can be offered by WSIL (Web Services Inspection Language), also known as WS-Inspection. The characteristics of WSIL are close to those of the Parlay Framework (see the Framework discussion in Section 11.4.5). The first and optional level of discovery, providing URLs of Parlay X Gateways, could be handled by a third party, e.g. implementing UDDI. In contrast, the second level of discovery provides access to the different Parlay X Web services themselves. WSIL files describe Web services, possibly in a hierarchical manner, while UDDI serves a centralized registration and service publication solution. UDDI or WSIL allow for Parlay X applications to discover published services. The UDDI or WSIL-driven registry, finally, contains information using which the Parlay X application can bind and activate the Parlay X service. Note that authorized personnel can extend the registry with more Parlay X services.

Besides discovery, registration, and publication, other common enablers also increase the appeal of Web services. As with the Parlay APIs discussed in the previous section, security, accounting, and assurance are important common enablers. The Parlay Group explicitly introduces the set of Framework APIs to address these issues on an application level. The Web services environment is different since several mature or existing technologies exist today. These can be the industry norm, and so can be leveraged to achieve the same result as UDDI or WSIL for discovery or retrieval of references to specific services.

Security in itself is not an enabler. Security consists of several aspects:

- *confidentiality*: keep the communication secret;
- *data integrity*: ensure the data is not modified;
- *authentication*: prove the identities of the party or parties involved;

- *authorization*: control who is allowed to do what;
- *non-repudiation*: provide undeniable evidence that an action has taken place.

Confidentiality can be addressed by HTTPS or having a VPN. HTTPS and VPN software typically encrypt the communication between participants. HTTPS and VPN also, typically, address data integrity through encrypted data. This is hard to modify without having access to the encryption algorithm and public and shared keys. When using HTTPS and a VPN we assume that the data integrity threat is highest during the transport phase. HTTPS and VPNs do not protect against threats compromising the data when it is decrypted at the VPN gateway or at hosts that receive the HTTPS traffic. Additionally, XML has support for digital signatures to sign the XML messages (W3C 2003). This way, proxies or intermediaries can verify the source of the message as well as inspecting the XML message before routing it or possibly even altering it. The message is signed before it leaves the proxy or intermediary.

HTTP, and hence HTTPS, as well as a VPN include authentication support at the transport layer. These authentication credentials can be used to verify the other end of the encrypted pipe. If we assume that this end of the pipe uses credentials that also authenticate the request, we can use these credentials to verify access rights. On the other hand, in the case where WS messages are routed through multiple proxies or intermediaries where each must inspect the message before routing it forward, authentication assertions on the XML message level are desired. SAML (Security Assertion Markup Language, SAML 2003) provides just that.

Based on the authentication information (e.g. user handle and password in the case of HTTPS) access to only the subscribed services can be enforced through dynamically generated WSIL files, ensuring authenticating access to subscribed services only. Alternatively, the XACML (eXtensible Access Control Markup Language) (XACML 2003) can be used, in conjunction with an authentication protocol like SAML, to specify access control lists. Hence, XACML addresses the authorization side of security. Finally, an XML signature can also be used to address the non-repudiation security aspect. In general, the solutions used to address certain aspects of security entirely depend on the threads one wishes to defend against, as well as the costs one can incur to introduce and use the protection. These are important observations as security solutions can have higher costs than the damage done through a security breach.

A good source for the exciting and fast moving field of XML security is XML-security (2003). It can be seen that standardizing common enablers is a very active field, best left to the respective experts in those fields (note that we have not even touched common enablers for accounting and assurance in this section!) Providing Web services for telecommunications enablers, however, is best left to the telecommunications domain experts, for example the Parlay X Group.

11.5.4 The Parlay X Web services

In 2003, Parlay X participating companies submitted contributions (Parlay Group 2003) that targeted Parlay X compatible definitions of Web services for: NI3PCC (Network Initiated Third Party Call Control), 3PCC (Third Party Call Control), Payment, Account Management, User Status, Terminal Location, and Messaging (e.g. SMS). 3PCC and NI3PCC Web services are motivated by the observation that applications that interface with telecommunications resources often initiate and receive voice calls. 3PCC supports initiation of voice calls (e.g. click-to-dial) and NI3PCC. The Account Management API allows consumers to increase the value of their pre-paid accounts and query payments made from the account. The payment API allows content providers to charge for certain types of content such that the billing is handled by the operator. Note that the Parlay X payment API is fully aligned with, and endorsed by, the PayCircle initiative (PayCircle 2003). Examples of content that can be charged for are downloadable ring tones or downloadable voicemail announcements. Next, the User Status and Terminal Location contribution focuses on presenting User Status (such as online, offline, or engaged) and Location (fine-grained as longitude and latitude or coarse as within or not within an area). Finally, the Messaging API is intended for sending messages, most notably SMSs (Short Message Service) and MMSs, to devices that can accept such messages.

11.5.5 Observations

It is hard to measure simplicity. Parlay X focuses on exposing capabilities through accepted and largely applied technologies by the IT industry. Furthermore, the IT industry is currently furthering the Web services architecture. Emerging standards for transactions and integrated security will make the use of Web services as a middleware solution even more attractive and will further reduce the complexity of creating telecommunications applications. Already, Parlay X is exploring the applicability of Web services middleware in constructing a programmable gateway. As outlined above, a number of contributions have been suggested and are currently being consolidated and processed for inclusion in the first release of the Parlay X APIs.

11.6 Conclusion

We have seen the history and current status of the standardized programmability effort with regard to network element capabilities such as call control. The work carried out on AIN resulted in 'local' access to network capabilities. 'Local' here means that network operators do not allow business entities outside their control to access and use the network capabilities, and for good reason! The network capabilities and the services provided through them to their subscriber base are the operator's 'bread and butter'. Any accidental erroneous impact on the capabilities' ability not to meet the Service Level Agreements established with customers or the regulators may have a far-reaching impact.

Additionally, AIN was not developed using contemporary programming languages, rather the network vendor has often designed their own, proprietary programming language or model. Worse still, network element vendors often delivered a development environment especially designed for that vendor's network element. As a result, the applications built on AIN services are not portable amongst different network element vendors' products and application developers that know the specifics of the environment are often employees with highly specialized capabilities. This combination of factors leads to a slow application development, deployment, and activation cycle. The network operators would have to synchronize the features of an application with the different network element providers, wait for their individual development and testing cycle to complete, integrate with and test the various pieces in the operator's network, and decide on an activation time.

The JAIN APIs seek to resolve this. The JAIN APIs must be accompanied by an associated TCK and RI. Hence, applications built against one JAIN API implementation ought to be portable to other JAIN API implementations provided that both implementations passed the TCK. Not only are the JAIN APIs defined in a popular and state-of-the-art programming language, Java, but also, the Java language comes with a host of tools and other APIs to use and integrate with. JAIN applications can now connect easily with another domain's capabilities or information sources such as the numerous databases or the WWW (World Wide Web). The chosen abstraction level and the chosen programming language ensures that a large base of developers – not necessarily telecommunications specialists or that particular vendor's network element specialists – can develop the new and innovative services of the future.

In parallel, The Parlay Group started defining their APIs in a programming language neutral way, largely to address the same objectives. The Parlay Group added one dimension, however, to enable third parties to not only develop new and innovative applications, but also to enable third parties to host them. Many Parlay use cases target enterprise operators as consumers of Parlay services, and hosts of Parlay applications. This added distributed, secure, and reliable access to the list of concerns. To tackle these issues, The Parlay Group defined the Framework. The Parlay Framework is a central capability that authenticates the third party, as well as brokering and authorizing access to the various services based on a service agreement. Additionally, the Framework is defined such that new third party services can be registered and brokered. Services can also be updated, and applications are made aware of the new services through one of the Framework interfaces. It can be seen that The Parlay Group offers a comprehensive and rigorously defined set of APIs that are endorsed by the 3GPP, 3GPP2, and the ETSI.

So what is in store for the future? Well, things that emerge in the near to medium future need to be 'in the pipeline' today. Commonly requested features for future programmability solutions include:

- More flexibility in selecting the capabilities that are to be made programmable. For example, if an operator has a particular, 'home-grown' trigger it sees a market for, then the operator might not be able to make use of the current JAIN or Parlay APIs. These APIs may be too rigid in their definitions to allow the operator to offer a programmable access to the trigger. In short, the operator cannot exploit their market differentiator.
- Effortless access for third parties to programmable capabilities. By embedding some of the framework functionality and value in the infrastructure, there can be improved tool support. Hence, application developers can exploit the capabilities in a more efficient and rapid manner. Additionally, it is anticipated that the most common applications will require just a few aspects of the service. In Parlay, an application often first has to interact with the service manager object, and subsequently with its service session object. It is a view shared by many that if a service aspect cannot be activated using one request/response sequence, then most applications will probably not use it.
- Use XML technology. It would be difficult to avoid the fact that recently there has been a very large interest in Web services, and indeed it is seen by many as the latest 'hot' technology.

The observations above are supported by the next wave of service capability APIs named Parlay X Web services. Stay tuned!

11.7 Further Reading

We recommend reading the 3GPP/ETSI specifications for further details of the Parlay APIs (3GPP 2000a to 3GPP 2000o, ETSI 2003a, ETSI 2003b, ETSI 2003c). Jain *et al.* (2004),

JAIN (2001, 2002a, 2002c, 2003a, 2003c, 2003d) provide more specification details of the JAIN initiative. Enabling Web services for Parlay X is a more recent activity, and details and specifications can be found in Bakker *et al.* (2002), Parlay Group (2002, 2003).

A reader who is intimidated by a stack of paper detailing the specifications mentioned above can download and interact with some software that implements these specifications. Some links to JAIN-compliant software are JTAPI (2003) and JAIN (2002b, 2003b). Parlay/OSA simulators can be found in Appium (2003), Ericsson (2002), Lucent (2003) and OpenAPI (2003).

12

Formal Methods for Services

Kenneth J. Turner
University of Stirling, UK

As will be seen, formal methods are precise languages and techniques for specifying and analyzing systems. In some applications, notably those that demand high standards of safety or integrity, the use of a formal method is highly desirable – and may even be mandated. Telecommunications services are expected to be very reliable and clearly defined. The services can be complex and time-critical. They may have to operate on a variety of equipment, and several parties may contribute to providing the service. Services may even be safety-critical, e.g. for handling emergency calls. A long-standing issue in service provision is that independently defined services may interfere with each other – the so-called feature interaction problem. For all these reasons, the rigor offered by a formal method is attractive.

However, there are major challenges to be met in applying formal methods to services. Some issues are purely technical, such as the expressive power of a language or the practicability of analyzing complex specifications. Other issues are more pragmatic, such as the usability of a method or the commercial support for it. This chapter gives a taste of formal methods in general, and explains how formal methods have been used on telecommunications services. Work on formal methods tends to be rather specialized, so this chapter aims to give a not too technical introduction. There are extensive references to allow specific techniques to be investigated further.

12.1 What is a Formal Method?

12.1.1 Nature

The word 'formal' refers to form and systematic treatment. In the context of system development, a *formal approach* is one that is mathematically based. A *formal language* has precise rules for writing down statements (its *syntax*). Thus the syntax of a programming language is often defined formally using a notation like BNF (Backus-Naur Form). A formal language may also have precise rules for interpreting statements (its *semantics*). Programmers are usually not given a formal semantics for the languages they use. Instead, they learn intuitively

Service Provision – Technologies for Next Generation Communications. Edited by Kenneth J. Turner, Evan H. Magill and David J. Marples
 ISBN: 0-470-85066-3

about how the language works, such as how to use conditional statements and loops. However, it is possible to give a formal definition to the meaning of programs.

A *formal method* is a systematic technique that makes use of a formal language and formally-defined rules for development with the language. For example, a formal method might define how a high-level description of a problem can be turned into a detailed design of the solution. A formal method will usually give systematic rules for analyzing statements made in the language. Thus, consistency and completeness might be investigated. Normally, the supporting theory will allow comparison of different specifications written in the language. Some approaches allow the properties of a specification to be checked. It is desirable, for example, to show that a low-level design satisfies high-level requirements.

The field of formal methods is large. Unfortunately, it is difficult for the lay reader to get a good understanding of it. This is partly because there is a wide variety of formal methods, partly because the terminology is specialized, and partly due to the abstract mathematical basis. This chapter aims to help with the first two difficulties. For the third, the interested reader can obtain technical background from books on *discrete mathematics*, e.g. Rosen (1995). This is the kind of mathematics, such as set theory and logic, used in computer science.

12.1.2 Benefits and costs

Natural language is the commonest method of specification. However, natural language does not lend itself to precision. Its main advantage is that everyone can use it, although with different degrees of accuracy. Extra difficulties can arise if readers are obliged to use specifications that are not in their native language (e.g. as may happen with an international standard). In general, formal methods allow reasoning and analysis in a way that is not possible with natural languages. Much of the analysis can be automated, though certain aspects such as proving properties or theorems require expert human intervention.

The pros and cons of formal methods have been fiercely debated at times. Bowen and Hinchey (1995a,b), Hall (1990) and Le Charnier and Flener (1998) provide some insight into possible misconceptions about formal methods and what they offer. Hill (1982) is an amusing introduction to the limitations of natural language for specification. However, as pointed out by Naur (1982) the value of formal methods should not be overstated.

The very act of writing a formal specification forces attention to problems that might otherwise be missed. Omissions and ambiguities are often discovered during formalization. Writing a formal specification commonly helps to clarify the structure of a system, which is also a benefit in itself. The process of formalization also helps to identify assumptions that might be otherwise hidden or fuzzy.

Formal specifications are usually abstract. This reflects the nature of the languages, but is almost inevitable because it is impracticable to address too much detail. This is positive because the specifier is obliged to concentrate on what is essential. An implementer can easily miss important points in the welter of detail.

A problem with formal specifications is that they are written in languages that the customer (or even engineer) is unlikely to know. Some specifiers claim that it is easy to learn their specification language. However, the abstractness and discipline of a formal method are often difficult barriers. Executable specifications are helpful because they can be simulated to demonstrate the consequences of a specification to others.

Formal inspection (Fagan 1979) and structured walkthroughs (Weinberg and Freedman 1984) are common practice in software engineering. These are systematic but not precise in the sense of a formal method. However, they can be helpful in explaining formal specifications to non-specialists. And, of course, formal specifications also need debugging.

An executable specification can be turned into a lower-level implementation (e.g. with a conventional programming language). Some specification languages can be compiled into reasonably efficient code. The method then offers the precision and analytic power of formalization, coupled with the practicality of (semi-)automated implementation.

Formal methods are mainly used in the early phases of development, such as requirements specification and high-level design. This is partly because the specifications at this point are relatively small and manageable. Experience in systems engineering is that errors early in development become very costly to correct if they permeate through to final implementation. It is therefore very desirable to make specifications as accurate as possible. Fortunately, this is the very point in development where formal methods are most applicable.

Testing is the other development stage where formal methods play an important role. In an ideal world, implementations would be derived automatically and systematically from specifications. However this is rarely practicable. Usually there is an intuitive gap during implementation when the design and realization occur. As a result, the implementation may not satisfy the specification. It is possible to use a formal specification to generate tests of a proposed implementation, aiming to demonstrate its conformity to the specification. Unfortunately, it is rarely practicable to test an implementation completely. It is possible to demonstrate the presence of errors, but usually infeasible to show that no further errors exist.

The most obvious cost of formal methods is the need to use a special-purpose language. This has implications for training and tools. Formal methods tend to be regarded as obscure because they use unfamiliar symbols and concepts. However, many engineers learn and apply programming languages that are more complex than many formal languages. It is certainly unnecessary to be a mathematician in order to use formal methods. A number of formal languages have graphical representations that make them more attractive to industry.

Formal languages require their own tools. Some of these are common to any language (such as an editor, a compiler, or a debugger), but others are particular to the formal world (such as a verifier or a model checker). Formal methods tend to have a research focus, so there is often little or no commercial support. There are notable exceptions to this, such as SDL (Specification and Description Language, ITU 2000h). In the field of hardware description, there are also a number of commercially supported tools.

In certain applications such as safety-critical or mission-critical systems, the costs of formal methods are almost irrelevant. The penalties for system flaws are so severe that it is worthwhile to make every effort to find problems. In less critical applications, there is a trade-off between the cost of design faults and the cost of formal methods in finding them. A significant stimulus to develop formal methods has come from the telecommunications standards community. (Safety-critical systems and hardware design have also been major influences.) Compatibility is crucial in telecommunications, as the equipment that must interwork is heterogeneous and distributed. Overall, there is a gradual but definite trend towards using formal methods in a wider variety of areas.

12.1.3 Concepts and terminology

Of necessity, a formal approach has a precise mathematical basis. At the other extreme, there are *informal* languages and methods. Natural language, for example, does not have an exact interpretation, although it is widely used for description. Diagrams are also usually informal.

Programming languages are normally regarded as informal since their semantics is usually not defined precisely. However, some graphical notations (e.g. flowcharts or other kinds of graph) and some computer languages are close to being formal. It is therefore useful to recognize a *semi-formal* category for certain languages and methods. Even a fully formal method may not be used with complete mathematical precision. A *rigorous* approach is systematic and formally-based. For practical reasons it does not employ strict reasoning, although this is still possible because there must be a proper formal foundation.

A *proof* applies the *axioms* (assumptions) and *inference rules* of a formal language to establish the correctness of some statement. Thus, it might be proved that a program will never accept incorrect data. A *theorem prover* is a tool used to support the activity of checking or proving theorems. The terms 'verification' and 'validation' are often used with different meanings. In software engineering, *verification* checks for 'building the product right', while *validation* checks for 'building the right product' (Boehm 1984). In formal methods, verification means proof of correctness, while validation means testing. *Analysis* is a generic term for any systematic investigation of a formal specification.

The terms 'specification' and 'implementation' tend to be used loosely in system development. In a formal context, a *specification* is a high-level and abstract description of requirements. An *implementation* is a low-level and detailed description of a design. It is possible to define a formal relationship such that a specification is satisfied in some sense by an implementation. The most rigorous way of passing from a specification to an implementation is by means of *refinement*. This requires each design step to be formally justified. In practice, an implementation is normally created using the designer's skills. However it is still possible to show formally that an implementation respects its specification. *Description* is a neutral term referring to a specification or to an implementation. An FDT (*Formal Description Technique*) is the term used for a standardized formal language and method.

Because of their mathematical basis, formal methods tend to be abstract: they define *models* of the system being specified. It follows that they tend to be used mainly for writing specifications. Specification languages fall into two broad classes called constructive and non-constructive. A *constructive* language gives an explicit (although abstract and mathematical) *model* of the system to be built. A constructive specification is generally executable, that is, it can be simulated to give an understanding of what the specification means. A *non-constructive* language focuses on the *properties* that a system must have.

A system might be expected to respect certain *assertions* (general statements about the system) and *invariants* (properties that must always be true). A *model checker* is a tool that establishes if a model (formal specification) satisfies certain properties. Consider an algorithm for sorting values into order. A constructive specification of this might define a particular procedure (such as QuickSort). A non-constructive specification might state that the algorithm permutes its inputs, and that the outputs are in order.

Specification languages may deal with systems that are *sequential* (one action after another) or *concurrent* (actions in parallel). Communication among parts of a system may be modeled in a *synchronous* (direct) or *asynchronous* (buffered) manner.

12.2 Classification of Formal Methods

Formal methods offer a rigorous approach to developing computer systems. There is a large range of methods, though they tend to fall into broad groups. There have been numerous surveys of formal methods such as Austin and Parkin (1993), Clarke and Wing (1996), Courtiat *et al.* (1995), Craigen *et al.* (1993a,b), Mitchell (1987) and Parkin and Austin (1994). An introduction to formal methods is given by Ford and Ford (1993). Turner (1993) concentrates on standardized formal methods. The use of formal methods in standards is discussed by Ruggles (1990). There have also been several discussions about the applicability of formal methods in industry (e.g. Fraser and Vaishnavi 1997, Parkin and Austin 1994, Vissers 1993).

Most formal languages are textual, though some are graphical or have alternative graphical syntaxes. Sometimes the languages make use of special symbols, though the majority aim to use a standard character set for ease of editing and portability. There are usually accompanying tools for parsing and analyzing specifications. These include tools for simulation, state space exploration, model checking, equivalence checking, and theorem proving. The tools may be built with meta-tools that can be adapted to support specific languages.

The following subsections give a brief classification of formal methods. This is approximate since it does not cover every variety, and since some methods combine the techniques of different categories. Space permits only a limited range of typical references. Further articles can be found in many journals such as *Acta Informatica*, *Science of Computer Programming*, *Theoretical Computer Science*, *Transactions on Programming Languages and Systems* and *Transactions on Software Engineering*. Relevant articles can also be found in the proceedings of many conferences, in fact almost too numerous to mention. Starting points could include AMAST (*Algebraic Methods and Software Technology*), CAV (*Computer-Aided Verification*), CONCUR (*Concurrency*), FM (*Formal Methods*), FORTE (*Formal Techniques for Networked and Distributed Systems*), *Foundations of Computer Science*, IFM (*Integrated Formal Methods*), LiCS (*Logic in Computer Science*), *SDL User Forum*, and TAPSOFT (Theory and Practice of Software Development).

12.2.1 Model-based

A *model-based* language aims to specify a system using simple mathematical objects like sets and functions as the building blocks. System data is represented by corresponding types. The operations on the system are then functions that change the system state. Within this broad category, B (Abrial 1996), VDM (Vienna Development Method, Jones 1990) and Z (Spivey 1992) are well known. Both B and Z have seen significant industrial usage. The term *axiomatic* is also used of such languages.

A variety of languages loosely follow the same style. *Relational languages* focus on relations between sets as the main modeling tool. This ties in with the entity-relation style of modeling. *Declarative languages* are very high-level programming languages that aim to be suitable for specification as well. They can be regarded as executable formal languages because of their strong mathematical basis. *Functional programming*, for example, models the behavior of a system using executable functions. LISP (List Processor) and SML (Standard Meta Language, Wikström 1987) are well-known examples. *Logic programming* achieves goals using a collection of facts and inference rules. Prolog (Programming in Logic, Clocksin and Mellish 1994) is the archetype.

12.2.2 Logic

Many variations of mathematical logic are used for specification. For example, Prolog uses a restricted logic that can be executed. *Propositional calculus* (simple statements about objects) is too limited for practical use. *Predicate calculus* (also called *first-order logic*) allows statements about sets of objects. This is typically enriched in a number of ways to obtain a practical specification language.

Modal logic deals with various modes of expression such as talking about necessity, belief, or permission. The commonest logic used for specification is a *temporal logic* that expresses how the system behavior evolves over time. It is usually possible to make statements about the next state of the system, or about future states. Some temporal logics have additional operators that describe behavior until some condition holds, or that describe past behavior. CTL (Computational Tree Logic, Clarke *et al.* 1986) and TLA (Temporal Logic of Actions, Lamport 1993) are popular kinds of temporal logic.

12.2.3 State-based

An FSM (*Finite State Machine*) or FSA (*Finite State Automaton*) is an abstract machine with a set of states and transitions between these. The machine changes state when there is some input, and may produce an output in response. The most practical languages use an EFSM (*Extended Finite State Machine*) that has state variables as well as a major state. For example, the state variables of a service might record the amount of data sent or the names of its users. The major state of a service might indicate whether it is in use, idle, or unavailable.

State machines are very widely used in design. The machine is often drawn as a *state transition diagram* that shows the movement between states. *State tables* are also a common way of showing how the machine reacts to inputs in each state. Computer support often makes use of state tables in memory. There are many variations on state machines. Among the best known are ESTELLE (Extended Finite State Machine Language, ISO/IEC 1997b), SDL (Specification and Description Language, ITU 2000h) and Statecharts (Harel and Gery 1996). The MSC (Message Sequence Chart, ITU 2000f) notation is frequently combined with SDL, though its notion of state may be implicit.

Nets can be seen as a development of state machines. Tokens are placed in one or more places of a net to mark its state, and these places are connected by transitions. If the input places of a transition have tokens, the transition may fire and transfer tokens to the output places. The original notion of a Petri net (Petri 1962) has been considerably extended by variants that allow for different kinds of tokens, conditions on transitions, timing, etc. A standard has been developed for high-level Petri nets (ISO/IEC 2000b).

12.2.4 Algebraic

Process algebras are used in many areas. The approach resembles that of state machines, except that state is usually implicit. The focus is on the actions that cause transitions between states. Process algebras model a system as a collection of communicating processes. Each process describes the order in which events can occur, including sequential or concurrent behavior. Communication among processes may be asynchronous or synchronous. Well-known process algebras include ACP (Algebra of Communicating Processes, Baeten and Weijland 1990), CCS (Calculus of Communicating Systems, Milner 1989), CSP (Communicating Sequential Processes, Hoare 1985), and LOTOS (Language Of Temporal Ordering Specification, ISO/IEC 1989).

An *algebraic method* focuses on the expressions in a language and how they can be manipulated (their laws). However, the term tends to be used of techniques that formally specify

ADTs (*Abstract Data Types*). The goal is to encapsulate the data held by a system and the operations on it. OBJ (Goguen *et al.* 1988) is a typical example. *(Constructive) type theory* emphasizes how types and operations are built.

12.2.5 Structural

Structural approaches define the organization of a system in terms of its components. SA (Structured Analysis, Yourdon 1989) and its variants are very common in software engineering. Structured methods are not formal in themselves, but are often accompanied by a formal underpinning. UCMs (Use Case Maps, Buhr and Casselman 1996) are a visual notation for representing sequences of actions and the causality among them.

An ADL (*Architecture Description Language*, Medvidovic and Taylor 1997, Shaw and Garlan 1996) is used to describe the high-level organization of a (software) system. Architecture description languages support the concepts of component, connector, and configuration as the means of expressing system structure. ACME (Garlan *et al.* 1997) and Rapide (Luckham *et al.* 1995) are examples that have been given formal support. The term *architectural semantics* is used for the relationship between architectural and formal concepts (Turner 2000b).

Object-based or *object-oriented* methods are, of course, well known as a means of modeling systems as a collection of interacting objects. They are not formal in their own right, but formal theories of objects have been developed. UML (Unified Modeling Language, e.g. Stevens and Pooley 2000) is the dominant method in object-oriented design. UML is a collection of multiple notations, some more formal than others. For example, UML state diagrams are similar to Statecharts, UML sequence diagrams to Message Sequence Charts, and UML activity diagrams to Use Case Maps. There has been significant work on relating SDL and MSCs to UML (ITU 2000g). Object concepts have been retrospectively added to most of the major formal languages (e.g. LOTOS, SDL, Z). Formal object-oriented methods have also been developed from scratch (Derrick and Bowman 2001).

12.3 Formal Methods for Communications Services

The term 'service' tends to be used rather loosely and in different ways depending on the application area. This section looks at formal methods for describing communications (data) services in networked and distributed systems. Space permits only a limited range of typical references. Further articles can be found in journals such as *Computer Networks*, *Computer Standards and Interfaces* and *Transactions on Communications*, as well as general computer science publications. Relevant articles can also be found in the proceedings of conferences such as FORTE (*Formal Techniques for Networked and Distributed Systems*) and PSTV (*Protocol Specification, Testing and Verification*).

12.3.1 OSI (Open Systems Interconnection)

OSI (Open Systems Interconnection, ISO/IEC 1994c) standardized the ways in which networked systems can communicate. OSI first gave prominence to the concept of service (Vissers and Logrippo 1986). An OSI service can be viewed as the boundary between protocol layers and also as the abstraction of all lower layers. A service therefore isolates layers so that they can be developed and changed independently.

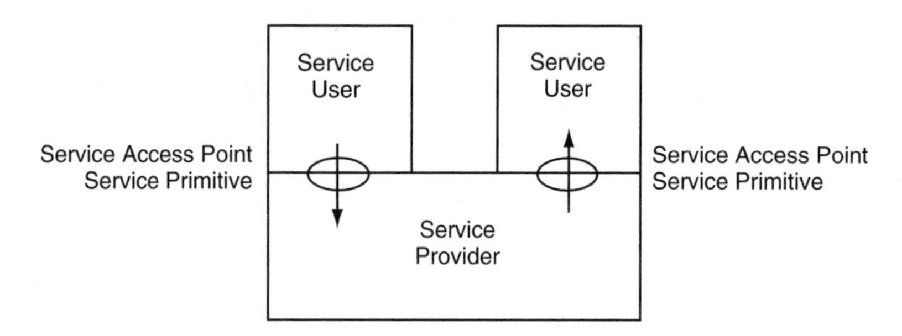

Figure 12.1 OSI service.

Figure 12.1 shows the main concepts of an OSI service. The ovals are SAPs (Service Access Points) that serve as abstract interfaces between a service user and the service provider. The arrows represent service primitives that are abstract calls across this interface, for example to establish a connection or to transfer data.

Much of the early work on formalizing services was undertaken on OSI. This was partly because OSI offered a sound architecture. However, formal specifications of services also proved useful in reasoning about protocols. Since a service is an abstraction of a protocol, it reflects its functionality in a compact way. It is therefore easier to reason about a service than about a protocol. As an example, it is desirable to show that a service is satisfied by a protocol and the underlying service that supports it (e.g. what the network offers). Conversely, a service specification can be treated as the requirements for a protocol. A number of protocols might thus be designed to fulfill the same service. The OSI Transport Protocol (ISO/IEC 1986a), for example, has five classes of protocol that can be used to realize the Transport Service (ISO/IEC 1986b).

Another motivation for formalizing OSI services was the existence of different FDTs (Formal Description Techniques): ESTELLE, LOTOS, and SDL were all used with OSI standards. There was therefore an issue of how specifications of protocols in different languages could be related and combined. Even in the same language, it was important to specify different protocol layers in a consistent manner.

A common, formal interpretation of services was therefore developed to tie these together. As part of the OSI work, an architectural semantics was defined to link architectural and language elements. There are two choices in such an approach. It is possible to show how every architectural concept can be represented in each formal language, but this leads to considerable work. It is therefore preferable to relate architectures and formal languages via an intermediate set of fundamental information processing concepts and their combinations (Turner 1997b).

Guidelines were produced for using FDTs on OSI standards (ISO/IEC 1990c, ITU 1992c). An introduction to the FDTs and their use with OSI-like services appears in Turner (1993). Detailed guidance was developed for these languages, and particularly for LOTOS (Turner and van Sinderen 1995) and SDL (Belina *et al.* 1992). The following gives a flavor of the LOTOS work.

A service user and the service provider are LOTOS processes that run in parallel. A service is specified by defining the behavior of the provider process. The service access points of the service correspond to a LOTOS event gate – an abstract interface where communication takes place with the provider. Service primitives are LOTOS events, i.e. communications between

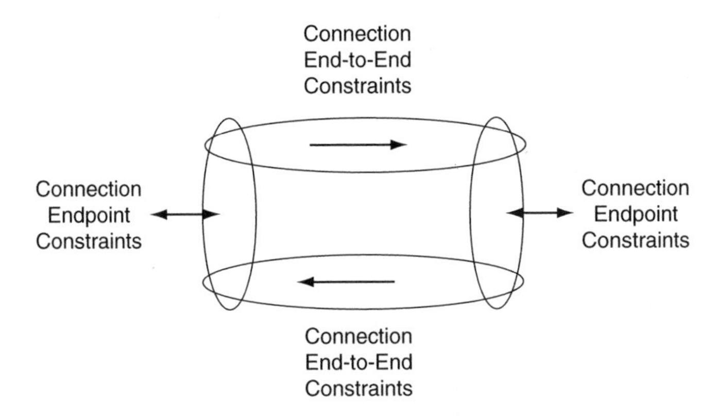

Figure 12.2 LOTOS constraint-oriented specification of a service connection.

a service user and the service provider. Individual service access points are identified by an address parameter in events.

LOTOS can be used to specify both abstract data types and process behavior. Static information about a service, such as addresses and titles (service user names), is given using types. The dynamic behavior of a service is defined by processes. This is usually broken down into individual protocol functions. For example, the techniques of multiplexing (combining data streams) and concatenation (combining protocol data units) can be specified separately in LOTOS. A complete protocol specification then consists of these building blocks in combination.

A constraint-oriented style was developed for LOTOS (Vissers *et al.* 1991). This allows separation of concerns, i.e. different aspects of a service or protocol can be specified relatively independently. Figure 12.2 shows how this can be applied to the description of a service connection. At each endpoint of a connection, there are constraints (rules) on behavior. For example, the service user is not allowed to send data before a connection has been established. There are also constraints on the transfer of data in each direction. For example, expedited (priority) data may catch up and overtake normal data. Each of the four constraints may be defined independently, though symmetry usually means that the endpoint constraints are the same and similarly the end-to-end constraints. Where the constraints overlap, the behavior must match. Thus, expedited and normal data cannot be sent until there is a connection. But once a connection has been established, expedited data may be sent and overtake normal data.

The application of formal methods to OSI services led to a variety of standards documents such as ISO/IEC (1990a) for the Session Service and ISO/IEC (1990b) for the Transport Service.

12.3.2 ODP (Open Distributed Processing)

ODP (Open Distributed Processing, ISO/IEC 1995, ITU 1996c) standardized a framework for distributed systems. Part 1 (Overview) of this is a general guide to the standard. Part 2 (Foundations) defines the concepts and architecture, while Part 3 (Architecture) specifies the characteristics of an open distributed system. Part 4 (Architectural Semantics) is a direct descendant of the work on architectural semantics for OSI. ODP describes systems from five complementary viewpoints: Enterprise (high-level statements of purpose and policy),

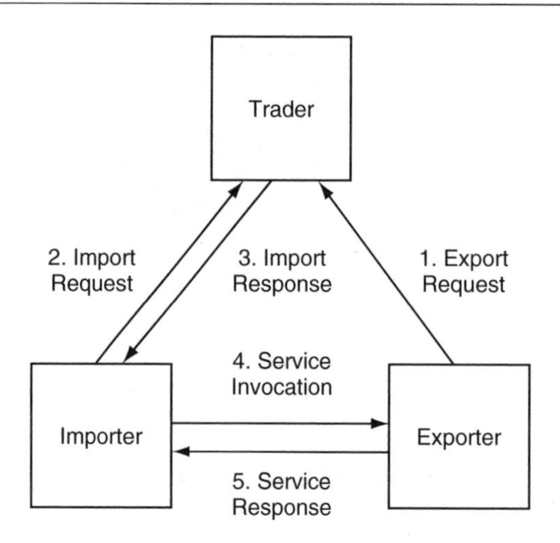

Figure 12.3 Service trading in Open Distributed Processing.

Information (processing and manipulation of information), Computational (functional decomposition), Engineering (infrastructure), and Technology (supporting technologies).

TINA (Telecommunications Intelligent Network Architecture, Kristiansen 1997) drew on ODP principles to create a well-developed service architecture. SDL has been used to describe and create service frameworks in the context of TINA (Kolberg *et al.* 1999b; Sinnott and Kolberg 1999a).

The architectural semantics work was particularly intensive for ODP. In fact it was possible to consider formalizations of architectural concepts while the reference model was being developed. This allowed the architecture to be tightened up during standardization, as well as guidelines for formal specification to be developed concurrently. Architectural semantics were developed for ESTELLE, LOTOS, SDL, and Z (ISO/IEC 1996b, ITU 1996d). This work concentrated on formalizing the ODP basic modeling concepts, the specification concepts, the Information viewpoint, and the Computational viewpoint. Sinnott and Turner (1995) report their experience with this work.

An ODP service is the functionality made available at an interface. This is similar to the concept of a service (set of related interface methods) in object-oriented design. ODP services are intimately bound up with the concept of trading (ISO/IEC 1996a). In a distributed system, it is necessary to discover what services are available and to bind to selected services. Such services might provide printing, document storage, image conversion, etc. An exporter is an object able to provide services. It provides a description of the services it offers. A service is characterized by a number of properties such as cost, quality, reliability, etc. An importer is an object that requires services. It sends out a request for the services its needs, qualifying these with the required properties. As shown in Figure 12.3, an ODP trader receives export and import requests. It attempts to match these, if necessary by consulting other traders. An import response indicates exporters with suitable services. The importer then interacts directly with its selected exporter.

Formal specifications were written of ODP services and service trading using LOTOS (Sinnott and Turner 1997) and Z (Dong and Duke 1993, Indulska *et al.* 1993). The ODP trader was subsequently used as an example to demonstrate the capabilities of E-LOTOS (Enhanced LOTOS, ISO/IEC 2001b).

The LOTOS specification treats the trader from the Enterprise, Information, and Computational viewpoints. The Engineering and Technology viewpoints are too low-level for them to be appropriate in a requirements specification. The Enterprise viewpoint is rather high-level. In fact it is not so easy to represent policies in LOTOS as the language is better suited to specifying definite behaviors. Policies discuss obligations, prohibitions, and permissions. Policies may, for example, deal with security considerations, service contracts, and resource consumption. These are represented by guards (conditions) in the LOTOS specification. These may prevent an action from occurring (prohibition) or may allow it to occur (permission). Obligation arises if only one action is permitted.

LOTOS lends itself rather better to specifying the Information viewpoint. Types are used to specify static aspects such as services, service properties, interface types, and service matching. The Computational viewpoint is also straightforward in LOTOS, as is the Engineering viewpoint (although with a lot more detail). The behavior of a trader is simply to accept export and import requests, marrying these up. The trader maintains a database of available services, and respects the policies imposed by the Enterprise viewpoint.

12.3.3 Other kinds of service

Multimedia services are commonplace. Blair *et al.* (1995) discuss general issues in specifying multimedia systems. The broad background to this topic appears in Blair *et al.* (1998), which advocates a method based on LOTOS for behavioral specification and a temporal logic for performance specification. A number of formal languages have been used to specify multimedia aspects, including LOTOS (Sinnott 1999a), SDL (Sinnott and Kolberg 1999b) and Z (Sinnott 1999b). Undesirable interactions among multimedia services have been investigated by Tsang *et al.* (1997).

Zave and Jackson (2000) use a DFC (Distributed Feature Composition) architecture that borrows ideas from Architecture Description Languages. The method has been applied to multimedia services, but also to mobile telecommunications services – an increasing growth area. Svensson and Andersson (1998) use a state-space method to discover feature interactions in mobile terminals, mainly in the user interface. Amyot *et al.* (1999a) report their experience of using formal methods to describe mobility standards.

Electronic mail is a well-established communications service. Hall (2000b) demonstrates a significant number of interactions among email features. For example, the ability to forward email could conflict with encryption and security.

Plath and Ryan (1998) deal with analysis and detection of feature interactions among conventional telecommunications services. However, the general ideas behind this work apply to other applications such as a lift (elevator) service.

12.4 Formal Methods for Telecommunications Services

This section looks at formal methods for describing telecommunications (voice) services in telephony and the IN (Intelligent Network, ITU 1993h). The services are thus oriented towards supporting voice calls. The emphasis has tended to be on the feature interaction problem (Calder *et al.* 2003, Cameron *et al.* 1993, Cameron and Lin 1998, Keck and Kühn 1998), since this is of considerable industrial importance. *Features* are the building blocks of services. Features may work satisfactorily in isolation. *Feature interaction* occurs when the behavior of several features interferes in some way. Usually interactions are undesirable, but they sometimes have positive benefits.

The following discussion follows the classification of formal methods given in Section 12.2. Techniques for dealing with feature interaction are classified into those that avoid interactions, those that detect interactions, and those that resolve interactions (Bouma and Velthuijsen 1994a). The latter two classes are split into offline and online methods. Formal methods have largely been applied to offline detection. However, there has been work on hybrid methods that combine formal analysis with online resolution (Calder *et al.* 1999).

Space permits only a limited range of typical references. The proceedings of FIW (the Feature Interaction Workshop) are a rich source of further material (Bouma and Velthuijsen 1994b, Calder and Magill 2000, Cheng and Ohta 1995, Dini *et al.* 1997, Kimbler and Bouma 1998, Velthuijsen *et al.* 1992). Further guidance is given by ISO/IEC (1990c), ITU (1992c), and Turner (1993) on the description of services using the FDTs ESTELLE, LOTOS, and SDL.

12.4.1 Model-based

Although B and Z have been widely used for specification in other domains, they do not figure so prominently for specifying telecommunications services. Bredereke (2000), uses an object-oriented variant of Z to define data-oriented aspects of telephony. These are used as additional constraints on the CSP-defined behavior.

Frappier *et al.* (1997) adopt a relational style of specification to detect interactions among IN-like features. Despite the name of the approach, its essence is similar to state-based specification.

Hall (1998, 2000a) separates a foreground model (the essential changes introduced by features) from a background model (the basic telephone service). A LISP-like (functional) language is then used to specify these models and to identify interactions. Erlang (Dacker 1995) was developed by Ericsson specifically to support telecommunications development.

12.4.2 Logic

Features are specified as constraints by Accorsi *et al.* (2000). Model checking is then used to determine the existence of interactions. Muller *et al.* (1992) make use of constraint satisfaction methods.

Temporal logic is employed in numerous methods for service specification. Blom *et al.* (1995) discover interactions by looking for states in which interaction-causing events occur. Lustre, a form of temporal logic, is used by du Bousquet *et al.* (1998, 2000). A Lustre specification of telephone services is exercised with test data, and an oracle judges whether an interaction takes place. Another variant of temporal logic is used by Cassez *et al.* (2001). Jonsson *et al.* (2000) exploit temporal logic to achieve an incremental specification of requirements for evolving systems. The Temporal Logic of Actions is used by Gibson *et al.* (2000) to define fair objects that constitute a telephone system. Capellmann *et al.* (1997) use both a goal model (based mainly on temporal logic) and a functional model (based mainly on SDL).

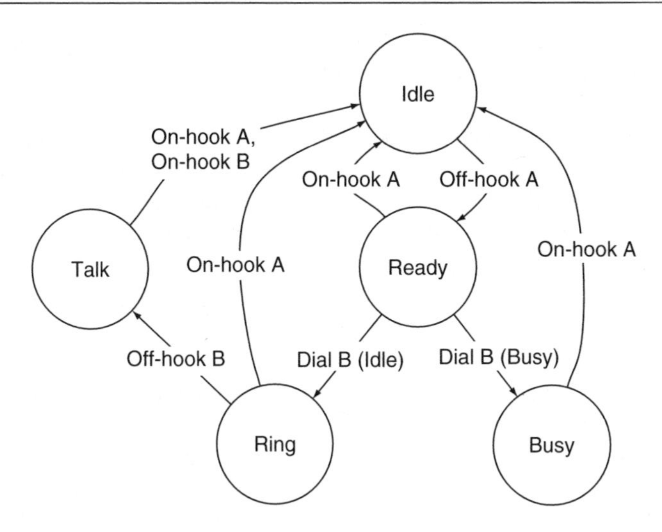

Figure 12.4 Simplified state transition diagram for a telephone call.

12.4.3 State-based

State-based methods are very widely used, although the notations vary considerably. Figure 12.4 is a simplified model of a telephone call, showing the state transitions in response to user actions. This diagram shows a complete call, but the calling and called state machines are often separated. SDL is a popular state-based language for specifying services since it is commonly used in telecommunications (Kolberg *et al.* 1999b, Sinnott and Kolberg 1999a).

Au and Atlee (1997) use their own state-based language to discover interactions through various techniques such as detecting resource contention or assertion violation. The state-based method used by Klein *et al.* (1997) focuses on the addition of features as refinements to an existing telephony specification. The STR (State Transition Rule) technique presented by Yoneda and Ohta (1998) adds priorities to state transitions. Interactions are detected when abnormal states or transitions arise, or when normal states or transitions become unreachable. The work of Dssouli *et al.* (1997) on services is unusual in using timed automata. This allows violation of timing constraints to be detected.

Petri nets are often used as the basis for verification, even for other techniques. As typical examples, Capellmann *et al.* (1997) and Yoneda and Ohta (1998) perform service verification using a variety of nets.

12.4.4 Algebraic

Among process algebras, LOTOS has seen most use on telecommunications services. Faci *et al.* (1997) show how to develop structured models of telephone systems using LOTOS. LOTOS is also used by Kamoun and Logrippo (1998), whose technique tries to find traces satisfying goals that reflect feature interactions. Fu *et al.* (2000) report on interaction detection using watchdogs and observers. Although Stepien and Logrippo (1995) also employ LOTOS, the focus is on the use of types to reflect user intentions. Thomas (1997) combines LOTOS and the modal μ-calculus to develop a theory of features, such as their relative priorities and interworking. Dahl and Najm (1994) use a LOTOS-based technique that looks for potential overlaps among features.

LOTOS is also used to support a number of other techniques. For example, Use Case Maps have been supplemented with LOTOS for analysis (Amyot *et al.* 2000). The building blocks of Turner (1997a) are translated into LOTOS specifications for validation of features (Turner 1998). The diagrammatic service notation used by Turner (2000a) is translated into LOTOS (and SDL) for formal analysis.

12.4.5 Structural

Van der Linden (1994) argues that a proper architecture is important when developing services. In particular, it is contended that separation (isolation) and substitutability (type compatibility) are important issues. Architecture Description Languages influenced Jackson and Zave (1998) and Zave (1998) to develop their DFC (Distributed Feature Composition) method. This adopts a 'pipe and filter' architecture whereby features can be chained to obtain the overall service. A similar building block approach was taken by Lin and Lin (1994), though the method for combining features is different. Turner (1997a, 1998) develops a service architecture based on the combination of elementary behaviors called 'features', though these are smaller capabilities (such as dialing) than is usual. It is claimed that this achieves more consistent services than the SIBs (Service Independent Building Blocks) used by the IN (ITU 1993g).

Formal, object-oriented methods have been applied to service specification. The usual interpretation of an object's service is adopted in the LOTOS formulation by Moreira and Clark (1996). In Lucidi *et al.* (1996), the object-oriented capabilities of SDL are used when modeling IN services. Kolberg *et al.* (1999b) and Sinnott and Kolberg (1999a) also use SDL to develop a framework for multimedia services. The concept of fair objects is formalized by Gibson *et al.* (2000) using the Temporal Logic of Actions, and is used to analyze feature interactions.

Several methods have a graphical emphasis. Use Case Maps have been used successfully to describe voice services (Amyot *et al.* 1999b). Figure 12.5, based on Amyot *et al.* (2000), illustrates the notation. Causal paths flow from the filled circles to the bars. Responsibilities (actions) are shown as crosses, and conditions are shown in brackets.

A dashed diamond is a dynamic stub that may be replaced by a plug-in submap. The left-hand stub in the figure is where outgoing calls are handled. Alternative plug-ins for this are shown above the figure. The first plug-in describes a normal call, where *Out2* is a dummy connector since no action is taken. The second plug-in performs screening on outgoing calls, either permitting or rejecting them. The right-hand stub is where terminating calls are handled. For simplicity, its alternative plug-ins are not shown in the figure.

A formal link has been provided from Use Case Maps to LOTOS, and techniques have been developed to filter out behavior that may be prone to interactions (Nakamura *et al.* 2000). Filtering interaction-prone behavior is also performed by Keck (1998), who uses static rather than dynamic properties of services to identify potential problems.

The CHISEL notation (Aho *et al.* 1998) is also graphical. It portrays sequences of actions taken from models of POTS (Plain Old Telephone Service) and SCP (Service Control Point) behavior. Turner (2000a) strengthens CHISEL by tightening up the notation, particularly for combination of features, and translating the graphical descriptions into LOTOS and SDL.

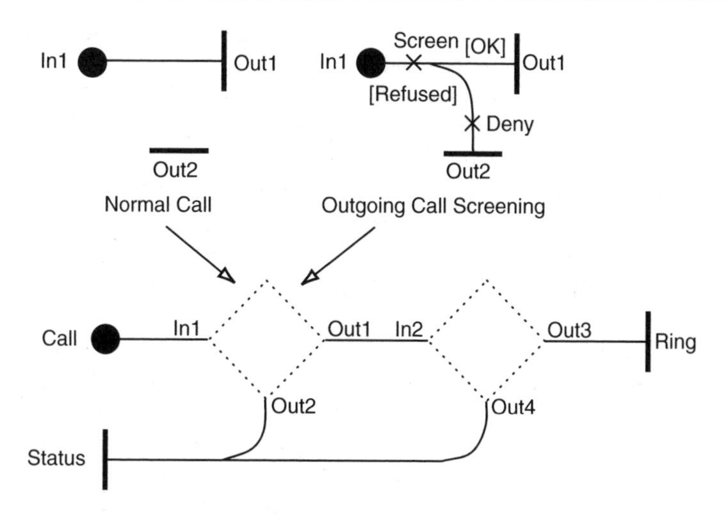

Figure 12.5 Outline Use Case Maps for a telephone call.

Agent-based systems have been used for negotiation of features (Griffeth and Velthuijsen 1994). This technique aims to resolve feature interactions at run-time. Buhr *et al.* (1998) combine agent techniques with Use Case Maps to handle feature interactions.

12.5 Evaluation of Formal Methods for Services

12.5.1 Results

Most surveys of formal methods provide an evaluation of their utility (e.g. Austin and Parkin 1993, Clarke and Wing 1996, Courtiat *et al.* 1995, Craigen *et al.* 1993a,b, Mitchell 1987, Parkin and Austin 1994). Specifically for services, Calder (1998) and Logrippo (2000) provide thoughtful insights into the applicability and value of formal methods.

It has been amply demonstrated that formal methods can be successfully used with a variety of services. Most of the work has been in (tele)communications, but the general principles apply to other kinds of service. Standards have provided a focus since they are international and industrially important.

Useful formal models have been created of services in various domains. The formalization has helped to clarify and make precise the exact nature of these services. More importantly, formal specifications have allowed rigorous analysis of services. The most significant results have been in detecting incompatibilities among services due to feature interactions. However, the ability to prototype services early in their lifecycle has been demonstrated. It has also been possible to use service specifications in creating operational code.

Voice services such as those drawn from the IN have been the most intensively researched. To a large extent, the formal methods community has been catching up with industry. Study of work on feature interaction, for example, will show that most of the services considered have been deployed for some time. Critics of formal methods might claim that little new has been discovered about these services. However, the IN has been a useful testbed for developing techniques that can analyze services. The fact that formal methods can rediscover well-known interactions should be taken as reassuring. When these methods are applied to new kinds of services, there can be confidence in the results.

Formal methods are mainly researched and applied by the academic community. However, it is heartening to see the number of joint studies undertaken with industry, and the number of industrially-oriented applications. Perhaps surprisingly there has not been very much integration of formal methods with conventional development methods, though there are encouraging trends. Techniques like CHISEL are unusual in being solidly based on industrial practice. Certainly there is more scope for integration of formal and conventional methods, particularly in dealing with legacy systems (which form a very substantial part of the telecommunications infrastructure).

The scope of service development in industry is very wide, including business planning, marketing, and customer support. The technical development, where formal methods come into play, is only part of a much larger whole. Formal methods also tend to be developed as isolated solutions. It is not so common for different formal methods to be combined to address a larger range of issues. There is a tendency to expect that formal methods will be used to support the entire development process. In industrial practice, this may not be practicable or cost-effective. As has been advocated elsewhere, selective use of formal methods on particular issues is more likely to be accepted in industry.

12.5.2 Prospects

IN-like services are well established. The thrust in the telecommunications industry is towards new kinds of services such as for multimedia, mobile communication, the Web, the Grid and Internet telephony. Studies of multimedia services have shown that formal methods can already play a role. The early work on formal analysis of mobile services has also proven to be beneficial.

Web services are a major growth area but have not yet seen much exposure to formal methods. It should be noted that nearly all formal work on services has concentrated on what might be termed call control. In a more general setting, such as the Web, a service is just about any functionality that might be offered to a user. A website might, for example, provide news services, financial services, or entertainment services. This is a much broader interpretation of service, though it is one that is common in the business world. The potential difficulty for formal methods is that these services are much harder to pin down and to formalize.

Internet telephony, for example as supported by SIP (Session Initiation Protocol, Rosenberg *et al.* 2002), is growing rapidly. SIP conforms to the Internet philosophy of small, lightweight protocols that are combined to achieve more complex goals. Perhaps inevitably, the companies developing SIP have initially reinvented IN-like services using the Internet as the communications medium. This allows them to provide the same kinds of service as users currently expect. In fact this recreates the kind of service environment in which formal methods have already been proven, so it is certain that existing techniques can be re-used successfully.

An issue that has become increasingly clear is that user intentions are very important when considering services. Many feature interactions can be traced to (mis)interpreting intentions. Resolving the conflict between Call Forwarding and Call Waiting, for example, needs knowledge of what the user really wants to do. Currently, such resolutions are often decided statically on the basis of assumed priorities. In fact, different users or differing circumstances may require alternative resolutions. With a wider range of services, it will be essential to capture (and formalize) the user's intentions.

There is a trend towards allowing users to define policies for how they wish their services to work. General work on policies for (distributed) systems will be useful here (Lupu and Sloman 1997). In fact, this moves the resolution of interactions out to the edge of the network, where it properly belongs. This will simplify future networks, but will complicate user support. It is also likely that users (or more likely their computerized agents) will wish to negotiate the characteristics of services that link them. Formal methods can be used to underpin this.

A major issue will arise from pushing service provision out to network endpoints. Industry initiatives like JAIN (Java API for Integrated Networks, Sun Microsystems 2003), Parlay (Parlay Group 2001) and TAPI (Telephony Application Programming Interface, Microsoft 2000) support third party service provision. Such services tend to be commercial in nature, such as in a call center. With Internet telephony it becomes easy, even natural, for users to create their own services. A user might, for example, decide to handle calls differently depending on who is calling and when. A major challenge, and opportunity, for formal methods is to come to grips with third party and end-user service description. Feature interaction in current networks can be managed because the services and the networks are under the control of only a few authorities. Services on the edge of the network will require more sophisticated definition, as well as run-time resolution of feature interaction.

13

Feature Interaction: Old Hat or Deadly New Menace?

Evan H. Magill

University of Stirling, UK

13.1 Overview

Consider the family automobile or car. It has a good alarm system that works well. It also has a crash protection system with airbags. They are designed as independent systems that may well have been designed by separate teams and indeed the teams may have been located on different sites.

A car thief approaches the car, and strikes the fender (bumper) with a two-pound (1 kg) hammer. The crash system unlocks all the doors. It has subverted the intentions of the security system. This is called *feature interaction*. Two systems (with a software component) sharing a common resource or component have interfered so that the behavior of one has altered the expected behavior of the other. (The thief also ran off with your favorite CD!)

Of course it can be fixed. Alter the security system so that it only responds to the thump if the car is doing more than 5 mph (8 kph). But it's an expensive mistake, and how many systems are there to potentially interfere with each other? Yet communications systems are more complicated.

Feature interaction has been a problem since telephony features were added to SPC (Stored Program Control) telephone switches (exchanges) and exhibits the same characteristics. Feature interaction is where at least two features, which operate successfully independently, interfere with each other when used together within a telephone call. The term *feature interaction* clearly betrays its origins, it is however a very contemporary problem seen not just in telephony, but in IP-based multimedia services. The term feature interaction has stuck, even if the interaction is between email features!

This chapter portrays feature interaction both from a traditional telephony angle, *and* from the perspective of advanced network architectures supporting multimedia services. As such, the chapter is suitable for those who have very little knowledge or experience of feature interaction, and those who perhaps only know of it within telephony. Also, while the focus is on

Service Provision – Technologies for Next Generation Communications. Edited by Kenneth J. Turner, Evan H. Magill and David J. Marples
 ISBN: 0-470-85066-3

incompatibilities within communication systems, the chapter addresses issues that impact any distributed system sharing resources that employ separate software development.

13.2 Introduction

The title, *Feature interaction: old hat or deadly new menace?* reveals a theme that is explored throughout the length of the chapter. However, it is first and foremost a tutorial on feature interaction to provide insight that readers can adopt to their own circumstances. The theme is employed to emphasize the changing nature of feature interaction and its continued relevance to modern services and features in communication networks. Again, the term *communications* is used to emphasize the move from voice services and telecommunications to multimedia services provided by numerous vendors across a variety of networks.

Before looking at feature interaction in any detail it is worthwhile setting it in perspective. It is only one of a number of concerns when a communications service or feature is produced. Indeed this book is testimony to that. Issues such as billing, QoS, testing, customer facing aspects, version control, and performance are just a few topics that need to be addressed to launch a service. For some the impetus to get a service onto the market means feature interaction is often dismissed as being of secondary importance. Yet many realize that the compatibility of the product with existing offerings is commercially important. Perhaps this omission occurs because the term *feature interaction* conjures up images of traditional telephony where many of the interactions are publicly documented, where one operator manages all the features, and where there is little growth in the number of features.

But this would be misleading. While feature interaction *is* an historical term, that *does* have its roots in telephony, it now has a much broader impact. Indeed, feature interaction is often reported outside the communications arena, however the chapter will focus on communications.

Now consider a scene at home set in the not-so-distant future:

> *You are at home expecting some messages and telephone calls while you sit watching a movie on TV. You are watching the movie using a* ***video-on-demand*** *service through a set-top box, which also hosts a* ***follow-me alert*** *service that routes any incoming communications to a window on the TV screen. This service handles email, voice, and instant messaging. With minimum disturbance you can watch the movie while waiting for an important telephone call, and an important message.*
>
> *Suddenly a window pops up to tell you that the important call is here. Using a simple menu you redirect the call to your cell (mobile) phone that acts as a cordless handset at home. It's a business call so you also instruct your extended voicemail service to record the call.*
>
> *Later, a message pops up to tell you that your partner has left the gym. Soon another message appears that was destined for the networked intelligent oven. It contains instructions for the oven to come on, and you ignore it. The oven knows from a* ***traffic-report*** *serrvice that the highway has 15 minute delays so it waits 15 minutes before turning on.*
>
> *You settle back into the movie.*

Perhaps this all seems rather far-fetched, but the technologies are in place to do this *now*. Of course there are still some hurdles to overcome, but this is *not* science fiction. This chapter focuses on one hurdle, the *feature interaction problem*. The term feature is used to explain the incremental change to the underlying telephone service. However, the same symptoms are observed between services and applications. As noted earlier, the historical term is used to explain the problem in a variety of situations.

Communications technology changes are not just limited to the home. The public and private networks used to transmit voice, music, pictures, and movies are also changing. For example, the merging of voice and data networks, with the introduction of protocols such as the ITU's H.323 and the IETF's SIP (Session Initiation Protocol) which allow for a rapid increase in the number of services. But technology is only one factor; deregulation will encourage a free competitive market with numerous companies vying for custom. Again, an important hurdle to overcome is guaranteeing the compatibility between these services, possibly between services offered by different companies.

Will such incompatibilities be a problem? The history of existing networks strongly suggests they will. Take the traditional voice network. In North America this is often referred to as POTS or the Plain Old Telephone Service, while in Europe the rather grander title of Public Switched Telephone Network, or PSTN has been adopted. These networks have experienced serious difficulties with incompatible software. It has required the careful choice of service bundles, and a lot of effort has been made to attack the problem. For example, some telephone switches now ensure that a particular line cannot be allocated incompatible features.

Consider what would happen if the *follow-me alert* service explained above was designed *not* to pass messages on to the original destination, in this case the oven! The behavior of the *follow-me alert* service has interfered with the expected behavior of the oven. This is *feature interaction* and it has ensured that the oven will not operate as expected. It is not a bug, simply a symptom of isolated or separate design and development.

This chapter will explain what feature interaction is, the difficulties it has caused to date, and what the vendors, operators, and researchers have been doing about it. It starts by considering POTS, but this is not enough. On the verge of a rapidly escalating number of new services, the later sections of the chapter look at the challenges ahead. Those familiar with feature interaction in POTS may wish to skip Section 13.3, and possibly 13.4. Also, although not essential, familiarity with the material described in Chapters 3 and 9 is desirable.

13.3 Feature Interaction in POTS

The control software in a traditional telephone switch that provides a Plain Old Telephone Service is often referred to as controlling the *basic call*. That is, it responds to off-hook and on-hooks of the telephone receiver, with responses such as dial-tone and ringing current to telephones. It also routes calls in response to dialed digits. Any additions to the basic call are called *features*. Software that re-routes a call if the called party is busy, for example, would be the *call forwarding busy* feature, CFB for short. This feature is not standalone, it cannot exist without the basic service. In other words the features provide increments to POTS.

In practice, although a switch vendor produced the software for their switch, the sheer scale of the effort resulted in multiple teams operating on a number of sites, often across a variety of countries. This led to features being developed by one team being incompatible with a feature developed by another. This feature interaction often being discovered during system tests. Incompatibilities between vendors are commonly handled by the network operator.

Feature interaction is said to occur when one telecommunications feature modifies or subverts the operation of another one. This is *not* a coding error, but a clash of design requirements that are developed in isolation.

This phenomenon is not unique to telecommunications systems, but can also occur in any large and distributed software system. There is no single characteristic that makes telecommunications unique, it is simply a blend of characteristics that have made it a focus for these concerns. Telecommunication systems are seen as being very large, changing rapidly, employing

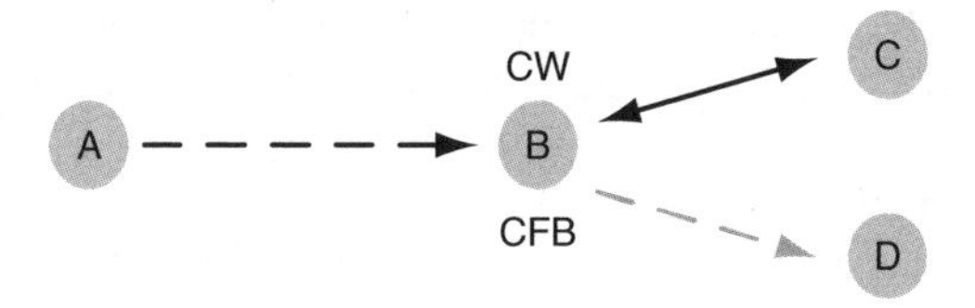

Figure 13.1 An interaction scenario with both CW and CFB features on the same phone.

distributed control and data, being real-time, and needing high reliability and availability. This is quite a cocktail!

Many feature interaction scenarios have been identified. One well-known example is between the feature CW (Call Waiting) and the feature CFB (Call Forwarding Busy). The scenario is played out diagrammatically in Figure 13.1, where User B has both features. A calls B, who is already busy talking to C. What happens next? Does CW provide User B with the call waiting tone, or does CFB forward the call to D? Well there is no clear answer, except to say that they cannot both have their way! Clearly it can be 'resolved' by giving one feature priority, or it can be 'avoided' by not allowing the two features to be on the same phone. However, this too implies deciding which feature has priority. Strictly there is no solution, they cannot be made to work together.

In the previous example, both features are on the same phone. This does not have to be the case, and indeed many cases exist where the features are on different phones within the same call. Consider the following scenario. Here A has the feature OCS (Originating Call Screening) and B has the feature CFU (Call Forwarding Unconditional). OCS blocks outgoing calls to particular directory numbers, while CFU forwards all incoming calls to the subscriber's line to a predetermined number. If User A has OCS with User C's number to be blocked, and User B has CFU to User C, then if A calls B, A is eventually connected to C. Clearly, this violates the OCS feature. This is shown diagrammatically in Figure 13.2.

These two scenarios illustrate interactions involving call control that affect the routing of connections. However, feature interaction in POTS is broader. Consider the case where a # can be used to start another credit card call, *or* it can be used to query your voicemail. If a user uses the *credit card call* feature to query a *voicemail* feature, and happens to operate the key sequence very rapidly without waiting for prompts, situations occur when it is not clear which feature should receive the #. The user's intentions are ambiguous. This results, in part, from a limited number of key symbols being shared between features. In other words, poor signaling to the CPE. This represents a very different form of feature interaction. In an attempt to capture this breadth, a taxonomy has been developed (Cameron *et al.* 1993, 1994).

Scenarios, such as that given in Figure 13.1, with both features associated with the same phone are referred to as *single user* feature interactions, whereas scenarios such as the second case are called *multiple user* feature interactions. Feature interactions are also categorized by

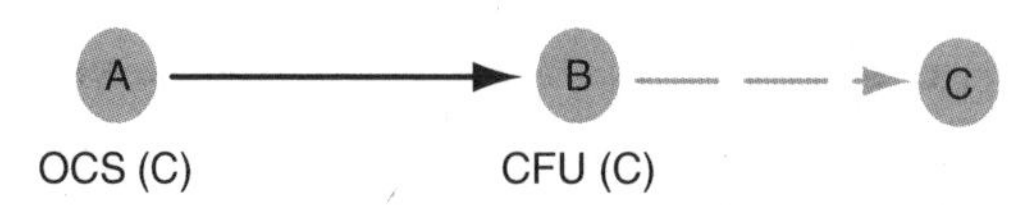

Figure 13.2 An interaction scenario with OCS and CFU features controlling separate phones.

Table 13.1 Feature interaction taxonomy

Acronym	Description
SUSC	Single user single component
MUSC	Multiple user single component
SUMC	Single user multiple component
MUMC	Multiple user multiple component

the number of components they use. So, if both features were hosted on the same telephone switch it would be a *single component* case, if the features were hosted on separate switches there would be *multiple component* interaction. Hence, there are single component and multiple component feature interactions. These are added to the number of users to form four categories in total, as illustrated in Table 13.1.

In practice, single user and multiple user characterization has proved useful, but the single and multiple component characterization has been less so. This results from features being quite distinct identities even within a single network component such as a switch in POTS, or an SCP within an IN. They cannot communicate directly so their colocation matters little. Rarely is this proximity exploited. So, while the two earlier examples show the difference between single user and multiple user scenarios, the number of components in either case is not a matter of concern. Consider the case when the feature interaction scenario is necessarily distributed across components. OCS and *operator services* are an example. In practice, operator services are implemented on network components separate from the remainder of the network. Yet if they were colocated, in their current form they would not cooperate. In contrast, features associated with separate directory numbers on a single (shared) physical line will still be implemented as separate components.

It is noteworthy that the number of components is also important in the application of some techniques to resolve feature interaction. An example is the application of an FIM (Feature Interaction Manager) within an SCP. The IN architecture allows for a FIM to detect and resolve feature interactions. The standards describe the location and environment, but do not specify how it should operate. Research in this area has developed techniques for a single SCP, but they do not port easily to the multiple SCP case (Marples and Magill 1998).

The taxonomy for feature interaction in Table 13.1, is often used for *customer* services and features. That is, they are supplied to a particular subscriber. This implies the feature is associated with a particular telephone line, and that the user is closely associated with the subscriber. OCS or CW is a typical example here. In contrast, *system features* such as 911 or 411 (999 in the UK) are associated with the system. Here the subscriber, a government agency, is separate from the individual users. When customer features and system features interact, reference is made to CUSY (Customer–System) feature interactions. Typical examples include calls with customer features interacting with a system service such as billing.

This feature interaction taxonomy is perhaps the most popular in the feature interaction community. While not regarded as ideal, and although other taxonomies have been presented, it has remained the most cited measure. Marples (2002) used triggers as a basis of classification. Essentially this is close to traditional switched-circuit technology where triggers such as *off-hook* are being used to activate call features. The call control is assumed to be event driven. Feature interactions are described as being in one of four categories: STIs (Shared

Trigger Interactions), SAIs (Sequential Action Interactions), looping interactions, and MTIs (Missed Trigger Interactions). Shared trigger interactions are between features that share a trigger event such as 'call attempt rejected – line busy' Often the interactions are undesirable. Sequential action interactions are where the behavior of one feature subsequently causes a second to be triggered. Here, the interaction may be expected, or undesirable. Looping interactions are those where a feature triggers another, which in turn triggers the original. Two call forwarding features is a common example of these undesirable interactions. This continuous loop may of course have more than two features. Finally, missed trigger interactions are where the behavior of a feature ensures that a second feature does not receive a trigger.

It has been noted that feature interactions are not errors, in the sense of a design or coding errors. The features work correctly by themselves. The problems occur because the features were not designed to interwork with other features. Many feature interactions occur because a second feature breaks assumptions about the first, be the assumptions implicit or explicit. In short, the features are designed and implemented in isolation.

It is possible to be more specific about the causes of feature interaction (Cameron *et al.* 1993, 1994). Generally it is accepted that three broad causes can be identified. First, the *breaking of assumptions* of one feature by another. The assumptions might be about call control as seen between OCS and CFU. But they can also be about naming, data, signaling, and administrative domains. Secondly, *limitations of network support* which can range from poor CPE signaling, seen in the *credit-call* and *voicemail* scenario given earlier, to limited resources within the network, such as the number of SCP processes allocated to a call. Thirdly, *intrinsic problems in distributed systems*. These are problems found in all distributed systems such as resource contention, race conditions, data consistency, and non-atomic operations. Again, the *credit-call* and *voicemail* scenario given above also occurs due to race conditions. So there may be more than one cause to any particular feature interaction scenario.

Breaking assumptions is a typical feature interaction cause and is the main focus of effort. It is fundamentally at the heart of the feature interaction problem, as management of the network cannot solve it, indeed it is independent of any particular network. The second cause can be viewed as having less and less influence. Improved CPE signaling and cheaper computing resources make this cause less likely. There is also the possibility of managing network resources to overcome some aspects. In addition, it is difficult to fix from a research perspective. Effort on the third cause relates to work on distributed systems in general, rather than feature interaction specifically, and so is not generally addressed with feature interaction research. In short, most feature interaction efforts have focused on the broken assumptions.

Given a taxonomy, has it been possible to provide a solution? A general solution has not, as yet, been identified, nor has it been possible to provide a set of specific solutions to each scenario *on any reasonable scale*. Consider the example above with OCS and CFU. OCS assumes it has total control of the call. It simply screens the dialed digits. No attempt is made to liaise with any call forwarding features such as CFU. It, or any feature, was not considered in its design. In *hindsight* of course, OCS could be improved to monitor the network signaling to ensure a connection is not made to a screened number. Indeed, some private systems do this and hence differentiate between OCS and ODS (Origination Dialed Screening). The latter making explicit a straightforward screening approach.

However, in general, it is not possible to retroactively alter the numerous existing features to correct such broken assumptions. There are too many, and new features are constantly being added. In fact the relationship between features grows with the *square* of the number of features, So with ten features, there are one hundred pairs of features, each pair often having more than one feature interaction scenario. Moreover, details of new features may not be released, as they are commercially sensitive in a competitive market. In short, to design or alter features to work with all other features is impracticable.

The chapter has deliberately limited the number of examples so as not to swamp the text. Yet there is a danger that this presents a view that feature interaction is a limited problem. It is not. Those wishing to seek a broader range of examples or indeed more detail on the taxonomy are referred to Cameron *et al.* (1994).

13.4 Response of the Communications Community

Feature interaction is an established problem in POTS, the IN, and in telephony generally. This has allowed time for a response. Three responses can be identified, that of the network operators, the equipment vendors, and the research community. While the operators' response is the closest for users, the concepts and techniques often originate in the research community. Hence this response is reported first, followed by the other two. It is hoped that this division will provide an insight from the experience of others that readers can modify to their own circumstances.

13.4.1 Response of researchers

As a research topic feature interaction is addressing a *problem*, it is not a single technique or the study of scientific principles. Hence, it focuses on technological pull, rather than scientific push. So it has drawn researchers from a broad range of topics and experience, and enjoys a healthy mix of engineers and computer scientists. But how can such a broad range of work be categorized? Grouping such as *online*, *offline*, *architectural*, *detection*, *resolution*, and *avoidance* are often used. The most recent survey of the research (Calder *et al.* 2003) splits the work up into three areas, *software engineering*, *formal methods*, and *online techniques*. Here the description is limited to these general approaches, for those seeking a deeper insight the relevant research surveys are recommended (Calder *et al.* 2003, Keck and Kühn 1998).

Although the research literature on feature interaction is well distributed, much of the work is captured in the proceedings of a series of international workshops on feature interaction. The series began in 1992 and has become the premier forum for feature interaction research. The proceedings are available from the second workshop onwards (Bouma and Velthuijsen 1994a,b, Cheng and Ohta 1995, Dini *et al.* 1997, Kimbler and Bouma 1998, Calder and Magill 2000, Amyot and Logrippo 2003).

The *software engineering* area captures the work that alters the creation process of features and services. This may be *direct*, in that new development processes are added to handle interactions. An example here is filtering (filtering identifies likely feature interaction scenarios, which are used to direct a subsequent manual effort). It may also be *indirect*, in that increased rigor introduced from (say) a new notation helps mitigate the potential for feature interactions.

While direct approaches may introduce new *processes*, such as filtering, that are specific to feature interaction, they can also *introduce techniques* employed more generally in software engineering. Use case maps are a candidate here. The techniques are often adapted to provide feature interaction detection capabilities.

Most of this work concentrates on the early stages of development. As a research area, this has probably offered the most to practice, and indeed it is this area of research that has a significant overlap with vendors and operators.

It is clear from the previous chapter that the *formal methods* area is a very active area, indeed it forms the largest body of work within the feature interaction community. By employing formal methods to model features, services, and sometimes networks, it is possible to reason about the potential for interactions.

Three broad formal method approaches have been identified. First, to capture the properties of features and to ensure the properties are consistent and can be satisfied. Secondly, the behavior of services and features are captured in an 'executable' way, say through automata. They can then be analyzed for deadlock and ambiguity. And finally, the two approaches are combined. The behavior acts as a model against which the properties can be checked. This is often referred to as *satisfaction on a model*. In contrast, the first approach is often called *satisfiability*.

These essentially *offline* approaches, in that they are not used within a running communications system, often require detailed knowledge of features and services and are often criticized for this. They also have scalability problems, as the number of comparisons grows with the square of the number of features. Moreover, as there is a move towards a more liberated regulatory environment, details of all the features may simply be unavailable. Either through sheer numbers of new offerings from numerous providers, or through propriety concerns. There are also some difficulties with the techniques themselves, for example, is the description complete?

As a result, some researchers have employed *online* methods where the approaches are built into the network to handle interactions as the features and services execute. Other perceived advantages of online approaches include support for a quicker time to market, feature operation in a natural environment, and a better resistance to future changes. In this environment it is assumed that features and services meet for the first time on the network and that knowledge of the other features may be limited, especially if another vendor provides them.

Two broad approaches have been identified, first using a *feature interaction manager*, and secondly using forms of *negotiation*. The former requires a point of control (the manager) that monitors and modifies the flow of signals between the features and services. This requires a set of rules or algorithm to follow. The rules can be set up in advance (*a priori*) or they may be acquired at run-time. The literature often gives case studies using *a priori* knowledge, but acquisition of rules at run-time is employed much less. Unfortunately, using *a priori* rules removes many of the advantages of employing online approaches! Another concern is how well feature interaction managers can cooperate once they are distributed across the network(s).

In contrast, negotiation offers a fundamentally distributed approach. Again, rules are required to produce a coherent negotiation, and again the rules have concentrated on *a priori* information. Although there is more potential for more abstract rules that are not service or feature specific.

A difficulty with run-time approaches is that detection alone is insufficient. The technique must also be able to resolve. Detailed *a priori* knowledge has been the most successful at resolution, but removes many of the advantages of online approaches. It has to be said that run-time approaches in general have proved difficult. The volume of work in this area has never been very large, and now appears to be dropping. Although this may change with increasing deregulation, and new network architectures.

It should also be remembered that with the exception of the software engineering area, the impact of feature interaction research on practice has proved frustratingly limited. Also, until recently much of the work concentrated on POTS. But with new technologies this all seems rather out of date. This is perhaps unfair, as to experiment on any form of interactions, in whatever type of network, there is a need to have information on *known* interactions. The feature interaction problem is not formally defined, so all experimentation is, by its nature, empirical. Hence, there must be a set of *known* problems, and the richest is in the POTS environment, although some multimedia type interactions have been reported (Tsang *et al.* 1996, 1997, Zave and Jackson 2000).

13.4.2 Response of operators

Although there is a widespread expectation of more deregulation of the communications market, at the moment most operators control both the network, and the services and features provided on that network. This provides a single unified 'authority' to oversee and control feature interactions. In addition, the operators have a limited number of services and features, which makes the problem manageable. Either the removal of a single controlling agency, or a large increase in the number of offerings would greatly exacerbate their problems.

In practice, many operators simply manage the problem manually. Equipped with knowledge of known interactions, especially for POTS features that are well documented, they can ensure features are kept apart. Pragmatic techniques such as filtering act as aids. Here, a straightforward filter looks for combinations of features that are *likely* to cause interactions. This may be as simple as *if feature A blocks, and feature B diverts, then mark the pair as a likely interaction candidate*. These candidates can then be processed manually. In short, the filter directs the manual effort. Other filter techniques simply use records of known interactions, others use more sophisticated approaches.

13.4.3 Response of vendors

In part, the response from vendors is similar to the operators. Again, the number of features are limited, and there is one controlling agency, albeit over a smaller set of equipment. However, in the future, while deregulation may cause a large surge in the number of features, it is unclear if this will be met by existing vendors or new entrants. It seems reasonable to assume that existing vendors will see an increase in the number of offerings. At present they simply ensure that all their offerings are compatible, and note those which are not. Indeed some vendors build in systems to ensure incompatible features cannot be placed on the same telephone line. The detection process is essentially part of the system test, and regression test processes.

However, vendors are able to go further than operators. They need not simply test for interactions in a reactive manner. They can modify their development process to take feature interaction into account during the early stages of development, and they can alter their service architecture to avoid interactions. Some have done so, most notably BellCore (Cameron *et al.* 1997, 1998). Here, they add feature interaction analysis into a modified service creation process and make strong use of tool support. There is potential for the introduction of new languages (Aho *et al.* 1998, Turner 2000a), and others have advocated building filtering into the early stages of the development cycle (Kimbler 1997). Some have service architectures that limit the potential for interaction at run-time (Lin and Lin 1994, Jackson and Zave 1998).

13.5 Regulatory Changes

The deregulation of the telecommunications market is having a major impact on feature interaction. Without the opening up of the services market much would remain unchanged from POTS, and indeed many of the technological advances that are discussed in the following section would be largely irrelevant without deregulation. In short, advances in network technology *and* deregulation are both required to provide an expanding services market. It is this synergy that exposes all the 'network players' to feature interactions.

Thus, the following section on technology advances implies a deregulated environment.

13.6 Technological Changes

Here we assume that readers are familiar with many of the technological changes current in communications. This section onwards concentrates on the *impact* of these changes to the feature interaction problem, and indeed on the changing *nature* of feature interaction. Noting the concentration of feature interaction studies within POTS, how does feature interaction stand in light of the rapid technological change in advanced communications networks? Elsewhere in this book there is a description of the merging of telecommunications and computing technology to produce an increasing emphasis on packet switching, object-oriented technology, the distribution of services and applications, and an increasing user control or programmability of networks.

In POTS, services and features are provided centrally and are strongly embedded within the network. The user buys the service from the operators and has little control over the functionality. In return, the operator is responsible for maintenance and support. The operator in turn is likely to subcontract the service creation and feature development to their switch vendors. With the advent of the *Intelligent Network* this relationship is largely unaltered, and the service is still within the network. In short, this is the traditional telecommunications philosophy to service provision. New protocols such as SIP, and new APIs such as Parlay (see Chapter 11), have a major impact on the location, content, and control of services. Services can now be located on the periphery of the network. The services can therefore be distributed both within the network, and around the outside of the network. In the latter case they can enjoy both control of the switching functions of the network, and be provided by a myriad of service providers. From a technical perspective they need not be provided by the network incumbent. Indeed, it is expected that *users* will be able to write and provision services. This implies a huge surge in the number of features, and the loss of any controlling body attempting to harmonize behavior. In other words, the telecommunications market will migrate to resemble that of the Internet. While this greatly increases the potential for service and feature interactions, there are also strong concerns about billing, QoS, and security. However, feature interaction remains the focus here.

In addition to an increase in the number of services and vendors, there is a more subtle implication of these technological changes. An implication of protocols such as SIP and associated notations is that users, either as individuals or enterprises, can provide their own services. But *how* services are provided by service vendors may also change. The vendor may no longer wish to support the service in the all encompassing manner provided by the telecommunication companies, but prefer to sell them in the manner of PC applications. That is, the ultimate responsibility moves to the user, albeit that separate support service could be purchased. It is already commonplace to find support for computer applications supplied free of charge through websites. This includes upgrades, software patches, and limited technical support. Comprehensive support tends to involve costs to the user. From a service interaction perspective, it will be the consumer who worries about feature interaction when two interacting communication services are provided by separate vendors; not the service vendors.

The use of circuit switching in the existing POTS network sets it apart from the packet switching used within the Internet. Circuit switching provides a single voice stream, a single type of stream. In contrast, packet switching allows a combination of stream types. Voice, picture, and video can be *combined* to provide exciting new services. Emerging network architectures and protocols such as RTP and SIP allow for voice services to be integrated with other types of media steam. The combination of media streams will act to increase the number of services, but it will also increase the complexity and diversity of service offerings. No longer will interactions simply involve one media stream, they will involve a number of streams that may *change* during the session.

But the move to packet switching has further implications. An essential element of any communications network is a signaling mechanism. A means of transporting control signals around the network. This is fixed in traditional telephony, employing specialized components, protocols, and transport for the purpose. The *type* of information that can be passed is static or at least changes very slowly. In contrast, with packet switching the same transport mechanism (IP) can be shared between control information and payload. This permits lightweight control protocols to be added relatively quickly for particular purposes; SIP to control session, RSVP to control bandwidth allocation. Moreover, it allows middleware such as CORBA (or at least specialized communication ORBs) to provide inter-application and service communication. This removes a great difficulty with feature interaction in POTS. The signaling does not allow interacting features to locate each other, or to communicate with each other, at best they may be aware of each other through the unexpected behavior of a shared resource. Removing this 'signaling bottleneck' offers a new paradigm to feature interaction detection and resolution. So while many technological changes exasperate the problem, this offers an exciting potential to attack the problem.

Also, a class of feature interaction may vanish. There are feature interaction scenarios involving features that use complex key sequences on a traditional handset. An earlier example described how if there is more than one feature active on the phone, a # key for example, may go to the 'wrong' feature. With improved terminals, this class of feature interaction is expected to go.

Technological changes outside the network also have a major impact on feature interaction. The introduction of increasingly sophisticated, and increasingly mobile, communication peripherals heralds a major impact for feature interaction. Apart from simply increasing the number of services, they are expected to operate in different 'contexts', moving from one context to another. Consider a PDA with wireless communication capabilities incorporating a powerful *do not disturb* service. This will operate differently at an office desk, to when in the boardroom, to when it's at home. It may receive ad hoc user control, and it may receive network control. Context sensitive services subject to rapid change will prove challenging to feature interaction detection and resolution.

Not only may the advanced mobile peripherals support different roles, but when used with advanced networks they can alter the network. Consider *Programmable Networks*. Here the network components may be controlled, for example by users, operators, and services; they are programmable. Typically from outside the network through APIs and protocols, but also from within where some of the packets transferred through the network form the programmable control. This latter approach is often referred to as an *Active Network*. It is also foreseen that this can be taken one stage further to form *ad hoc* networks. Here, Intelligent Peripherals will form themselves into cohesive (logical) networks on an as-required basis. As the name suggests, it is predicted that these sessions will be rather dynamic. At a recent research meeting (OPENSIG 2001) it was stated that the number of peripherals is expected to number in the *trillions* opening up the potential for a huge number of dynamically changing sessions. The changing nature and the scale here is very significant to feature interaction. Solutions cannot afford to be heavyweight, rather they need to be efficient and require few network resources.

Changes to software technologies also play an important role. Earlier there was a description of the application of service architectures and of notations to provide a development infrastructure to avoid interactions. It is predicted that this supportive role will develop further. The increasing move to object-oriented technology allows a separation of concerns in constructing services. That is, the access to a service is separated from the service control, which is separated from the connection control. Again, this rigor helps avoid interactions.

While many changes to the *software* technology employed in creating services can be expected to help interactions, others are expected to exasperate the problem. Traditionally, services were developed in a vertical manner, that is, they were written with a particular network, or network component, in mind. In an attempt to provide a more flexible alternative, a horizontal paradigm is being investigated. Here the service is developed independently of the network on which it will operate. This allows the development of services that are independent of the network with all the productivity gains this implies. It also implies services operating with a number of different APIs. Indeed this has been taken to the stage of constructing services from a number of components in both a horizontal and vertical manner. This has already driven the IEEE P1520 standard (Biswas *et al.* 1998), and can only be expected to grow with the maturity of component technology. Software technology is going from a monolithic construction of services and features to a 'building block' approach. The implication for feature interaction is interesting. No longer is it simply a matter of looking at interactions between services, but also within service compositions.

The software technology behind the APIs also increases the potential for feature interactions. Typically the APIs offer two aspects, not only the API itself, but also a set of service capabilities. It is predicted that a third party service could suffer feature interactions with the API service capabilities. Presumably care would also be required to ensure the set of service capabilities did not have feature interactions.

So imagine a picture where a network will have embedded services (both built-in and (possibly) through Active Network technology), advanced peripheral devices with embedded services, third party services (composed of a number of components), user-constructed services, and service functionality from API capabilities. Of course there is likely to be more than a single third party service, and more than one network. It is a very frightening prospect! The following sections consider this more closely.

13.7 Future Services, Old Interactions

The chapter has now come full circle. At the start of the chapter, services in the home of the not-so-distant future were described. What sort of incompatibilities, what sort of feature interactions, are expected? Earlier it was noted what might happen if the instructions were not passed on to the oven. But the scenario is fertile for many others too, for example, if the home PC hosted a *terminating call screening* service, the *follow-me alert* service would accept calls that might otherwise be screened.

First, note that many of the feature interactions seen with POTS will still exist! Some of course will not. Feature interactions, such as *credit card calling* and *voicemail* from traditional handsets with their inherent signal ambiguities, are removed, there is no longer any constraint from the traditional telephone handset. Also, some features such as *call waiting* are no longer required, again because there is no longer a requirement for a simple telephone handset to *control* the call. Any terminal supporting a number of call terminations (such as a key-set or a desktop PC) does not require call waiting.

However, features that bar (screen) connections combined with features that redirect connections are expected to act in their normal manner. Many of the feature interaction scenarios found in POTS contain a feature that blocks calls (such as *origination call screening* or *terminating call screening*), and a feature that redirects calls (such as *call forwarding busy* or *answer phone*). Indeed the example given in Figure 13.2 at the start of the chapter is just such a case. The media stream may no longer be simply voice, but the potential for feature interaction remains. This is examined more fully in Tsang *et al.* (1996, 1997) and Zave and Jackson (2000).

Consider the case of voice over IP employing SIP (Session Initiation Protocol). Assume that the 'originator' has a feature barring invites to certain identifiers, and that the destination address redirects the invite to a third party that the 'originator' barred. The result is identical to *originating call screening* and *call forwarding* in POTS. The *cause* in both cases is the same, the forwarding aspect breaks the assumptions of the screening aspect. Of course, unlike any *multiple user* case in public POTS networks, once the incompatibility is identified it potentially can be solved. For example, the barring feature can check the returned OK message from the far end; although this is non-trivial as the 'originating proxy' is stateless.

But of course the real issue in feature interactions is the new *unknown* feature combinations, especially with rapid growth in service provision from a large variety of sources.

13.8 Future Services, Future Interactions

Familiar interactions are a potential hazard, but so too are new feature interactions. Traditional POTS feature interactions can be extended to the new services expected on emerging networks. This is as true for voice over IP as it is for multimedia services. With an enormous operational experience of POTS, the feature interaction community are well-rehearsed in possible interaction scenarios. However, with the possible exception of email, there is not as much experience with new styles of service. Nonetheless there are many potential difficulties (Zave and Jackson 2000).

Services embedded in peripheral devices: Increasingly there are more sophisticated features built into handsets and other, often portable, peripheral devices. The potential here for interactions is large. Consider the Parlay demonstration feature that in effect gives group ringing. An important customer-facing employee may have a number of landline and mobile terminal devices that may have a range of 'numbers'. If any one number is called, all the terminal devices in the group 'ring'. Yet if the call is answered using one particular device, and within the group there is an 'unanswered' mobile handset, its *internal features* will inform the user that there is a missing call. The owner of the handset may in time return the call, especially if it is an important client. The client will of course be confused as they had their call answered – and hopefully their query handled!

Resource conflict: Networks are engineered to provide adequate services for the vast majority of users, the vast majority of the time. As a shared resource a useful 'statistical gain' can often be achieved. This is not true for access to networks from individual users at home. Historically, the access satisfies the requirements of one voice channel, but if the user tries to use it to carry more demanding traffic, it can offer a poor service. This is true of broadband access too. If there is 2 Mbytes per second broadband access and there are two services that *each* use 2 Mbs – then they cannot be used at the same time. There is an interaction. Remember feature interactions are about the inability of features and services to share resources successfully. As services and features carry more multimedia traffic, with an increasing need of access bandwidth, it is expected there will be an increase in this form of feature interaction.

Of course the resource need not just be access bandwidth. It could be any resource with a finite capacity, such as shared recording devices.

Interaction modes in other media: No longer will communication media be used to simply carry voice. Other media streams will be carried too (video, text, pictures) and these can introduce feature interactions that are associated with a particular medium. Perhaps the best documented is email (Hall 2000b). This study revealed 26 feature interactions between email features. Many had characteristics seen in traditional telephony. For example *forward message* and *filter message*. A user forwards their email to another domain where (unknown to them) it is filtered out. The forwarded messages simply vanish.

Other email interactions do not have a traditional telephony flavor, *encrypt message* and *autoresponder* are examples. The email autoresponder can reveal the subject line of the encrypted message to others.

The email study is pertinent as it reveals new types of interaction from new types of feature. It also emphasizes the ease with which feature interaction can happen where the features are developed separately and are employed across a range of responsibilities. Twenty-six[1] interactions were detected from ten features!

Swapping and mixing media: This is expected to be an important aspect of services in the future. Already it is expected that SIP will allow voice call attempts to be routed to Web pages. In other words, the media is changed by a recipient of the invite message. But what if the 'originator' does not want to do so, or perhaps has a terminal device that cannot support Web information? Again there is feature interaction.

Another documented scenario is mixing voice and text media (Zave and Jackson 2000). It is not uncommon for many to prefer to use an instant text message in preference to the telephone, indeed it is not unreasonable to expect users to use text to check availability for a telephone conversation. Often email is used for this too. Some users may wish to subscribe to a service that bars calls unless they have received a text message seeking guidance on availability. Also, imagine that the potential originator of a call is using a dedicated multimedia phone that has been designed to establish a voice connection before any other media can begin. Once again there is feature interaction.

[1] 26 are given in the paper proceedings, 27 were presented at the workshop itself.

13.9 Future Services, More Complex Interactions

Future services and features will suffer from both well-known and new types of interaction. Yet this says little about the *volume* and *complexity* of feature interactions in the future. There are many regulatory and technical changes that point to a much larger volume of complex feature interaction scenarios.

Service mobility: In POTS, features and services are associated with a fixed telephone line. It is expected that there will be services that are associated with a user. So whatever handset they use they obtain the same set of features. This is not to be confused with mobile handsets where a feature set for the handset is fixed, rather like a traditional landline.

In addition to associating features with individuals, it is expected that feature sets will be associated with roles. Here, an individual may have a number of roles, perhaps a business role and a private role. Of course there will still be sets of features associated with individual handsets. This creates the potential for a huge dynamic mixing of features and hence inevitably feature interaction.

Feature packaging: As more service sets are purchased from different service providers, the potential for feature interaction occurs. You may have purchased a video-on-demand service that incorporates a screening feature to ensure that 'adult material' is barred. However, if a user has a videoconference service from another vendor, it is unlikely to be able to access the screening feature and so access to unsuitable material is possible.

An alternative scenario is buying video-on-demand at domestic rates, and using video-call-forwarding to supply video to a video telephony service in commercial premises. A single vendor is unlikely to supply commercial and domestic services within a single package, yet they cannot stop such abuse.

Markets: The lesson from the PC and Internet market is that services at the periphery of the IP network are huge and growing. In contrast, the embedded telephony services have been moribund. It is in this latter market that feature interaction has been observed and managed to date. All this is set to change, and the sheer number of service offerings will be vast. The market model for communication services is expected to gradually move towards that of the Internet, and many analysts expect a large surge in the number of services from about 2003. Although the latest downturn in the market may delay this, it is still inevitable. The potential for consumer disappointment is of course great, although it can be argued that the PC software market has been very tolerant.

13.10 New Challenges, New Opportunities

A case has been developed that the threat from feature interaction will grow in the next few years, yet the freedom from signaling constraints may make this an easier task to overcome. Rather than features and services interworking without explicit communications, services are able to manage their interworking directly.

Within feature interactions, *interworking* and *incompatibility* have clear interpretations. Components must interwork to share a (communications) resource. The components may interwork *explicitly* through an exchange of information with each other, or *implicitly* through this shared resource. In the latter case, the shared resource, such as a telephone or call session, links the controlling components together. The components often do not know that the other exists, however they may be aware of another controlling influence from the behavior of the resource. Explicit interworking is possible in advanced session control architectures and was first seen with TINA (Kolberg *et al.* 1999a). The interworking is controlled and expected. Implicit interworking is common in traditional telephony and occurs simply because the components (often unknowingly) attempt to share a resource.

When components interwork to share communication resources, they are *compatible* if the joint behavior of the resource is acceptable. Typically, this would require that the behavior of the resource did not break the expectations of all its controlling components. Here, compatibility does *not* refer to simple coding errors, nor to the adherence of interfaces or protocols, but to the adequate behavior of a resource under the joint control of interworking components.

The ability to have explicit controlling communications between services encourages online methods of feature interaction resolution. The volume of new offerings from a wide range of providers will limit offline approaches to checks between services from a single service vendor. Here, software engineering will be a powerful tool in helping detect and resolve interactions.

In the future, expect considerable effort on how services can exchange information to handle interactions at run-time. And, more importantly, what will that information be?

13.11 Summary

While the feature interaction problem may be considered an old problem, it is set to become increasingly important. The deregulation of the telecommunications market is expected to rapidly increase the number of services deployed in the telecommunications networks that aim to control the set-up of connections and the establishment of streams. The increase in numbers is facilitated by initiatives such as JAIN and Parlay. For the first time, third party service providers, independent from the network operators, will be able to provide services. This allows for much more competition in the telecommunications industry. The speed with which services can be implemented and deployed will be an important business differentiator.

In sharp contrast to these new business opportunities, service creation is still a very costly and long process. One of the reasons for this is that extensive validation and testing activities are required. This has partly been addressed by developing new service architectures and APIs as well as service creation approaches. In spite of all these activities, the feature interaction problem remains important when services interwork.

Part III

The Future of Services

This section very much looks forward, and so we cannot be certain that everything we write is correct. The best that we can do is to make predictions based on where we find ourselves today. We draw on the technologies, protocols, and approaches currently in our research laboratories and universities to establish observable trends. Here, we see an array of competing and synergetic advances, yet it can be difficult to predict how they will interact. Not only is the outcome uncertain, but new drivers will always be impinging. No one can be certain, but the next three chapters provide a cognitive reflection on future possibilities.

Chapter 14 (Advances in services) argues that a miscellany of services and infrastructure will only find coherence through convergence. Indeed, it argues that convergence is a 'mega-trend' with a deep influence on communications standards and markets. Yet how can such a divergence of technology converge? It is strongly argued that *context* will be an essential element. The chapter describes what context is within communications, and explores the technologies and protocols that support it within communication systems. This 'glue', it is argued, will provide an almost limitless array of new, exciting services. Telephony, wireless devices, home networks, and data services will become a seamless whole. Media stream services and content services will amalgamate to provide a rich source of services, and of course revenue. The chapter is sprinkled with concrete examples to provide glimpses of an exciting future.

In contrast, Chapter 15 (Evolving service technology) looks deeper into the techniques and technologies that will be applicable in this new world, as we move from procedural event–response models of system operation to much more collaborative environments. Here, Software Agents operate asynchronously on the behalf of users, performing their mundane, repetitive, or time consuming tasks and generally representing them in the electronic world as their 'proxy'. In this chapter, Constraint Satisfaction, Artificial Neural Networks and Genetic Algorithms are all discussed as ways of finding solutions to problems which are not amenable to straightforward procedural programming. As the communication environments in which we operate become more complex, with more actors and more variables, these techniques will, by necessity, become much more important in ensuring that users' real requirements are met.

Finally, Chapter 16 (Prospects) takes a broad look at the future for services. In addition to a rounded look at technological changes (with some emphasis on devices), the chapter includes a look at the commercial and market environments. These are important drivers that have an impact on the uptake of technological advances. Advances that may in turn cause significant societal changes. The changes due to mobile telephony being a case in point! In short, the chapter considers a more automated future.

14

Advances in Services

James T. Smith

Consultant, USA

14.1 Introduction

The chapter starts with consideration of the *mega-trend* forces behind the explosion of new services that the communications services industry is now witnessing across all communications media (wireless, broadband). This is on a scale as great as, if not greater than, the recent explosion of Web-based Internet-enabled services that were introduced during the late 1990s and early 2000s.

First, those 'mega-trends' that are exerting the greatest impact on the direction that new service development is taking will be considered. Answers will be proposed to such questions as to how and why such 'mega-trend' changes are expected.

The discussion begins with an extrapolation from the already observable consequences of convergence on communications architectures and infrastructures, and leads to their impact on communications services. Hence, a formal definition will be offered of what constitutes service convergence and how it is characterized. In particular, the primary role that context management is expected to play in the design and the deployment of converging services will be examined.

Key technologies needed for the successful development of converged communications services will be identified. They will be discussed from the perspective of how they contribute to the development of converged communications services. In particular, their role in the provision of context management is considered. Elements of their technical details are addressed in other chapters.

Examples of how these technologies are already being applied to transform the communications services industry, will be examined in the effort to elucidate the underlying design principles that developers of 21st century communications services must master. In particular, the emerging market of SOHO (Small Office-Home Office) networks provides a rich exemplary environment for how services that exemplify the principles discussed here will be developed.

The discussion of communications services would not be complete without also considering the corresponding evolution of communications platforms that is accompanying the development

Service Provision – Technologies for Next Generation Communications. Edited by Kenneth J. Turner, Evan H. Magill and David J. Marples
 ISBN: 0-470-85066-3

of new services. The telecommunications services industry has witnessed tremendous improvements from the days of the standard black phone, and later the Princess handset, and now is on the verge of a multitude of newly emerging possibilities.

The much discussed and contemplated concept of the Internet appliance is prototypical of this new communications platform development. In like fashion, this platform development is exhibiting the consequences of service convergence. In some cases, the communications platform is assuming new roles, typified by the cell phone that also has messaging capabilities, a built-in calculator, and a day-timer. On the other hand, devices not generally thought of as communications appliances are gaining such capabilities, typified by PDAs and even wristwatches that have various forms of integrated communications support.

How well newly proposed services address these considerations will exert a profound influence in determining their potential for success. Hopefully, this higher level perspective of what constitutes successful 21st century services will prove invaluable to the student of communications service development.

14.2 Service Convergence

Several *futurist* writers have proffered their lists of century-defining mega-trends. The classic list perhaps is that provided by Donald Tapscott in his book, *The Digital Economy: Promise and Peril in the Age of Networked Intelligence*. (Tapscott 1996). Of these, the one mega-trend most apropos to communications is that of *convergence*.

The mega-trend of convergence is accomplished by the delicate balancing of the two forces of mass production and mass customization, together with a balancing of the *pull* of customer demand in juxtaposition with the *push* of technology enablement.

The element of *mass production* implies the economies of ubiquity, commonality, consolidation, standardization, and interoperability, etc. In contrast, the element of *mass customization* implies dynamism, adaptability, and personalization, etc. Because of the many recent technological advances, the service developer no longer has to choose one of these forces at the expense of the other. On the contrary, this counterbalance demands equal consideration be given both to service ubiquity (the manifestation of mass production), and to service personalization (the manifestation of mass customization).

The effects of this convergence take two general forms: *sustaining convergence* that leverages the resources of mass production, and *disruptive convergence* that leverages the creative potential of mass customization. Both of these must be considered to assess fully the impact of convergence on 21st century communications services.

These two forms of convergence, directly traceable to the general concepts of sustaining and disruptive technologies, are manifested at every level of the communications hierarchy from the physical layer to the applications layer.

Consequences of the *convergence mega-trend* have been transforming all levels and all aspects of the communications industry. They are now beginning to impact communications services as the highest level of the communications value chain. The focal point of this service convergence ultimately is the customer. The customer figures strategically into the core of all the subthemes that define the *convergence* mega-trend.

To set the stage for a discussion of how convergence will profoundly influence the characterization of 21st century communications services, we begin with a formal definition of converged services, one that reflects the general characteristics of convergence just presented above.

Converged services are services that so leverage each other that the value of the resulting whole is greater than the sum of its parts. This convergence enables capabilities that the standalone component services could never deliver simply by being offered, or packaged, as independent components.

Recent examples of this service convergence phenomenon include the increased use of multimedia presentations that incorporate not only rich graphics, but also animations and sound effects, and the development of unified messaging services that attempt to manage email, voicemail, faxes, and call control as an integrated package.

Nuances of this definition will be explored in the sections that follow. First, consideration will be given to the impact of convergence on other layers of the communications hierarchy. Here the consequences are already more noticeable as they are affecting aspects such as the communications infrastructure and technology adoption.

14.2.1 Metcalfe's Law

The convergence mega-trend is more or less a direct consequence of Metcalfe's Law (Boyd 2001) which states that 'the usefulness, or utility, of a network equals the square of the number of users.'

Metcalfe's Law in its simplest form is a statement regarding the interconnectivity of a fully connected N-node network. There exist an order of N-squared one-to-one connections, and hence N-squared potential reasons why one could justify participation in such a network. As a network expands in geographic reach and size (or number of nodes, number of connections), the opportunities existing for potentially profitable communication and interaction grow quadratically.

Some consequences of this convergence have already been realized at the lower layers of the communications hierarchy. Technology has enabled tremendous growth in the Internet, to the extent that other networks are being relegated to niche applications. Indeed, specialized proprietary networks are become increasingly uncompetitive and difficult to justify.

Exceptions to this pattern still exist under particular circumstances, e.g. extreme security requirements in the military establishment. Still, even such special cases will adopt core Internet technologies and capabilities to benefit from Metcalfe's Law.

Such effects of convergence as these are due to the consolidation aspects of convergence. For many discussions, the concept of convergence is viewed as a process of consolidation. This consolidation includes both technologies, say the choice of protocol standards and physical media, as well as the aggregation of businesses, and the transition or re-invention of ISPs (Internet Service Providers) into ASPs (Application Service Providers).

14.2.2 Sustaining service convergence

In most of these instances of convergence the consolidation aspects of convergence are of a *sustaining nature*. This results in incremental improvements to core products and services so they may, for example, operate in a manner that is faster, cheaper, and quieter. Otherwise, the fundamental assumptions remain valid. Existing products and services continue to operate and contribute to the value chain as they always have. Existing business models remain intact, and actors such as the incumbent vendors and service providers are very happy for matters to do so.

One immediate consequence of *network-level convergence* is that existing services, previously available only within particular networks, realize a degree of increased potential value purely from their ubiquity with many other potentially complementary services that could be leveraged. This consequence is compatible with a view of *sustaining technology*.

One example of *sustaining service convergence* with which everyone is familiar is typified by the universal adoption of SMTP (Simple Mail Transfer Protocol) as the preferred email protocol for interoperability over other protocols, e.g. X.400. Proprietary email frameworks continue to exist (Lotus Notes and Microsoft Exchange readily come to mind) but the ubiquitous availability of email to anyone anywhere has become a *must-have* capability. This is an obvious manifestation of Metcalfe's Law applied to the services communications layer.

14.2.3 Disruptive service convergence

In contrast to the sustaining convergence typified by the above described consolidation, there exists another and arguably more exciting perspective of how convergence already has begun affecting communications products and services. This alternate perspective is due to those deeper consequences of Metcalfe's Law as a corollary:

> *Until a critical mass of users is reached, a change in technology only affects the technology. But once critical mass is attained, social, political, and economic systems change.*

This last observation is what Downes and Mui (1998) have termed the *Law of Disruption*. The first sentence of the quotation describes the world of sustaining technology, while the latter situation is characteristic of disruptive technology.

Newly emerging communications technologies and support structures enable the user to utilize and interact with their services in previously unanticipated ways, and in some cases, almost unimaginable ones. Such a serendipitous view of service creation is aligned with the characterization of *disruptive technology*.

Interestingly, recently emerging messaging services also provide examples of disruptive service convergence: IM (Instant Messaging) from the Internet, and SMS (Short Message Service) from the wireless world. Each has formed the basis for a multitude of new value-added service capabilities by merging the immediacy of IM and SMS functionality with their own core messaging functionality.

14.2.4 Finding common ground

Both forms of service convergence, *sustaining* and *disruptive*, depend upon the recognition of *common ground* that is shared by a variety of individual services. At the lower levels of the communications hierarchy (the physical, transport, and network planes), common elements around which a case for convergence can be made are straightforward to identify.

At each layer of the communications hierarchy from the physical layer to the application or services layer, the problem space is searched for what could provide that common basis for convergence. As the focus moves up through the layers, the role of *mass customization* (customer focused) becomes increasingly pronounced relative to that of *mass production* (device and infrastructure focused).

On the one hand, the wireless industry and the PC industry will sell millions of devices and appliances, so mass production of services certainly remains an important factor. On the other hand, the multiplicity of distinct devices, applications, and services with which the user may interact is rapidly ballooning.

Identification of one overarching entity in common to all of them becomes increasingly difficult. What possibly can provide the required common ground on which to establish a new framework for convergence? It is not transport, nor is it a protocol, or a software interface.

14.3 Context: Putting Communications in Perspective

The last great frontier remaining to be conquered before a ubiquitous all encompassing service convergence framework can be achieved is the identification of that entity capable of binding all services. The entity that could provide that binding, or *common ground*, for all communications services is outside the realm of communications.

The identification of that mystery entity is based on the realization that the *act of communicating* is always part of a larger *context*. The customer would be better served if each communication were adequately connected to (i.e. associated with) the context that it supports and that gives it purpose. The synergism so derived would greatly enhance the value of such communications. On the basis of their support of a common context, any given set of services has the potential to become a *converged service*, the whole providing more than the sum of the parts.

14.3.1 One simple example

With today's technology, an Internet-enabled vendor can implement an e-commerce Web page containing product offerings, along with an embedded icon (e.g. a small telephone handset) by which the customer can initiate a *separate* phone call to discuss further the contents of that Web page with a customer representative. Supposedly, this phone call could be delivered via the same PC using some form of *net-to-phone* telephony that could possibly be based upon H.323 or SIP (Session Initiation Protocol).

Now, what if I am upstairs on the game room PC and would like to bring my wife into the discussion? I believe she is downstairs in the kitchen, or perhaps she is now in her car traveling to pick up the kids. In any case, I do not wish to be burdened with wondering which network (home, wireline, or wireless) should be used, or wondering about the right sequence of menu clicks or keypad strokes to make the conferencing happen.

On the other hand, I want the information under discussion that provides the *context* of the phone conversation to be provided to all involved parties: to myself, to my wife, and to the customer representative. For myself, the information probably would be presented via the PC.

Perhaps, if she is in the kitchen, my wife could receive this information displayed on a wireless voice-enabled webpad, or on the flat panel display built into the face of the refrigerator. If she is mobile, then perhaps the information, in a summarized format of course, is displayed on her Web-enabled cell phone. If she is driving her automobile, perhaps she must interact verbally by means of a hands-free interface.

The primary point being made by this contrived example is that the *act of communicating* is always part of a larger *context*. Ultimately, communications services will come to exhibit what some might describe as a *self-awareness*: a sense of why they are being used, of what task is being supported, of the goal to be accomplished, to which it must contribute as a team member.

The technology to enable the integration of this level of intelligence within communications services may appear currently to be out of our reach. However, a reasonable first pass at providing such intelligence is now possible. Indeed, service building blocks that could support this purpose already exist or are under development. As later sections show, the customer already has demonstrated a profound appreciation of such context.

14.3.2 Where is this thing called context?

Most of today's telephony communications services could be characterized as *context free*, providing no real-time context regarding the purpose or the circumstance of a phone call that one is receiving. Until now, this has been the state of affairs, one tolerated by most users, with no reasonable alternatives available. However, there is more than sufficient reason to believe that the user would like to have such information, and use it to manage the use of the phone or other communications services.

Traditionally, voice services have not been able to address the issue of context except in limited special circumstances. Perhaps this situation is due to the historical background of telecommunications. For decades, telecommunications services were very intelligent, with human operators in the loop. The operator not only knew who you wanted, but often other contextual information critical to a call's completion. For example, the party being called may have left a note to be called at the office. The operator, however, knew that in fact the party was at another location, because she had just completed another call for them.

One important example where context matters is the emergency E911 service now widely deployed in the United States. The placing of an E911 call automatically implies that the call's context involves some ongoing emergency. In this case, related location information about the source of the call is presented to the agency handling it. The wireless industry currently is facing new mandates to improve the effectiveness of the contextual information provided to support E911.

Many other examples less obvious than E911 also exist. For the want of a better solution, many customers attempt to compensate for this general lack of context by employing an *ad hoc* combination of *caller ID* or *calling name delivery* service, together with a home-based answering machine, as a way to capture additional call context prior to answering a call. The gathering of sufficient context to judge appropriate treatment of an incoming call is what *call screening* is supposed to address!

In many other cases, customers have been left to devise their own means to work around and to compensate for this general lack of contextual support. Alternatively, they choose to decline new services because they lack sufficient contextual capability. In particular, the inability of currently deployed network-based voicemail services to provide sufficient real-time call context is a primary reason why many customers continue to use a home-based answering machine, even though some inbound calls may be missed, and in spite of the burden of maintaining such devices!

14.3.3 A context logistics problem

The context of a single communication, such as a phone call or conversation, is important. As complex as this problem may seem, it is a subset of the larger problem of preserving contextual continuity from one communication to the next.

To make the general context management problem more concrete, the original example of Section 14.3.1 (the Web page with the phone icon embedded in it) is again considered. Suppose that I am not satisfied with the quality or depth of the information exchanged during my first conversation after clicking the phone icon, i.e. *antecedent context*.

Perhaps, both the agent and I must do our homework and a little background search. For example, I may need to take a measurement, while the agent needs to seek clarification of some inadequately documented product feature. We agree to 'get back to each other' with the additional information to conclude the original purpose of the Web page visit, and the supporting phone call it generated, i.e. *consequential context*.

In the interim, between the initial and the follow-up conversations, where is the context of the discussion *parked*? Today, that context is necessarily fragmented into separately managed pieces. In this example there is *no* convergence! All major telephone carriers offer services by which one can 'park' the call, that is, place it on hold if the hold time is not too long. With today's telephony services, I could *park a call* if I needed to pause, say to gather more information, or to await the actions of, or interaction with, someone else.

However, *any ancillary contextual information must be captured separately from the call processing*. As for capturing this additional documentation, I can make notes, write myself a post-it, or perhaps print a snapshot of the Web page.

The issue to be addressed is how to capture enough contextual information to enable a coherent 'multi-session' communication. This must be achieved without retracing our steps, or indeed starting again. The example demonstrates *contextual disconnect*, where the contextual chain that should connect a sequence of communications events is broken, that is disconnected, at the end of each communications session.

In contrast, with the use of *context-enabled services*, I should be able to *park the total context*, of which any phone conversations would be but a part. We should be able to resume the activity with its total context where we left off!

The tracking of context in such bits and pieces is the norm for today's services. But why must it be the status quo? Context-enabled services should manage my communications session as this one context, from my initial Web page viewing, to the associated phone call exchanges until, finally, I am able to conclude the matter for which the communications were initiated in the first place.

For the communications industry to achieve a recognizable level of *converged services*, where the whole is decidedly more valuable than the parts, the source of the synergism that holds the parts together must be determined and addressed. That source is non-other than the *context* those parts already share in reality, and so ought to share at the service level.

14.3.4 Context management: the basics

The general solution to context management, as typified in the above example, is not realizable within current service infrastructures. However, much of the fundamental functionality needed to enable such services is under development. Efforts that will enable the integration of such context-based features with telephony services already have been established. This section will examine those foundations.

The formal dictionary definition of *context* has two general meanings:

- *Cause–effect*: the influences and events that helped cause a particular event or situation to happen; and
- *Explanatory*: the text or speech that comes immediately before and after a particular phrase or piece of text and helps to explain its meaning.

To propose *context management* of the surrounding circumstance of a phone call, as an integral part of the phone call's *connection management*, represents an extraordinarily aggressive goal. To specify and capture all the information that possibly could impinge on a given call is beyond the scope of any system, now or ever.

However, one narrow and fairly well-defined component of context with considerable utility to call management is the *setting*. The 'setting' embodies parameters such as timing, places, and involved parties. This should be much easier to describe and to manage than the total context. Currently, telephony services manage the *literal connection*, for example, endpoints, circuits, and packet routing. They also manage simple aspects of the parties involved, to ensure the intended parties are properly identified and connected.

The issue to be resolved is how best to proceed. How are telephony services to derive the most value from the incorporation of setting management? Perhaps the parties in the call should determine extra aspects of the setting and context; both in selecting the appropriate aspects, and describing how the aspects should influence the 'call' progress. So, where are the means to allow the parties to participate?

From the above definition, the *causal dimension* of context refers to why the call is made by the *caller*, and received by the *callee*. The effect, or *consequence*, refers to the results or the benefits of the call. In other words, what was accomplished, and what could or should come next in the chain of events?

14.3.5 Context management: a historical perspective

The addition of contextual support to call management is not novel. Several papers on the subject have been published, such as Smith (1995). As previously described, the customer has employed various *ad hoc* methods such as caller ID, call-blocking, and a home-based answering machine. This is an effort to integrate some degree of *context management* into her *call management*.

In the mid-1990s, a number of AIN (Advanced Intelligent Network) based services were developed by the telecommunications carriers that permitted the customer to control a phone call's disposition. This was based on parameters such as the date and time of the call, and the identity (or at least the caller ID) of the calling party.

The new AIN service model represented significant improvements in the telecommunications industry's ability to accelerate the diversity of services. In the theme of sustaining service convergence, much was done to create standard methods to be employed by all the telecommunications industry to provide highly portable, readily deployable advanced services.

However, the AIN approach exhibited significant shortcomings. General areas in which shortcomings can be identified include the following:

- the user management interface;
- 'disconnectedness' of service from other context management tools and services; and
- inability to integrate the wishes of all parties (both the *caller* and the *called*).

Initially the default user management interface was an IVR (Interactive Voice Response) system. However, while it was rather complex and sophisticated, it proved very limited and trying for all but the most persistent users. It did not find an appropriate trade-off between flexibility (e.g. how many time slots in a diary schedule), and ease of use.

Later, GUI management interfaces were developed so that one could manage a schedule via the Web. Such GUIs were a definite improvement, but they too suffered from the 'disconnectedness' shortcoming. Consequently, the information to be entered was often a replication of the same information that had already been entered into other non-telephony systems, such as a company-provided planner. No one appreciated the requirement to maintain synchronization across multiple sets of systems containing the same information. Especially when issues such as multiple formats and data organizations were involved.

The underlying problem was the absence of *context sharing*. Constrained by the general AIN architecture the services were one-sided, as they were designed from the perspective of only one of the parties involved in the call. So the customer could configure a service to control certain aspects of call progress. Describing conditions such as when terminating calls would be answered.

However, no feedback mechanism existed by which the calling party could ascertain that information, either before or during the call's set-up. No means were provided through which the involved parties could *negotiate* the terms and conditions of the call.

An often overlooked but very important value of context is its *negotiation* value. This limited state of affairs has also made all but the simplest feature interaction problems intractable. No means existed by which the involved parties could negotiate a course of action from among ambiguous or equally plausible alternatives. Feature interaction is described in some detail in Chapter 13.

14.3.6 The end game: converged services through context

Exactly what steps and actions must be taken to achieve the quest for converged services is still an open question. Fortunately, the communications industry now recognizes the need, the challenge, the opportunity, and the urgency this quest forebodes.

John Jainschigg is editor of the former *Computer Telephony* (recently re-launched as *Communications Convergence* in recognition of the need to respond to the *disruptive technology*-driven Internet). He has described the situation thus (Jainschigg 2001):

> 'Next-gen communications applications – be they vertical or horizontal in nature – require context integration before they can deliver return on investment. And context integration – adapting infrastructure and applications to drive process and express strategy – is a complicated business, requiring great technical facility, coupled with new forms of business savvy (in forms both vertical and horizontal).'
>
> '... it involves knowing how complex communications systems, workers, partners, and markets interact, both in the horizontal context of general productivity and in the relevant vertical domain.'
>
> 'Today, our agents of change are building the rule-book for context integration. Because communications convergence demands this next step forward.'

Note how many times the term *context* was used in the above quotation to explain communications convergence. In particular, the last statement has specifically identified *context integration* as necessary to the *communications convergence* that will characterize the 21st century.

So, the next major convergence milestone of the communications and information services industries is now clearly discernable. On the one hand, the *communications services industry* must ensure context is shared across a multitude of applications that require it to perform their tasks. On the other hand, the *information services industry* must manage the context, and make it available for any communications that are required.

The convergence of the two, communications and information services, is much more intertwined. In particular, context serves a critical dualistic purpose, simultaneously being both the enabler of, and the purpose for, all communications:

- Context must be communicated as part of, or in support of, a given communication's purpose, be it a phone conversation about a document being developed, or an appointment being set up.
- Context is also necessary to support an increasingly sophisticated communications negotiation process for that particular conversation, from low-level call set-up, to sophisticated intelligent-agent anticipation of the customer's immediate preference.

Key here is the utilization of personal context.

14.4 Context: the next killer service feature

Despite the shortcomings of previous efforts to integrate context management with telephony services, the concept is now more viable than ever. In particular, the previous lack of converged context management has now been addressed. However, it is debatable with the integration of context management and telephony services, which of the two will be embedded within (and so subsumed by) the other.

Examples of emerging communications services that help illuminate the convergence of communications and information services around context management will be drawn from various service sectors. These include mobile services, service location and discovery, networked appliances, Home Network services, and unified communications. Each example cuts across these service sectors, such is the nature of converged services.

Each example in the follow sections is considered from the perspective of how the service can be enhanced by, and could contribute to, the management of context. Context can be *momentary*, 'Who's on the phone?' one might spontaneously ask. Context can be *temporal*, that is, cumulative over time, so 'Call me before you leave today, so we can discuss the matter' might be the request. Sometimes context is incomplete, as in, 'I guess I missed your call.'

14.4.1 Presence: personalized context management

One such context-rich technology now under development in the Internet is *presence management*. Jonathan Rosenberg, Chief Scientist at DynamicSoft, has provided the following definition:

> 'Presence is the dynamically changing set of means, willingness, capabilities, characteristics, and ability for users to communicate and interact with each other.'

Presence could be considered an appropriate label for that body of information that characterizes a person, that is, it is a person's *personal context*. As such, *presence* represents a critical first step in the quest for an ultimate comprehensive treatment of context.

Interestingly, the communications industry's current interest in *presence management* is not due as much to developments in telephony, as to rapid advances in the area known as IM (Instant Messaging).

With the arrival of IM on the Internet, the consumer's vision of messaging has been vastly enhanced and extended. Internet-based chat programs such as ICQ have been around for some time. AOL and others have been very successful in offering their own IM products, being able to add many new subscribers.

In its most basic form IM combines the functions of email, that is, the ability to send a text message, with the immediacy of a phone call, as delivery is in real time. This combines the two-way interactive nature of voice calls with the asynchronous nature of email, along with other feature-rich email-like capabilities such as message archival and retrieval, and multimedia attachments.

A number of special interest groups have been organized around a shared commitment to the integration of *presence management* as a core component of communications processing. The PAM Forum (PAM Forum 2001) is one such group:

> 'The PAM Forum is an independent non-profit consortium with members from the voice, data, and wireless networking services and applications community. It is dedicated to establishing and promoting the PAM (Presence and Availability Management) as an industry standard. PAM is a new, open software platform specification essential to creating advanced communications and messaging services that operate seamlessly across various telephony and IP (Internet Protocol) technologies.'

The adoption of this standard by the industry will be a boon to everyone. Service providers will be able to share critical data necessary to deliver advanced customizable services across a range of network architectures. This includes cable, wireline, fixed wireless, and mobile networks covering a variety of geographical ranges such as LANs, WANs, and PANs. The end-users will benefit from such services by being able to define, under a single identity, their personal preferences for all (participating) communication systems including email, telephony, wireless, and instant messaging.

Once we have presence awareness, we can expect to offer location awareness for those interested in us. This leads to the obvious need to control the dissemination of information about ourselves to those without a valid need. Developing policy and processes to control the dissemination of this information is a difficult problem.

Working under the auspices of the IETF is the IMPP Workgroup, another effort focused on the definition of protocols and data formats necessary to build on an Internet-scale, end-user presence awareness, notification, and instant messaging system. Its initial task is to determine specific design goals and requirements for such a service. Current IMPP proposals under consideration are: APEX (Application Exchange, also known as IMXP or Instant Messaging Exchange Protocol), PRIM (Presence and Instant Messaging) and SIMPLE (SIP for Instant Messaging and Presence Leveraging Extensions) (Disabatino 2001).

Of particular note among this group of proposals is SIMPLE. Both Microsoft and AOL have committed to the adoption of SIMPLE, essentially making it the *de facto* standard. The market momentum for SIMPLE, with its integration with SIP, is a reflection of the market's view that instant messaging is more than just another isolated application. The advocates of SIMPLE strongly envision *presence* and *instant messaging* becoming critical components of a broader suite of *integrated communications services* that includes telephone calls, voicemail and Web conferencing.

Because it is SIP-based, SIMPLE provides a common infrastructure from which to develop a *negotiation metaphor* that incorporates the capabilities of voice communications and messaging.

14.4.2 Wireless meets presence management

Additionally, the convergence of such Internet-based initiatives with the messaging capabilities of the wireless world is now a given, as typified by wireless SMS. Ericsson, Motorola, and Nokia, all major wireless equipment vendors, have collaborated on an initiative focused on the definition and the promotion of universal specifications for the exchange of messages and presence information between mobile devices, mobile services, and Internet instant messaging services.

Their Wireless Village Mobile Instant Messaging and Presence initiative, will deliver architectural and protocol specifications, test specifications, and tools for mobile IMPS (Instant Messaging and Presence Service). This group also will define procedures and tools for testing conformance and interoperability of IMPS. The emergence of such groups as the Wireless Village with their pre-emptive and defensive behavior illustrates an underlying concern of many service providers. They are aware that presence management represents a 'sticky service', in that it can provide the inertia to maintain customers that adopt such services.

Originally a text-based service, instant messaging has evolved to include rich multimedia content, such as audio, video clips, and images. *Presence services* provide a system for sharing personal information about the user's status (online, offline, busy), location (home, work, traveling), and the moods of their friends and colleagues (happy, angry). Presence services will allow users to 'subscribe to presence' and so access, for example, listings of friends and colleagues that are currently online. Additionally, these services will allow users to participate in private or public chat rooms with search capabilities.

Ultimately, network operators will be able to provide meeting and conferencing services with shared content. The convergence of voice and messaging services is about to attain a new realm of functional possibilities.

The instant messaging specification will be based on prevalent bearer protocols and other well-adapted standards, such as SMS (Short Message Service), MMS (Multimedia Messaging Service), WAP (Wireless Application Protocol), SIP (Session Initiation Protocol), and XML (Extensible Markup Language). This service will include security capabilities for user authentication, secure message transfer and access control. Operators will find these specifications applicable to existing 2G, and the new 2.5G (e.g. GPRS), as well as emerging 3G wireless network technologies.

According to Janice Webb (AFX News 2001), Senior Vice President and General Manager of Motorola's Internet Software and Content Group:

> 'Instant messaging and presence services are proving to be among the most exciting areas in today's wireless and wired world and initial signs are that this market is set to expand massively over the next few months and years.'

Presence could be considered an appropriate moniker for that body of information that characterizes a person – it is the person's *personal context*. As such, *presence* represents a critical first step in the quest for an ultimate comprehensive treatment of context.

14.5 Sharing Service Context

In spite of the recent infatuation of many business strategists with Web-enabled e-commerce, the most widely used and deployed Internet service has been, and continues to be, messaging with plain old asynchronous email! Similarly, the greatest use of the telecommunications networks is still POTS (Plain Old Telephone Service), not Centrex.

Perhaps an even more provocative trend to note is the relative utilitarian value that businesses now place on voice and messaging services. As early as the summer of 1998, the AMA (American Management Association), in conjunction with Ernst & Young LLP, conducted a survey to determine such preferences. Among their findings (Ernst & Young 1998), the Annual Human Resources Conference survey found that email has overtaken the telephone as the most frequently used communications tool among HR executives.

This observation is even more telling and indicative of what is to come when one considers that the currently expanding significance of instant messaging in business was then still over the horizon.

Advances in messaging, with its generation of new requirements and capabilities, such as presence management, will be the 'tail that wags the dog,' rather than adding one more nuance on Centrex. Messaging services provide the user with the means to share their personal context and their presence with others as they see fit. The user is thereby enabled to take call control negotiation to levels never imagined by the telecommunications industry.

14.5.1 Messaging: a historical perspective

The telecommunications industry has been focused for some time now on just how to integrate the two lucrative services of telephony and messaging. One such approach is referred to by the catch phrase UM (Unified Messaging) (International Messaging Consortium 2001).

> 'Unified messaging is the integration of several different communications media, such that users will be able to retrieve and send voice, fax, and email messages from a single interface, whether it be a wireline phone, wireless phone, PC, or Internet-enabled PC.'

Another related effort has focused on the interoperability of voicemail systems with each other, and with the Internet. They initially developed a non-Internet based approach that never became widely deployed. More recently, the EMA has proposed VPIM (Voice Profile for Internet Mail) which was approved by the IETF and published as RFC 2421 in September 1998. Now to gain the participation of a wider community, the technical work on VPIM has been transferred to the IETF with the creation of the VPIM Work Group in early 2000.

Most of the major incumbent voicemail vendors have embraced this vision, and most carriers have attempted to introduce UM products into their markets. Web-based interfaces for their UM products have been developed. The strategists have gone on the record predicting major growth opportunities (Hicks 1999):

> 'In fact, market research company Ovum Inc. believes that unified messaging will become almost universal in the business market, but not until 2006, with three-fourths of companies either using a service or deploying the equipment themselves. For unified messaging services alone, Ovum predicts that the number of users will increase from less than a half-million this year to 151.9 million in 2006, creating a $31 billion services market.'

The future of UM projected by the incumbents will supposedly be bright and shiny. The standards needed to support their UM vision are in place, and the incumbent players (the vendors and the service providers) know their parts.

But such a vision is not to be. As is the case with *disruptive technologies*, (Christensen 1997), the presumed playing field (market, technologies, major participants) is already experiencing the Internet's disruptive forces. Now the UM vision must be revisited.

Just as the Internet has forced the re-invention of industry segments such as EDI (Electronic Data Interchange), instant messaging and presence management now are forcing the re-invention of the incumbent *status quo* vision of unified messaging, and much more!

14.5.2 Messaging in all its flavors

This disruptive impact of the Internet on the telecom incumbent's vision of *messaging* is typified by comments such as the opening statement of an article appearing in *Wireless Week* (Smith 2001):

> 'First there was voice mail, then e-mail, unified messaging, and unified communications. Now there's adaptive communications.'

A plethora of messaging services have been defined in this rapidly evolving segment of communications:

- *Unified messaging* bundles voice, fax and email, allowing all three to be stored and retrieved together.
- *Integrated messaging*, a subset of UM, provides the same integration but with a different interface.
- *Unified communications* provides the same bundle as UM, but gives the users real-time access. In other words, users can create a profile with preferences regarding when and where they may be reached, i.e. a user-defined 'follow-me' capability.
- *Adaptive communications* is a new variation of unified communications that learns from and adapts to a user's habits.

The above description of *unified communications* deals with users creating a profile on a network with preferences as to when and where they may be reached. This sounds like instant messaging has arrived, and that many features previously offered as part of AIN telephony services now have been re-invented from a *communications-via-messaging* perspective.

The ADC Telecommunications announcement of *adaptive communications* introduces the concept of an *intelligent* presence agent that adapts to a person's habits and preferences. The agent is subject to a user's direct management through a user interface, which also allows an initial seeding of a user's preferences.

In its most basic implementation, IM combines the core functionality of email (the ability to send a text message) with the immediacy or spontaneity of a phone call. The interaction is essentially occurring in real time.

IM is well on its way to becoming the *de facto* point of convergence where the *two-way interactive immediate person-to-person* nature of voice communications is combined with the *asynchronous pause-and-start person-to-group* nature of email, along with other such feature-rich *data management capabilities* as message archival and retrieval. Furthermore, the degree of flexibility with which the user can manage and configure his use of IM already far exceeds that of the traditional telephony or voicemail services.

As previously noted, currently proposed IM standards support the use of SIP, as well as MIME (Multi-Purpose Internet E-Mail Extensions), LDAP (Lightweight Directory Access Protocol) and other Internet standards. Clearly, the environment now is ripe for the development of new classes of service that will converge the features of VoIP (Voice over Internet Protocol), IM, and more.

14.5.3 Intelligent messaging

The *adaptive communications* proposal of ADC Communications provides each user with her own personal HAL. This intelligent agent from *'2001: A Space Odyssey'* may seem a little premature both from a technology perspective as well as from consideration of the customer's readiness to accept such an agent. However, one could consider another recent announcement that leverages the IM *metaphor* and technologies.

As previously discussed, the use of SIP in general, and IM in particular, as a technology is not restricted to communications between two people. Using instant messages to interact with a computer-based intelligent agent *in the network*, rather than with another person, is an equally plausible application of this technology. In fact, new services that exploit this modified model of IM already are emerging.

One example of such a new service, called *ActiveBuddy*, is presented by Courtney (2001). This service enables the user to pose a question via instant messaging to a computer, or more specifically to an intelligent agent functioning as the customer's own *personal gopher*. The agent then responds via the customer's IM client with the answer.

For example, one might request a stock quote, a sports score, or any other piece of information. In the process of determining the answer, *ActiveBuddy* might send a referral URL where more information is found, or perhaps invoke a follow-up program.

The long-term implications of this new application of IM will be quite profound. *Active-Buddy* is the precursor of what will be a variety of new services that are enabled through instant messaging. Take a moment to grasp the significance that this expansion of IM-like functionality represents. Your PDA-enabled cell phone may come to be the *embodiment* of (or at least an access point to) your own *personal agent*. That is, an agent that in turn could interact on your behalf with yet another IM-enabled agent. Your personal agent may as well have access to its *own ActiveBuddy* client!

As noted at the beginning of this section, email gets credit for being the Internet's first *killer application*, with the Web being a strong second. Now services such as *ActiveBuddy* are attempting to combine the best aspects of both. Combining personal communication with fast and easy information retrieval.

For the immediate future, while we wait for the ultimate realization of that grand unified convergence vision, much can be accomplished with today's resources. The necessary resources for this first convergence wave are PIM (Presence and Instant Messaging), LBS (Location-Based Services), and SIP (Session Initiation Protocol). Particularly if PIM and LBS are coupled with the recently announced DAML (DARPA Agent Markup Language) effort to make the content of the WWW self-explanatory (Berners-Lee *et al.* 2001). This allows intelligent agents readily to process the WWW content. All provide critical resources to address *context-communications integration* and so facilitate true data communications convergence.

14.6 SIP: The Oil that makes Context flow

The ultimate strategic value of SIP (Session Initiation Protocol) is the ease with which it can be applied to a broad range of application domains. Certainly, SIP can be used to initiate, manage, and terminate *interactive sessions* between two or more users on the Internet. However, these *interactive sessions* need not be restricted to voice and video services, nor even to exchanges such as messages between two or more people!

More generally, and this is why SIP is such a strategic technology, SIP is an appropriate protocol by which two, or more, *endpoints* (people, objects, processes, intelligent agents, or some combination) engage in a *conversation* (an interchange of *messages*) for practically any purpose. That purpose can be for almost anything, including:

- voice and video;
- instant messaging;
- monitoring;
- information retrieval and exchange;
- command-n-control of applications;
- operation of devices;
- feature negotiation among intelligent entities.

As previously noted, SIP already provides the basis for SIMPLE as the proposed IM specification that has essentially already become a *de facto* standard, with its adoption by Microsoft and AOL. SIP also is finding its way into many other application domains. Some of these are very communications focused, and some would not have been imagined until recently.

The wireless carriers stand to gain just as much extra value from the adoption of SIP as do the wireline carriers. 3GPP (Third Generation Partnership Project) is dedicated to using SIP for call control (3GPP 2001). SIP is to be used from the terminal to the network, and between the network service nodes. In other words, all IP Multimedia call signaling will be performed via SIP. The value of the addressing in SIP is critical to its success.

A major consequence of the decision to employ SIP as a call control method is a complete decoupling of the services provided, from the network in which the user is currently operating. No longer will the limitations of the visited network have an impact on the services available to the user. Furthermore, this same decoupling effect will be realized in the SIP-enabled wireline-based network.

Another step in the direction of SIP extensions is the recently submitted IETF RFC-3087 (IETF 2001c) on Control of Service Context using SIP Request-URI. This proposal would extend the application of SIP beyond the current focus on call set-up and control to serve as a *bridge*. This provides a link between information needed to initiate and manage a communication, and additional *contextual* information that might be used by an application. For example, a help-desk application that operates in conjunction, or *in sync*, with a call.

The RFC-3087 provides as its prototypical example the description of a voicemail service. Going beyond the previously discussed messaging services currently in fashion, this example could readily be extended to formulate a generalized approach to the convergence of voice services with messaging services. In particular, the concepts presented in RFC-3087 are suggestive that SIP could be employed as the means by which various applications could share context at some level. For example, telephony services and e-commerce services would be able to collaborate by each sharing its context as part of a converged service.

Yet another example demonstrating the potential breadth and diversity of SIP is typified by another IETF draft on Pico SIP (IETF 2001d) proposed in February 2001. That IETF draft describes how the SIP protocol could be applied to *ad hoc* networks, or to Pico Networks, such as one might encounter in the home environment and in mobile scenarios. This proposal leverages the direct, client-to-client, peer-to-peer session initiation capabilities of SIP.

Finally, the *meta-negotiation* capabilities of SIP are now being explored as the basis for providing a generic feature negotiation capability. This is typified by the recent IETF draft on the SIP Negotiate Method (IETF 2002e). In this draft, the need is identified for the development of a generic approach to negotiation. One that would be applicable across all protocols used in the Internet world. This has been an area of active academic research since the Contract Net Protocol (Smith 1980) and the *negotiation metaphor* (Davis and Smith 1983). The authors of this SIP negotiation RFC draft have attempted to attack a more immediate problem in the SIP arena and establish a generic negotiation mechanism.

14.7 Discovering Service Context

To reiterate, *context* is key to the successful development of converged communications services in the 21st century. When known, context serves as that *common ground* described in Section 14.2.4 that binds together for a *common purpose* the multiple communications capabilities of the converged service. Next we consider the case where:

- not all the supporting or interacting components and participants of the service are known, be they devices or people;
- the context that binds them is incomplete or missing. For example, this could include personal presence, or it could include details of the task that motivated the communications.

Then, the missing components and contextual information must be *discovered*. Or alternately, the means and mechanisms to address incomplete situations must be determined, i.e. *negotiated*. Both of these cases are considered in the following sections.

Fortunately, much has been done to develop the *bootstrap* processes and protocols to effect this discovery. This is not a new concept. The classical case is the development in the United States of 411 directory services that support determining the telephone number of someone we wish to call. This once meant speaking human-to-human to request and receive the needed information. Now, this directory access has been automated with IVR front-ends that permit the calling party to speak the name of the called party and either have the number spoken, or even have it dialed directly.

14.7.1 Discovery characterized

As with the 411 directory service, discovery-oriented protocols and processes have been developed for each of the described communications environments. The consideration of what needs to be discovered may be open-ended. For example, is it a physical device, a logical entity, perhaps some incomplete information, a location, or a map of several locations? In one instance, a service may need access to a specific type of device, (say) an Epson color printer. In another, any available device with printer functionality may suffice.

Likewise, the discovery process is fundamentally an open-ended one. What is determined – *discovered* – at each stage of the discovery process may well raise yet other unanswered questions. Can multiple discovery paths be pursued simultaneously? The meta-search engines found on the Internet are an example of services that use this approach.

Ultimately, a service may have to proceed forward in a best effort with incomplete context. A classic example of this is found in the connectionless IP world of the Internet where there are no promises or guarantees that a packet will ultimately arrive where it was intended. For this reason, the more flexibility a service is given in how it may proceed, the more likely a service (instance) will operate successfully.

Often multiple entities may be involved, each with its own parochial perspective and agenda. This situation leads to the necessity for negotiation. The process that one's modem executes when it attempts to establish an Internet dial-up connection, is a simple example of negotiation. As the complexity, sophistication, and intelligence of services increase, negotiation will play an increasingly relevant role in a service's execution.

14.7.2 Device discovery pot pourri

With the arrival of increasingly specialized information appliances, a number of discovery architectures addressing mobile and specialized devices have been proposed recently. These architectures are essentially coordination frameworks that propose certain ways and means of device interaction with the ultimate aim of simple, seamless and scaleable device interoperability.

Among the better known proposals are UPnP (Universal Plug and Play), Jini, and Salutation, each of which are proposed by a major industry segment. Other minor efforts include the IETF's SLP, Lucent's Inferno, and CalTech's Infospheres Project.

Jini, developed by Sun Microsystems, provides a federation coordination framework evolved and adapted from the academic research of David Gelernter (Yale University) and Nick Carriero, tailored specifically to Sun's Java.

Universal Plug and Play, is pushed primarily by Microsoft. It is a framework defined to work primarily at the lower layer IP network protocols, rather than at the application level. It provides a set of defined network protocols for which device manufacturers may build their own APIs. The APIs may be implemented in whatever language or platform they choose.

Salutation, developed by device manufacturers, is drawn from research on intelligent agents and treads a middle way between device autonomy and standardization. This enables vendors readily to adapt many of their legacy products to interoperate with one another.

Generally speaking, *device coordination* essentially means providing a subset of the following capabilities to a device:

- ability to announce its presence to the network;
- automatic discovery of devices in the neighborhood and even those located remotely;
- ability to describe its capabilities, and to query and understand the capabilities of other devices;
- self-configuration without administrative intervention;
- seamless interoperability with other devices wherever meaningful.

14.7.3 Discovering and interpreting location

One specialized type of discovery service that is particularly relevant to mobile entities is *location*. For devices attached to a tethered network, such as the office LAN or the wireline telecommunications networks, the location problem is solved statically when an item is installed on the network. Its static location is noted and recorded in appropriate databases, e.g. the carrier's phone directory. However, the world of communications is becoming increasingly mobile, with the cellular industry being the obvious area of growth. Mobility means location is now a dynamic entity that must constantly be re-evaluated.

The location problem is particularly critical to services, such as E911, that must support emergency contexts where response time is of the utmost importance. Determination of where someone making an E911 call is located under all conditions and circumstances is a very difficult problem. For example, what accuracy is required? This can range from the nearest cell tower, through to the nearest intersection, to within a few meters.

Secondary concerns include such issues as privacy considerations. Should the wireless network *passively* provide the mobile party with the resources to locate themselves? Then, the network only knows about the party's location when that party chooses to reveal it. Should the network actively monitor the party's location within the provision for some agreed upon level of confidentiality?

Various technologies are now being proposed for determining location. Current technology to locate wireless phones takes several forms: GPS (Global Positioning System) devices, overlay triangulation technologies, and cell/sector information. Emerging approaches include UWB (Ultra-Wide Band) technology that is already in use for radar-based search and rescue applications. The application of pico-cell technologies, such as *Bluetooth*, is also being considered.

LBS (Location-Based Services) are an obvious strategic capability. In addition to emergency considerations and general tracking services, a multitude of other value-added opportunities based on LBS are being pursued. *Location-based services* are characterized by their blend of location information with other useful content. These provide relevant, timely, and intensively localized information. This can be passed directly to the consumer, on which the consumer may then act. Or, the information can be passed to other services that operate on behalf of the consumer.

As described, presence management services seek to integrate an individual's LBS information with their presence profile. For example, a set of LBS location coordinates may be interpreted to determine that someone is at home, work, or elsewhere. Hence various personal states, such as busy, resting, or eating, can be inferred.

14.7.4 Discovery not limited to physical entities

The need for a discovery process is not limited to devices. Sometimes, information also needs to be discovered. Fortunately, methods that would facilitate such information discovery are now under development.

The W3C (World Wide Web Consortium) is studying WS-Inspection (Web Services Inspection) to allow companies to post standardized directories of the kinds of Web services they have available. This specification would complement the already established UDDI (Universal Description, Discovery and Integration) standard. UDDI presents standard interfaces that companies can use to build directories of Web services that can be queried by type. UDDI performs similarly to a telephone Yellow Pages, such that services can be queried by business function.

Another much more ambitious effort is the Semantic Web program being pursued by the W3C and DARPA. They have started the development of DAML (DARPA Agent Markup Language). The concept is to enable a World Wide Web that not only links documents to each other, but also *recognizes the meaning* of the information found in those documents. This is an enormous undertaking.

The first step involves the establishment of standards that allow users to add explicit descriptive tags, or *meta-data*, to Web content, and thereby label the content. Also to be developed are methods and protocols that enable various programs to relate and to share meta-data between different websites. From this foundation, advanced capabilities can be developed, such as the ability to infer additional facts from the ones that are given. As a result, searches will be more accurate and thorough, data entry will be streamlined and the truthfulness of information will be easier to verify.

14.7.5 Negotiation extends discovery

A constrained negotiation framework is fundamental to each of these discovery frameworks. However, a general negotiation framework is apropos to far more than just the discovery of devices. The situation where multiple service entities are involved, each with its own parochial perspective and agenda, will almost invariably lead to the necessity of negotiation. As the complexity, sophistication, and intelligence of services increase, negotiation will play an increasingly relevant role in a service's execution.

The previously discussed proposal (Section 14.6) to develop a SIP-enabled negotiation framework is therefore quite relevant. Another recent IETF draft on Applying Contract Net Protocol to Mobile Handover State Transfer (IETF 2000d) proposes some ways that CNP (Contract Net Protocol) could be used to enhance macro and/or micro mobility protocols. Specifically, the draft focuses on how CNP could be used for AAA and QoS (Quality of Service) state transfer and handover negotiation in mobile IP networks.

Feature negotiation is not a new or difficult capability to understand. Similar functionality, for example, is now provided in specific and constrained domains such as *smart* modems. Here, the modem supports facilities such as the auto-negotiation of multiple bandwidths, error correction, data encryption, and data compression schemes.

Negotiation is an iterative process of *announcement-bid-award* in which an entity:

1. Poses its need (its requirements or simply its preferences) to other entities that might be able to satisfy the request;
2. Receives offers from responders; and
3. Evaluates and accepts the best offer of assistance.

Variations in the negotiation process include the possibility of iterative offers and counter-offers, and of teaming arrangements by multiple responders.

In each instance of negotiation, three fundamental components may be identified:

- *exchange of information* by, for example, requests and responses;
- *evaluation of exchanged information* by each entity from its own local perspective; and
- *final agreement* (the *contract*) by mutual selection.

In preparation for a solution by negotiation, there are *preparatory activities* which each involved party should complete. These include such items as the clarification from each negotiator's perspective of:

- *absolute bounds* of this give-n-take effort;
- *compromise fall-back positions*, multiple if possible; and
- *evaluation criteria* by which to judge any and all offers.

In support of the negotiation process, mechanisms, such as languages, data structures, and protocols, must be in place to enable negotiation rules. Hence, each party may *accurately* and *succinctly* communicate requests and bids.

Under the assumption that the involved parties are willing to cooperate, negotiation provides a reasonable vehicle for the identification of *common ground*, i.e. a *global* view of the problem's solution that is *locally* acceptable to each party involved.

In general, the system's negotiation mechanisms should facilitate the customer's indication of acceptable *fall-back* service behaviors. Much of multi-customer feature negotiation management is determining an acceptable level of service delivery *common* to the involved parties.

14.8 The New Converged Home Network

The networked home is expected to yield a plethora of opportunities for the development of new communications services. For many, the concept of the *networked home* has referred to a home LAN of networked PCs, perhaps with some attached printers and fax machines. Now, thanks to the new technologies being developed and the price points being realized, that vision has broadened to encompass many appliances. This ranges from refrigerators and microwaves in the kitchen, to the washer and dryer in the laundry, to the home infrastructure to include air-conditioning, lighting, and security. Home-based e-commerce, today restricted to a PC or possibly a cell phone, will also become embedded in every activity of the home.

Every service in the home is being re-evaluated for the additional value to be added if that service were network-enabled. For example, if your lawn sprinkler system were network-enabled, then perhaps it could be managed automatically by some network-hosted weather-knowledgeable service to only water when needed. This could, for example, adapt to watering moratoriums during droughts. Perhaps the homeowner would like to check the contents of the refrigerator before leaving the office, or indeed outsource the stocking of the refrigerator to a network-enabled grocery service. No doubt other network-enabled services that have no current standalone (non-networked) equivalent will be created. Controlling access to services in a SOHO, or allowing interaction from without, is a big issue.

14.8.1 The current state of affairs

Before our expectations begin to soar in anticipation of such possibilities, we first need to step back for a moment to consider the current state of affairs. The interoperability that convergence is to bring has many hurdles to overcome in the home environment. How many and what types of network should be deployed? What management strategy should be established?

Our daily lives in our homes are touched by a multitude of distinct networks from standalone pico-like networks to the Internet. They more or less operate as ships passing in the night.

They include not only the PSTNs, which are different for the United States, Europe, and throughout the world. They also include various wireless networks (such as AMPS, TDMA, CDMA, GSM, 802.11, and Bluetooth) as well as a multitude of other explicit and implicit networks, which are often proprietary and non-interoperable. Traditional radio and television (again different in Europe and America) are ubiquitous broadcast networks. The wide availability of cable-based and of global satellite-based broadcast networks, xDSL-based networks, and telephony networks is pervasive for many areas of the world. High-definition TV and digital radio are now being deployed.

In addition to a variety of networks, a person typically has to interact with a large miscellany of devices. There is a combination of wireline POTS phones, cordless phones, ISDN phones, many different types of cell phone, fax machines, and pagers. A person also has to interact with various radio and infrared-enabled devices (garage door openers and keyless car doors) and appliances (TVs and VCRs).

This multitude of often proprietary, explicit, and implicit networks are not interoperable in terms of either the underlying communications protocols, or the information that would use those protocols. For example, information from one source, say a speed calling list stored in my cell phone handset, cannot readily be transferred between, or synchronized with, other information sources such as a PIM (Personal Information Manager) on a PC or handheld PDA (Personal Digital Assistant); or indeed with the local telephone company's directory service.

Across the end of my own coffee table in the family room lie six different remote controls for various multimedia appliances: the TV, VCR, CD changer, DVD, amplifier-tuner, and a cable set-top box. The remote controls for my garage doors and those for my automobiles (power door locks and trunk release) are not interoperable with each other, nor do they interoperate with my home's security system.

I keep within reach my cell phone that I carry with me even when at home, and a cordless phone to access my wireline POTS service. I have ten distinct voicemail boxes, one on each of my family's six cell phones, one on my work phone, and four shared by my multiple wirelines.

I have to maintain two different remote access configurations for each of my home PCs: one for access to the company's remote-access network, and the other for access to the general Internet via a commercial ISP. My cable modem provides shared Internet access for a home LAN of seven PCs.

Clearly, the opportunities and the hurdles for achieving a networked home of converged services is very great. In the process, the installed base of non-interoperability such as those typified above must also be addressed. Retrofit and migration strategies will be required.

14.8.2 Great opportunities for service convergence

Many developers contemplate the vision of the *networked home* broadened to encompass just about every appliance one could imagine. The question then is what are the services that this new infrastructure might enable? While a comprehensive list would clearly be a moving target with frequent additions, the following potential application scenarios are frequently mentioned:

- home automation and security;
- home entertainment;
- energy management;

- remote monitoring and control;
- computer–telephony–CE integration;
- e-commerce.

A multitude of approaches are now being proposed by various special interest groups such as appliance manufacturers, entertainment providers, and utility companies. Until there is some consolidation of efforts and there is a convergence of approach, this will be quite a 'horse race'. Of course, each of these special interest groups has formed a corresponding organization to work at defining and marketing its vision, architecture, infrastructure, and standards, for the networked home. Each group approaches the networking of the home from its own particular perspective: the wireless industry, the appliance industry, the multimedia industry, and so on.

Some of the major organizations (alliances, consortia, and forums) announced thus far include:

- *Bluetooth*: http://www.bluetooth.com/index.asp;
- *Home RF*: Home Radio Frequency Work Group http://www.homerf.org/;
- *Home PNA*: Home Phoneline Networking Alliance, http://www.homepna.org/;
- *HomePlug*: Home Powerline Alliance, http://www.homeplug.org/;
- *Home API*: http://www.homeapi.org/;
- *ETI (Embed The Internet)*: http://www.emware/eti;
- *HAVi (Home Audio–Video interoperability)*: http://www.havi.org/;
- *AMIC (Automotive Multimedia Interface Consortium)*;
- *TSC (Telematics Suppliers Consortium)*: http://www.telematics-suppliers.org;
- *UPnP (Universal Plug and Play)*: http://www.upnp.org;
- *OSGi (Open Services Gateway initiative)*: http://www.osgi.org/osgi_html/osgi.html.

Bluetooth and Home RF are focused on defining a wireless network infrastructure for the home. Bluetooth also addresses mobile *ad hoc* peer-to-peer networks such as two cell phones swapping address book entries. Various wireless 802.11 IEEE workgroups also are considering how best to support this market opportunity.

Home PNA focuses on a network overlay of the phone wire already installed in the home, and is compatible with the ADSL-lite (splitter-less ADSL) broadband technology being delivered by the telephone companies and their re-sellers. In contrast, HomePlug focuses on a network overlay that utilizes the powerline running throughout the home.

Attempting to provide appropriate middleware for Home Network servers and appliances are Home API, Universal Plug-n-Play, and OSGi. These groups are dedicated to broadening the market for home automation by establishing an open industry specification. This defines a standard set of middleware software services and application programming interfaces that enable software applications to monitor, control, and otherwise interact with home devices.

These groups are dominated by software and systems vendors, and are dedicated to defining standard sets of software services and application programming interfaces. In particular, they allow both existing and future Home Network technologies such as HAVi, Home PNA, Home RF, CEBus, Lonworks, and X-10 to be more easily utilized. Furthermore, it should also be possible to integrate control of existing audio/video devices (using infrared control, for example) into one system.

In contrast, other groups with more of a hardware and component focus have offered lower-level appliance-based solutions. ETI is focused on the hardware devices that are to be made network intelligent. HAVi seeks to provide network interoperability to the multimedia devices of the home such as the VCR, TV, and stereo. AMIC is defining standards for an embedded automobile network.

Other fruitful areas besides the home where effort is underway to embed network interoperability, include the automobile. The main players in this area are AMIC (Automotive Multimedia Interface Consortium), SAE (Society for Automotive Engineers), and TSC (Telematics Suppliers Consortium).

AMIC has announced support for the ITS Data Bus, which is an emerging hardware specification. In collaboration with home-focused groups such as OSGi, it is seeking to create a common way for various electronic products to be plugged into different cars, while retaining their ability to work together. For example, a navigation system, PDA, pager, and other products could share a single screen in a vehicle, with data from one item driving a response from another.

The technical foundation for a common hardware interface based on the IEEE 488 specification has been under way for some time by the SAE. In liaison with AMIC, TSC is developing open standards from the vehicle out towards telematics services. Telematics is an emerging market of automotive communications technology that combines wireless, voice, and data to provide location-specific security and information services to drivers.

In addition to these consortium-led efforts, major appliance manufacturers (such as Seiko, Toshiba, and Sony) are forming partnerships with high-tech startups in their search for *light-weight* approaches to embedding Internet functionality. The focus of their efforts is to develop Internet-ready appliance components that are completely PC and OS-independent, as they are implemented entirely in hardware.

One such example is iReady, a start-up company that is attracting much interest (Yoshida and Hara 1998) with its Internet-ready LCD called the *Internet tuner*. The small iReady LCD panels feature a chip-on-flex (COF) module with built-in network, email and Web-browsing capabilities.

Embedding Internet-ready functionality into an appliance has the potential to facilitate applications that have little to do with accessing email, schedules, or websites. These devices, for example, could use Internet protocols not necessarily to search Web pages, but to download specific types of information available on a certain network.

The degree of service convergence that will be enabled once a home appliance is networked within the home, is almost limitless. For example, once the refrigerator is networked, a multitude of e-commerce services are enabled. This can range from the dynamic and instantaneous monitoring of what is in the refrigerator, to the auto-replenishing from the local grocery supplier. More provocative opportunities for converged services also are enabled, as typified by medical-related services such as assisting with assuring that someone's elderly parents have opened their refrigerator daily and whether they've been eating properly.

Researchers at Telcordia have proposed that the *generalized* use of SIP is in *appliance control*. As an ongoing follow-up to this effort, a group within the IETF is now exploring this venue for SIP at the site. Related drafts deal with 'SIP Extensions for Communicating with Networked Appliances' (Greenhouse 2000a), and a 'Framework Draft for Networked Appliances Using the Session Initiation Protocol' (Greenhouse 2000b). Figure 14.1 depicts the functional integration of a SIP-enabled smart home.

SIP could support a variety of *converged sessions*. Besides call control, with SIP one could also control, program, and monitor appliances. An example of a *converged service* would be

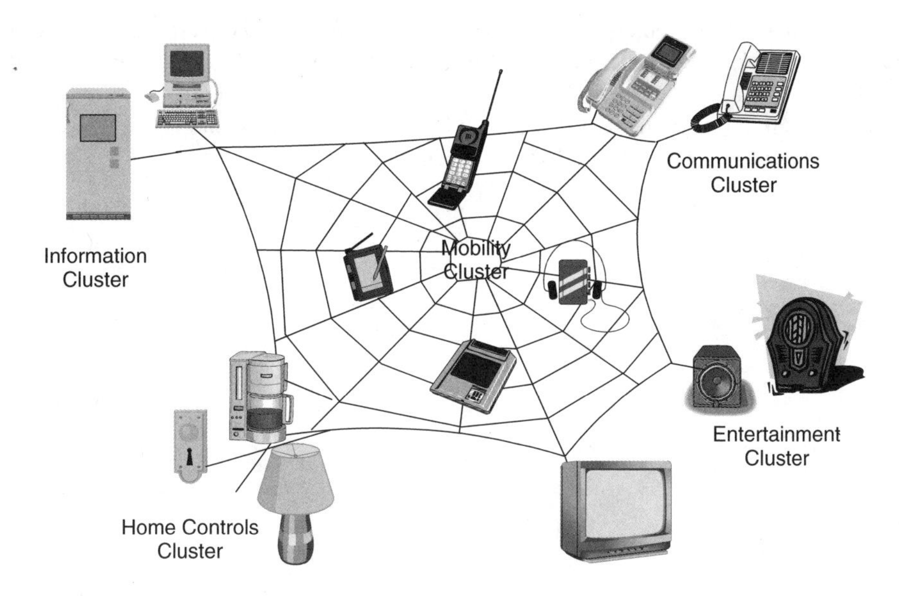

Figure 14.1 The SIP-enabled networked home.

to turn on the bedside lamp automatically and to mute the background music when the bedside phone rings in the middle of the night.

In addition to home appliances that generally are fixed in their location (e.g. refrigerator, dishwasher), another class of communications-enabled mobile appliance is emerging. One example of how the convergence of appliances and communications is proceeding is typified by the move of the watch industry to incorporate technologies to their products that allow users to communicate while on the move. They can find directions while hiking, or perhaps access their PCs.

First generation versions of such devices often have employed proprietary technologies to communicate with special tethered docking stations. The generation of watches now appearing on the market (e.g. Citizen's Wrist Browser wristwatch) are adopting standards such as Bluetooth to give a peer-to-peer, pico-cell wireless technology. This will enable their interoperability, for example, with a variety of other communications-enabled devices. Witness the new generation of Bluetooth-enabled cell phones, Bluetooth-enabled cash registers, vending machines, and Bluetooth-enabled VCRs.

How many of the population will opt for Dick Tracy wrist phones? In contrast, how many will choose to wear cell phone-friendly wristwatches, or some other approach to service and device convergence? Only time will tell.

14.9 Conclusion

This chapter has attempted to provide the reader with a panoramic view of the almost limitless possibilities that the convergence mega-trend is bringing to communications services. It has highlighted the *out-of-the-box* thinking that will be required of developers who wish to take full advantage of the emerging technologies. The resulting convergence has been described and meaningful examples have been presented from which the developer can build and extend.

But that is not all, there are many other possibilities that will enable and leverage converging communications services. We are living in exciting times!

15

Evolving Service Technology

Peter Martin[1] and Stephen Corley[2]
[1]Marconi plc, UK [2]British Telecommunications plc, UK

15.1 Introduction

Preceding chapters have described technologies that are currently being used in mainstream service engineering. As the diversity and complexity of networks and services increases, so more powerful and flexible solutions will be required. This chapter looks at some of these newer, still evolving software technologies that may play a central or supporting role in service engineering in the future. The technologies looked at in this chapter are: Software Agents, Constraint Satisfaction, Artificial Neural Networks, and Genetic Programming.

Knowing which technologies will be important in the future is highly speculative. However, the technologies described in this chapter have been selected because evidence from experiments so far has shown promise. Each technology is introduced and an explanation given of how it can be applied usefully to service engineering. Issues arising from each technology are discussed, and overall conclusions given.

15.2 Software Agents

15.2.1 Introduction

The term 'agent' in the context of software engineering (EURESCOM 2001a, Genesereth and Ketchpel 1994, Wooldridge and Jennings 1995, Bradshaw 1997) has become very overused in recent years such that there is no universally agreed definition. The broadest possible definition is 'software which acts on behalf of another entity'. So, just like a travel agent a Software Agent carries out a task or a role on behalf of a client. In a computing environment the client may be a user or another software entity. This definition, however, is not very useful because almost any piece of software could be viewed in this way.

Although there is no universal agreement, a popular definition (Wooldridge and Jennings 1995) describes an agent as a software entity that has the characteristics of autonomy (acts independently), proactivity (goal-based), reactivity (responds in a timely fashion to events),

Service Provision – Technologies for Next Generation Communications. Edited by Kenneth J. Turner, Evan H. Magill and David J. Marples
 ISBN: 0-470-85066-3

and social ability (communicates with other agents to achieve goals collaboratively). Other characteristics frequently quoted include mobility (the ability to move from one host to another) and learning (the ability to improve performance over time based on previous experiences).

Software with the above characteristics offers the possibility of systems which can lower the cost and improve the performance of business operations by: automating mundane tasks; enabling users/customers to complete tasks that would otherwise be very difficult, time consuming, costly, or just impossible; and adapting to unexpected events or changes in the environment automatically.

Of course it may be possible to achieve cost saving and performance boosting solutions without agents, but agent technology provides a natural model of the real world (i.e. a community of entities each with their own goals, communicating and often working together to achieve mutual benefit) compared to existing software paradigms such as object orientation.

Furthermore, agent technology consolidates and builds upon a number of important computing technologies (object orientation, distributed computing, parallel processing, mobile code, symbolic processing) and research results from other disciplines (artificial intelligence, biology, mathematics). In this way, agent technology offers a way to unify and simplify the use of the wide range of software technologies available today.

Agent solutions can be broadly categorized into four types:

Personal Agents

These agents act on behalf of a user. The agent has a profile of the user that records specific information about users such as their interests, modes of operation, skill level and likes/dislikes. Personal Agents are typically deployed as assistant services, that is they provide support to the user in much the same way as a human assistant.

Multi Agents

Here a community of agents acts in cooperation with each other to achieve an overall system goal. Typically, each agent in the system will have specific capabilities that can be made available to other agents. Agents can be added and removed without disrupting operation.

Ant-like Agents

These agents are lightweight and mobile with little intelligence. Ant-like systems typically comprise numerous mobile agents of several types, each behaving according to a small number of rules. The power of Ant-like Agents comes from the overall system behavior which emerges from the community of the individuals. This type of intelligence is often referred to as 'swarm intelligence'.

Hybrid Agents

These are agents which combine one or more of the previous characteristics, for example a multi-personal agent system.

15.2.2 Software Agents for service engineering

Agents can be applied in a wide range of contexts within service engineering. Personal Agents can be incorporated into the service itself to provide intelligent assistance to the user. For example, the agent could act as an intelligent answering service for answering calls, taking and filtering messages, and notifying the user of important ones. This model could be extended to personal number and unified messaging solutions to provide an enhanced service to help manage all of their communications, covering calls and messages. In EURESCOM project 712 (EURESCOM 2001b), each user has a Message Handling Agent that monitors and scans incoming messages to detect potential problems. If a problem is detected, the agent attempts to resolve the problem automatically. In the project's experiments, the problem of unrecognized email attachment types was the focus; see Figure 15.1.

In the event of email arriving with an attachment that cannot be opened and read, the following solutions are possible:

- install in the user's computer an appropriate reader (e.g. Adobe Acrobat, Ghost View);
- convert the attachment to a format that can be read;
- request the sender to re-send the email with the attachment in another format.

The important requirements for the solution are: domain knowledge about email attachments, problems that can occur and suitable solutions; and a profile covering capabilities of the user's working machine and knowledge of the user's preferred working formats.

The third solution (request re-send of email) requires a Multi Agent solution with the MHAs (Message Handling Agents) being able to communicate to agree a format which is acceptable to both sender and receiver. The project successfully demonstrated such a solution using JIAC (Java Intelligent Agent Componentware). Expanding this solution to a one-to-many (i.e. broadcast) situation would provide a significant increase in convenience to all users, such as when attachments are sent to mailing lists.

Other problems that could be handled in the messaging domain could include managing oversized message or attachments, overloading of inboxes, virus and security checking, and support for list maintenance.

Personal Agents can also be used to support the creation and management of services. Here, the users are service designers, architects and technicians, rather than the ultimate users of the service. Typically, SCE (Service Creation Environment) solutions address the implementation and validation of new services. In EURESCOM project 815 (EURESCOM 2001c) a Personal Agent solution was used to extend the SCE to support the service creation process

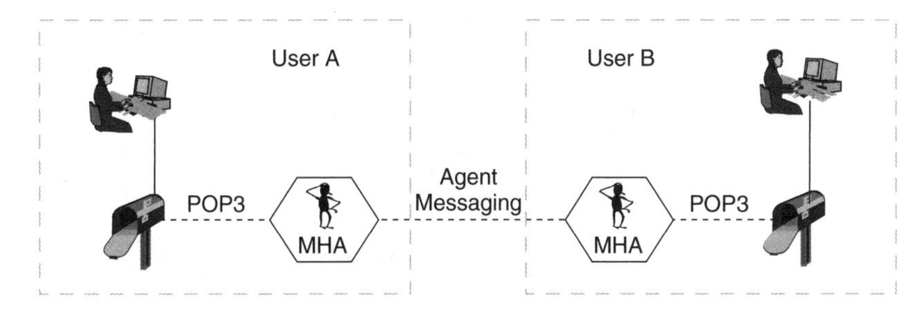

Figure 15.1 Personal Message Handling Agents (MHAs) for managing email.

across the complete lifecycle of an IN (Intelligent Network) service creation project. The project manager had a Personal Agent that supported the set-up and management of the project (e.g. defining tasks, allocating resources, monitoring progress, and generating reports). Service developers were supported by their own Personal Agents that helped them to know which tasks to work on, to monitor deadlines, to prepare reports, to negotiate deadlines with the project manager, to monitor other user activities, and to collaborate with users.

An example of how agents can be applied to service management is the support of fault and problem management. EURESCOM (2001b) used a combination of Personal and Mobile Agents to provide support to helpdesk technicians, as illustrated in Figure 15.2. When a problem is reported by a user, a technician determines the symptoms from the user and supplies this information to their Personal Agent. The agent searches its knowledge base to find solutions that have been used before. In the case of an exact match the solution is offered to the technician. In the case of a partial match, it may be that the solution can be adapted to the current problem. When a solution is ultimately found, this new knowledge is recorded in the knowledge base. In this way the performance of the agent improves over time. In the case where the technicians are distributed across geographical locations, then agents can provide an enhanced level of support. If a solution cannot be found in the agent's knowledge base, it can broadcast to other agents to ascertain if they have a solution. This communication could be carried out using

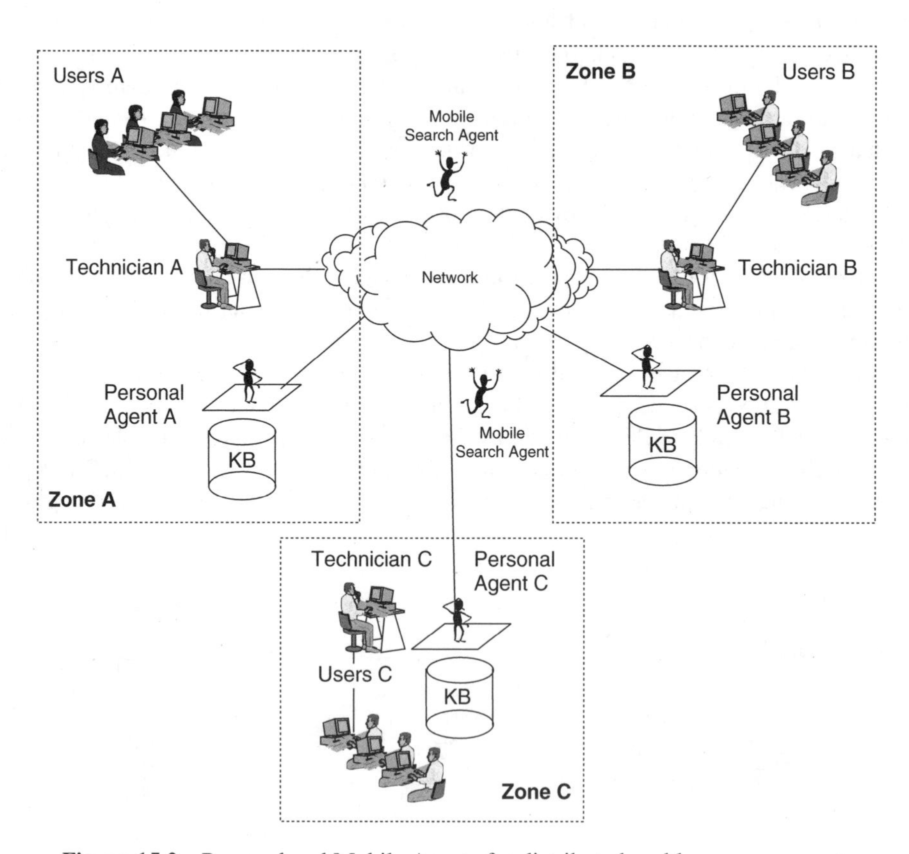

Figure 15.2 Personal and Mobile Agents for distributed problem management.

messages or Mobile Agents. In either case, a solution may or may not be found. If only a slightly matching solution is found, the Personal Agents could arrange a teleconference (or other interactive session, such as text chat) between the technician and the technician who handled the past problem. The Personal Agents have access to their technician's diary and so would be able to arrange a convenient time. In the event of a convenient time not being available, the problem can be escalated automatically to the next line of support, again with the agents arranging a convenient time for the technicians to talk.

Fully autonomous agents can be embedded in the service infrastructure. Typically, these agents carry out a particular role on behalf of the organizations providing the service. Examples could be Network Monitoring Agents, Call Routing Agents, Billing Agents, or Fraud Detection Agents. Multi Agent systems are able to provide a further level of sophistication due to their ability to interact with each other. For example in EURESCOM (2001b), Service Provider Agents interact with Network Provider Agents to negotiate the best deal for network resources. These two types of agent are considered next.

SPA (Service Provider Agent)

The SPA represents the interests of a telecommunications service provider and supports the provisioning of telecommunication services to customers. Its two roles are: client of network services supplied by the NPAs (Network Provider Agents), and manager of various telecommunication services to end customers.

The key functions performed by the SPA during service provisioning are to:

- authenticate the user;
- determine component network service requirements;
- identify secure NPAs for the supply of component network services;
- negotiate with NPAs for component network services specifying Quality of Service, bandwidth, endpoints, and period required.

The SPA aims to find the optimal solution in terms of Quality of Service and cost for providing the service to the end-user.

NPA (Network Provider Agent)

The NPA represents a network domain. It is responsible for provisioning network connectivity on request from the SPA. The NPA interacts with the network management system controlling the network resources and with other NPAs representing other network domains.

The key functions performed by the NPA during service provisioning are to:

- negotiate with the SPA regarding the cost, quality, and bandwidth for the delivery of the network service;
- negotiate with other NPAs regarding terms and conditions for external connection segments in order to provide end-to-end connectivity;
- reserve the resources;
- control the activation and deactivation of network resources according to the agreed contract.

In the solution proposed by EURESCOM (2001b), a Personal Communications Agent (PCA) exists. This is another example of a personal user agent. The PCA acts on behalf of the user for provision of the service. It elicits requirements from the user and then negotiates with SPAs to obtain the best deal. The PCA learns which service providers provided good service in the past and will give them preferential consideration in future.

The applications discussed so far have used agents that can be considered complex in that they utilize one or more of user profiling, deep domain knowledge, sophisticated inter-agent communications capabilities, and powerful reasoning to make decisions. Applications of 'Ant-like Agents' (or swarm intelligence) draw upon the emergence of intelligent behavior generated from the interaction of many simple agents following a few rudimentary rules. This approach has been applied to a number of service engineering contexts. Early work in this area (Appleby and Steward 1994) proposed that teams of small Mobile Agents could be generated at trouble spots in telephone networks where they would reconfigure routing tables. Real ants lay down pheromones to communicate with other ants, for example to find the shortest paths to food sources. As the ants travel they lay down the pheromone for others to follow. The pheromone evaporates after a time so the paths that have the highest level of pheromones will be the ones that are most traveled, indicating the most successful for obtaining food. In this way good paths are reinforced while bad paths are discouraged. This behavior has been modeled and applied to the routing of data packets around a network, avoiding and adapting to congestion and failure conditions (Schoonderwoerd *et al.* 1997, Bonabeau *et al.* 1998).

15.2.3 Discussion

In order for agent technology to be adopted in mainstream service engineering, a number of issues need to be addressed. These are discussed below.

The major barrier to the uptake of Mobile Agents is that of security. There have been many viruses in recent times, e.g. in 2003 there were several high-profile viruses and worms including Mimail, Klez, Yaha, and BugBear. For many IT managers the risk of allowing executable code to roam from machine to machine between organizations is just too risky (Chess 1998). Mobile Agents would certainly provide an alternative mechanism for unleashing viruses. There may be hope with the integration of high-strength security techniques and protocols into the Mobile Agent platforms. NIST (the US National Institute of Standards and Technology) has a project focused on the security of Mobile Agents (NIST 2003).

The problem-solving abilities of ant-like systems are attractive. However, the fundamental barrier is that the emergent behavior is generally not understood, and can be unpredictable. Further research is required before telecommunications players will let them loose in their services and networks.

One of the strengths of the agent-oriented approach is that it can help solve complex distributed problems. In practice this means that a number of different systems will need to interact to provide the whole solution. For this interoperability to happen as smoothly as possible, standards are required. The current leading agent standards activities are being carried out by The Foundation for Intelligent Physical Agents (FIPA, http://www.fipa.org/). FIPA was formed in 1996 and has a membership of 50 organizations from academia and industry. The focus is on intercommunication between agent systems. FIPA is evolving specifications for agent communications languages, protocols, public ontologies, and agent management architectures. With the backing of IT companies (e.g. Sun, Toshiba, Fujitsu) and telecommunications companies (e.g. British Telecommunications, France Telecom, Nippon Telegraph and Telephone), the future for agent technology looks promising.

15.3 Constraint Satisfaction

15.3.1 Introduction

Constraint Satisfaction technologies model the system of concern as a set of variables, values and constraints. Each variable has a set of associated domain values. The constraints are sets of allowable domain values that define stable or consistent states of the system. Generally, solutions in constraint-based problem solving are those assignments of values to variables that do not cause constraints to be violated (i.e. the constraints are satisfied).

Classical Constraint Satisfaction problems have fixed variables, values, and constraints. In systems that are changing continuously, such as telecommunications systems, classical constraint solutions are not adequate. To deal with these types of problem the classical techniques have been extended to create what is known as Dynamic Constraint Satisfaction.

15.3.2 Constraint Satisfaction for service engineering

This section focuses on the work of Elfe *et al.* (1998) that applied dynamic constraint satisfaction to the feature interaction problem (see Chapter 13). The constraint satisfaction approach to this problem is particularly interesting because success has been achieved in the three basic areas of avoidance, detection, and resolution. Some examples are given below.

Feature ordering

Interactions can occur when features are triggered in the wrong order. For example, consider speed dialing and outgoing call screening. SD (Speed Dialing) allows a shorthand number to be associated with a full number, so saving in dialed digits. OCS (Outgoing Call Screening) prevents calls being connected to certain numbers. If the OCS is applied to a dialed number first then it is possible that an invalid number will pass. This is because OCS will check the untranslated (short) number rather than the actual number it represents. If a constraint-based system is monitoring and controlling the operation then it is possible for this situation to be avoided. In a well-designed system, OCS should be the last feature to manipulate the dialed number, blocking any further changes to the terminating number register. If SD follows OCS then the attempted write by SD to the terminating number register will cause an inconsistency in the constraint model. This inconsistency indicates an interaction. Rather than simply detecting the problem, the constraint system can backtrack to find a valid sequence to avoid the problem, in this case activating SD before OCS.

Resource contention

Resource contention interactions occur when two or more features attempt to manipulate the same resource at the same time. The example of CW (Call Waiting) and CFB (Call Forwarding Busy) can be used to illustrate this situation. If a called party has both features activated, then an incoming call when that party is busy will cause both features to be triggered. The feature that processes the signal first will create a new session and set its type. When the second feature attempts to process the signal, it finds the new session has already been created with an incompatible type. This will cause a conflict in the constraints, so detecting the interaction.

To resolve the interactions in these circumstances would generally require some change to the performance of one or more features. Constraints can model the allowable changes such that when an interaction is detected, a new configuration is found that removes the conflict. This approach is described by Mittal and Falkenhainer (1990). A simple example for the CW and CFB case would be for one to relinquish control to the other. A more useful solution could incorporate the ordering solution described above. In this case, CFB would defer to CW, with CFB being triggered only if CW failed to complete the call. Note that the alternative order of CW deferring to CFB would not make sense since once the call had been diverted there would be no role for CW. This can be modeled by having constraints that ensure that the ordering of features is such that each feature would not progress the call beyond the target session. Many more complicated scenarios may be possible. The strategy might be to take the simplest solution or the one that requires the fewest changes to normal operation. One approach based on minimal diagnosis is described by Sabin *et al.* (1995).

Context violation

In a normal caller to callee connection, each party takes on a specific role and for each role there is a specific context: the caller is the initiating party and the callee is the terminating party within the context of an originating and a terminating number. In the presence of features, each party can take on multiple roles, each in a different context. As an example, consider CFB and Figure 15.3.

If caller *B* is already in a call and a new call arrives, then caller *B* takes on the role of initiating party and terminating party simultaneously. When CFB is triggered, a new call leg is set up with yet another role and context being created for party *B*. Furthermore, a new session is created for the *A* to *C* connection. The problem is that features are not designed to record the various levels of context information. This can lead to context violations that cause confusing and unexpected behaviors from the user's point of view. For example, if *C* is on *A*'s outgoing screening list, then the call from *A* to *B* (forwarded to *C*) would fail. This is confusing to

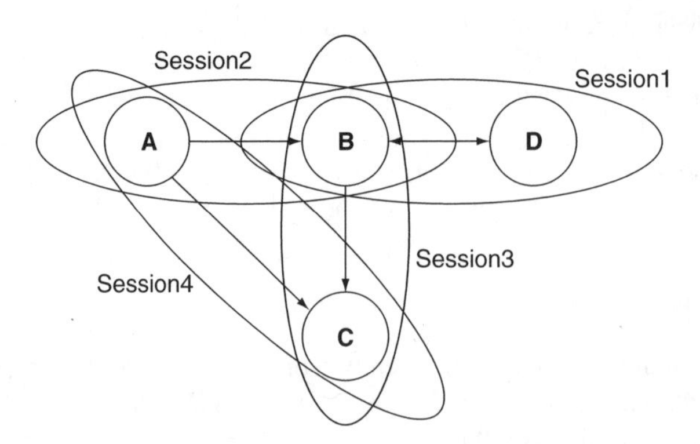

Figure 15.3 Maintenance of session context in constraint-based feature interaction management.

A because it was *B* that was called, not *C*. By modeling the roles and contexts using a constraint-based system, not only can violations be detected, but because the context information is available then explanations can be provided and solutions determined. For the example here, information could be given to party *A* that the call was being diverted to *C* with an explanation why (i.e. because *B* was busy). Party *A* could be given the option of allowing the call to go ahead.

Constraint Satisfaction has been applied to planning (Smith *et al.* 2000), network management (Frei and Faltings 2000, Frühwirth and Brisset 1997), diagnosis (Santos 1993), and network protocol verification (Riese 1993). Constraint Satisfaction techniques could be applied to other configuration type situations, for example in Service Creation Environments where the service components have constraints that model their allowable connections and valid behavior. Invalid constructions could be flagged dynamically to the user during construction. The constraints would support simulation and testing by detecting behavior outside that intended or expected.

15.3.3 Discussion

Constraint Satisfaction enables problems to be modeled and solved in a declarative manner, i.e. by specifying *what* relationships need to be maintained rather than *how* they should be maintained. The advantage for the programmer is that programs are very expressive, leading to short and easy to understand code. Constraint systems are also flexible in their ability to model a wide range of problems and to enable prototypes to be prepared and modified quickly.

Although constraint technology has been successfully applied to a range of problem domains, a number of challenges remain before constraint development and runtime systems are likely to become mainstream. A fundamental usability aspect is support for the modeling phase. Graphical specification tools such as Visual CHIP (Simonis 1997) have been used, whereby the constraints are generated from pictorial representations of the problem domain. Specialized languages have been developed such as CML (Andersson and Hjerpe 1998) to translate natural language descriptions into constraints. Machine learning has been used to automatically generate the constraints, for example Multi-tac (Minton 1996, Rossi and Sperduti 1998).

Another area requiring attention is the debugging of constraint programs. This can be particularly problematic because minor changes to the constraints can lead to unpredicted and significant changes to the overall system behavior. Graphical approaches (Carro and Hernenegildo 1998, Schulte 1997) have proven successful in enabling the programmer to visualize and hence understand how the constraint system carries out the search for solutions.

Soft constraints are among the other variations of constraints that have been introduced to enable real-world situations to be modeled more accurately. Soft constraints enable preferences and uncertainties to be modeled. Soft constraints also enable over-constrained problems to be solved (Jampel 1996). When no solution is possible, some constraints are relaxed. The solution chosen is the assignment of variables which provides the 'best' value according to some criteria, and could involve user input to arrive at the solution. The trade-off of more flexible types of constraint is that the implementation of the solving mechanisms becomes more complicated, leading to less efficient solutions.

Constraint technology is able to solve real-world problems now but is still very much an active research area.

15.4 Artificial Neural Networks

15.4.1 Introduction

The study of ANNs (Artificial Neural Networks) is a branch of computer science that was originally aimed at modeling the network of real neurons in the brain. ANNs have a long history in computer science, McCulloch and Pitts (1943) proposed a simple model of a neuron as a binary threshold unit. The output y is dependent on the weighted input values $x_1, x_2, \ldots, x_n$ and is determined by the transfer function $f(x)$. The function of interest to McCulloch and Pitts was a threshold function that gave the result 1 or 0. This is illustrated in Figure 15.4.

Following on from this there was much work to develop the ideas, but further real progress was not made until Rosenblatt (1957) proposed the perceptron, a pattern recognition device with learning capabilities. By the mid-1980s the hierarchical neural network was established as the most common form of ANN. A hierarchical neural network is one that links multiple neurons, as shown in Figure 15.5.

Each of the connections has a weight associated with it that modulates the signal arriving at the target node. A signal arriving at the input nodes $i_1, i_2, \ldots, i_n$ is processed by each node using the transfer function, and the result is propagated to the next node. This process is repeated until the signal reaches the output layer where the result is available as signals $o_1, o_2, \ldots, o_n$, thus completing the processing of the input signals. The connections between

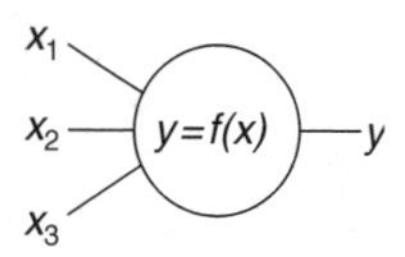

Figure 15.4 Simple model of a neuron as proposed by McCulloch and Pitts.

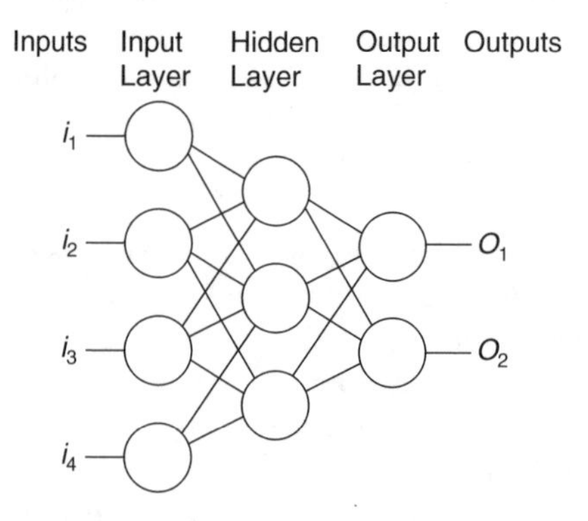

Figure 15.5 Hierarchical neural network.

the layers, the threshold parameters, and the weights associated with each connection determine the output signal. The principal dificulty with this network is arriving at an optimum set of weights and thresholds, so a learning method is employed to train the weights. This is achieved by creating a network with all the parameter values selected at random. This initial network is then used to perform input-to-output transformations for problems for which the result is known. This is called the training set. The correct final parameters are obtained by changing the values of parameters in accordance with the errors that the network makes in the process. There are several learning methods that have been proposed. Probably the best known of these is the error back propagation learning method proposed by Rumelhart *et al.* (1986). Several other alternative architectures have been proposed for ANNs, including self-organizing networks, of which Adaptive Resonance Theory (Carpenter and Grossberg 1988) is one popular example. Today, ANNs are being applied to an increasing number of real-world problems of considerable complexity, including financial risk assessment, protein structure analysis, game playing, and speech, character, and signal recognition. They are good pattern recognition engines and robust classifiers, with the ability to generalize in making decisions about imprecise input data. They have also been applied to functional prediction and system modeling problems where the physical processes are not always well understood or are highly non-linear. ANNs may also be applied to control problems, where the input variables are measurements used to drive an output actuator, and the network learns the control function.

15.4.2 Artificial Neural Networks for service engineering

Location information plays a central role in cellular networks. The physical location of a mobile handset is monitored constantly by the base stations in order to manage hand-off between cells. This location information has great potential for enabling a new breed of services. This is gradually happening as cellular operators are agreeing deals with service providers for the use of this information. Services that would benefit from location information include logistics and fleet management, security and safety services, map and direction services for tourists, personalized information services based on location, and multi-player gaming. See Chapter 6 for more information on cellular networks.

There are many techniques for estimating the location of a mobile handset (Caffery and Stuber 1998) with the most popular based on signal strength, angle of arrival of the signal, and time of arrival of the signal. The data is combined with reference to several base stations to arrive at an estimation of the location which could be accurate to better than 100 m. The accuracy depends upon many factors such as density of cells and obstructions such as hills and buildings. ANNs could be used as an alternative mechanism for determining location. Such an approach is described by Ahmed *et al.* (2001) where an ANN is used to learn locations of handsets based on signal strength. A four base station scenario was used for the experiments. The ANN consisted of a multi-layer back propagation network with four input nodes (one for each signal strength measurement received at each base station), 15 hidden nodes, and two output nodes (the X and Y location of the handset). Once trained, the ANN gave reasonable results with improved accuracy requiring increased training examples. The training phase was found to be computationally intensive and time consuming. However, the operational performance of estimation location was found to be very fast, since only a few computations are required by the ANN to generate the result. In current signal strength location estimation techniques, much effort is required to model the effects of multi-path propagation and shadowing. ANNs offer a potential method of modeling these effects implicitly as part of the training process, rather than having to create a model of specific terrains and urban building layouts.

Experiments in the application of ANNs to overcoming co-channel interference in digital systems have shown very good promise. The Decision Feedback Functional Link Equalizer proposed by Hussain *et al.* (1997) incorporates an ANN with 41 input nodes (values extracted from the distorted channel), 42 hidden layer nodes, and one output node (for the estimated output signal). This network has been shown through simulation to provide significantly enhanced BER (Bit Error Rate) performance and reduced computational complexity at moderate to high signal to interference ratios. These results compare favorably with the feed forward technique (Gan *et al.* 1992) and radial bias function technique (Chen and Mulgrew 1992). Furthermore, the equalizer gives better BER performance compared to conventional Decision Feedback Equalizer and Linear Transversal Equalizer approaches, but at the expense of increased computational complexity.

ANNs have also been applied to problems that would be very useful for new and advanced services. For example, incorporation of handwriting, speech, and face recognition would help to provide a range of interesting future services.

15.4.3 Discussion

The attractiveness of ANNs is due to their ability to model problems that would be impossible or impracticable to tackle manually, particularly involving data that is noisy and complex. Although training may be computationally expensive and time consuming, the real-time operation is generally extremely fast. The most significant downside to ANNs is that the resulting model is not explicit to humans, so it is not easy to obtain an explanation about why a particular result is given by an ANN (a current area of research). However, for particular kinds of problems, ANNs look very promising in a number of areas of service engineering, both in the management of services and networks and in the services themselves.

15.5 Genetic Programming for Service Creation

15.5.1 Introduction

Chapter 9 describes the current state of the art in service creation methods, showing that the majority of services today are developed using toolsets supplied by switch vendors. In many cases these tools have a graphical element that allows the service creator to express the service logic in the form of graphs or trees. Other methods of service creation of course use standard APIs (Application Programming Interfaces), some of which are described in Chapter 11. In all these cases, the service creator is required to go through the standard software engineering phases of requirements capture, analysis, design, code, and test, with possibly several iterations through this process as the requirements change.

This section presents a technique called GP (Genetic Programming) that has been shown to allow a high-level requirement to be translated into a computer program, without the direct intervention of a designer. Using a GP system, services for Intelligent Networks can potentially be automatically evolved.

Genetic Programming is an evolutionary technique inspired by the ideas of natural selection (Darwin 1859) and genetic inheritance. It uses an evolutionary approach to solving problems. By starting with a pool of randomly created programs and applying the principles of natural selection, programs evolve towards satisfying the high-level requirements of the problem

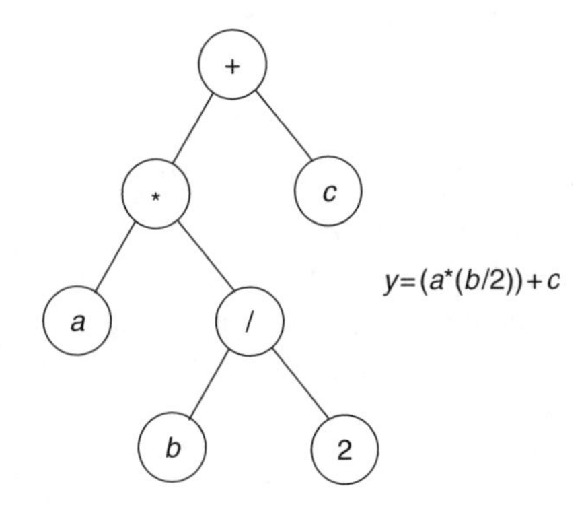

Figure 15.6 Tree representation of a program.

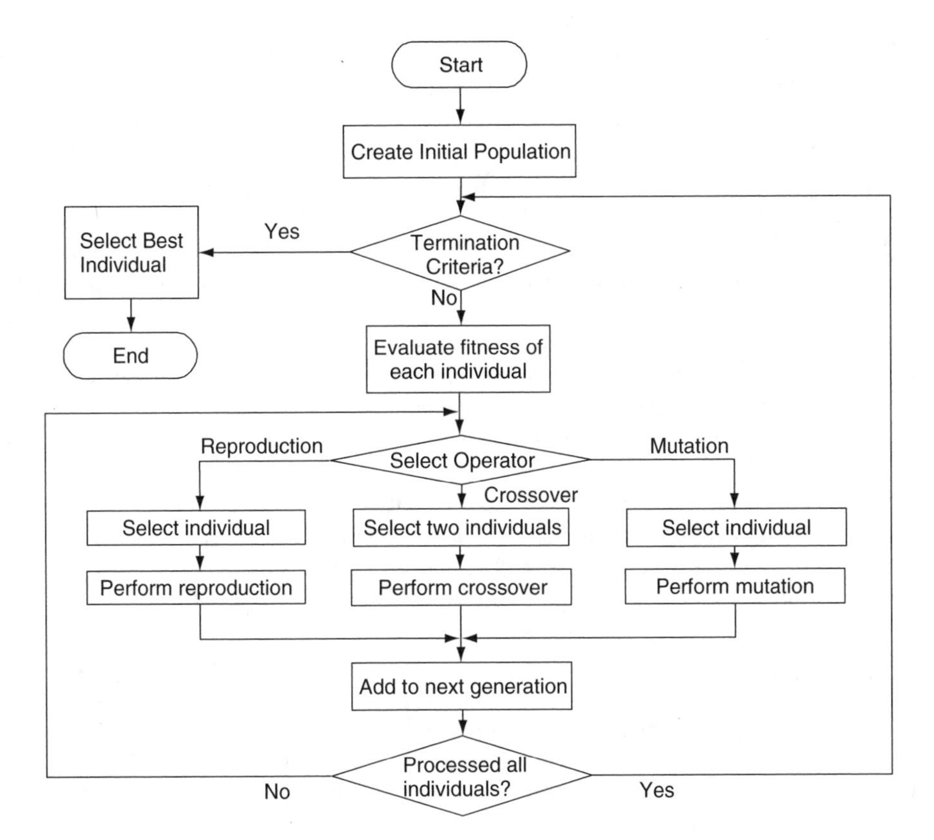

Figure 15.7 Flowchart showing the operation of Genetic Programming.

being solved. It is important to note that this automatic creation of programs is achieved without direct human intervention.

GP was first proposed by Koza (1992) as an extension of GAs (Genetic Algorithms, Holland 1975). In GAs the individuals that make up a population are fixed-length, limited-alphabet strings. In contrast, GP uses variable-length structures that represent programs. The structures are typically trees that describe the program, though other structures are possible such as linear representations (Banzhaf 1993, Nordin 1997) or directed graphs (Kantschik *et al.* 1999). Figure 15.6 shows a typical tree representation of a program and its corresponding symbolic expression.

GP uses five steps to solve a problem. First, a set of functions and variables is chosen from which to build a program. Secondly, individuals (programs) are randomly created from the chosen set of functions and variables; this is the initial population. Thirdly, the programs are evaluated (executed or interpreted) for fitness, and a fitness value is assigned to each individual. Fourthly, the population of individuals is used to form the next generation by selective breeding between the individuals in the population. The third and fourth steps are repeated until either a pre-determined number of generations has been processed or an individual meets a pre-determined level of fitness. Lastly, the fittest individual is selected as the output from the system. These steps are shown in the flowchart of Figure 15.7.

Choosing the functions and variables

In classic tree-based GP, each genetic program consists of one or more nodes chosen from one of two sets. The non-leaf nodes are known as the function set F. All nodes in F have arity (that is, can take a number of arguments) one or greater. The leaf nodes are the terminal set T. Nodes in T have arity of zero. Variables are members of T.

If the members of T are considered as functions with arity zero, then the total set of nodes is $C = F \cup T$, and the search space is the set of all possible compositions of the members of C. So that GP can function, this set must exhibit two properties: closure and sufficiency.

Closure requires each member of C to accept as its arguments any other member in C. This property is required in order to guarantee that programs can operate without run-time errors being generated. The common example cited is that of protecting the division operator to prevent division by zero errors, but also extends to data types used when calling functions and accessing terminal types.

The sufficiency property requires that the set of functions in C be sufficient to express a program capable of solving the problem under consideration. In general this is a problem-specific property and must be resolved before any GP can be evolved. This, together with determining a suitable fitness test, requires the most effort by a user of GP.

Creating the initial population

The initial population is created by randomly building tree structures from the function and terminal sets already described. The depth to which trees are created and the symmetry of the structures is a tunable parameter that has a significant impact on the overall performance. A key property of an initial population is that it should possess a high degree of diversity.

Evaluating the population

Each individual in the population is tested against a fitness function to determine how well it meets the criteria specified to solve the given problem. In most GP systems this is done by evaluating each program using a problem-specific interpreter. After evaluation, each individual is given a score indicating its fitness.

Once the individuals in a generation have been evaluated for fitness, the fittest individuals need to be identified by means of a selection process. The two main methods of selecting individuals from a generation are fitness proportionate and tournament. When using fitness proportionate, all individuals are ranked according to their relative fitness values and the best selected. A refinement on this is rank selection which reduces the influence of single highly fit individuals. In tournament selection, *n* individuals are selected and the best one in the selection is propagated to the next generation. The value of *n* can be any number greater than one.

Reproduction operators (how new generations are created)

The third step required when running a GP system is to apply a reproduction operator to the current generation to build the next generation. The three principal operators are asexual reproduction, sexual reproduction, and mutation. Asexual reproduction is the simplest operator. It describes the straight copying of one individual from one generation to the next without any changes to its structure.

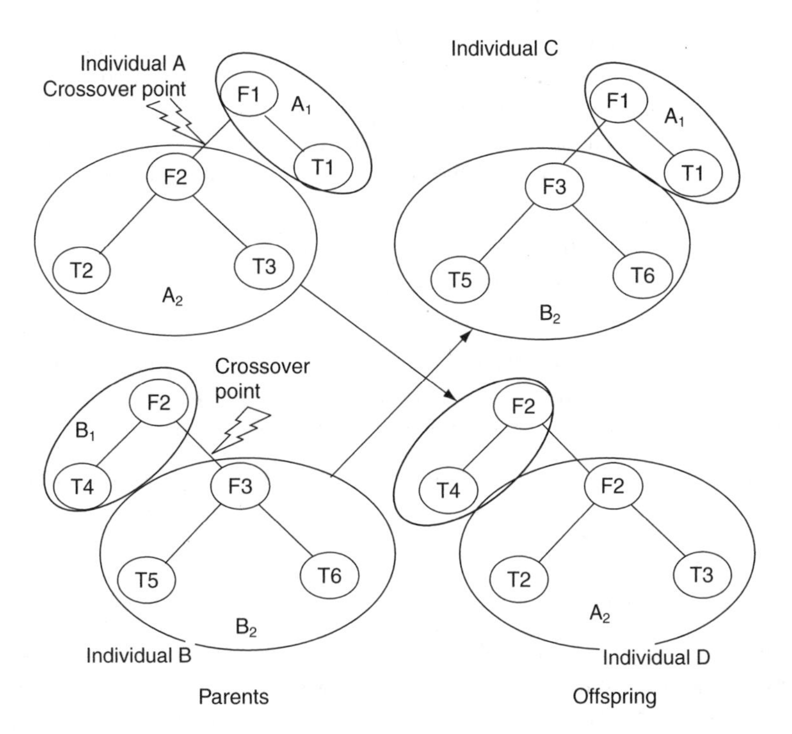

Figure 15.8 The crossover operator applied to two trees.

Sexual reproduction is commonly implemented by taking two individual program trees, say *A* and *B*, and splitting them both into two parts, selecting a random point at which to split them. This gives rise to four subtrees A_1, A_2, B_1 and B_2. These are then re-combined to form two new individuals *C* and *D* where $C=A_1+B_2$ and $D=B_1+A_2$. This is shown diagrammatically in Figure 15.8.

The main effect of crossover is to potentially combine a useful subtree from *A* with another useful subtree from *B* to make a fitter offspring.

Finally, mutation can be applied when performing both sexual and asexual reproduction. It is most often implemented by changing a single node from one function/terminal to another.

Genetic Programming has been applied to a number of applications related to telecommunication services and networks (Aiyarak *et al.* 1997, Lukschandl *et al.* 1999, Shami *et al.* 1997).

15.5.2 Genetic Programming for service creation

The important steps required to tailor GP to a particular domain are the selection of the function and terminal set, and the creation of a fitness function. Considering the functions and terminals first, we have already seen that the set of functions must satisfy the sufficiency property. That is, they must be rich enough to allow an evolving program to satisfy the functional requirements. For instance, a requirement for a program to generate messages would need one or more functions to support this, or sufficient primitives to allow GP to synthesize such a function.

Choosing functions and terminals for services

Various levels of abstraction can be selected for GP. When considering Intelligent Network service creation, the spectrum ranges from SIBs (Service Independent Building Blocks) to operating system primitives that send/receive messages, manipulate strings, perform arithmetic and so forth. Examples from both ends of this have been shown to be feasible (Martin 2000), but in the interest of brevity only the higher-level abstraction is described here.

FSTART takes two arguments. It accepts an Initial Detection Point message from the Service Switching Point, and the Called Dialed Number value is stored at the location returned as a result of evaluating its first parameter. The second argument is evaluated and the result of this is returned.

FDBREAD takes three arguments. This reads a database, using the value of the first argument as a key and placing the result in the location returned as a result of evaluating its second parameter. Finally the third argument is evaluated and the result returned.

FROUTE takes two arguments. Evaluates the first argument which is used to furnish the new routed number for the connection message.

FEND sends a pre-arranged end to the SSP.

STRSUB performs a simple string shift operation to simulate real-life number manipulations.

Terminals may have side effects and/or yield data. For this work, the functions were chosen to perform all external operations, while the terminals were chosen to yield data. Since most IN services require some state information to be stored between message transfer points, as well as per-state variables, the terminals are used to access variables stored in memory that is persistent for the lifetime of a program. In this way both state information and per-state data can be implemented.

The fitness function for service creation

A natural point to measure the fitness of the GP is at the external INAP interface (Intelligent Network Application Part) since this is a standardized external interface as described in Q.1211 (ITU 1995a), and allows the specification of services to be performed at the network level.

The BCSM (Basic Call State Machine) described in Q.1214 (ITU 1995b) is simplified, and termed an SCSM (Simple Call State Model) in order to focus on the GP technique rather than being distracted by the complexities of the BCSM.

In order to simplify the system, the execution of the system was driven by the service logic, so that when evaluating fitness the service logic is executed directly. It then makes requests to the SCSM simulator as required. This inversion of roles removes the problems of detecting unresponsive service logic programs and simplifies the initial debugging and verification.

When running a fitness test, two related problem-specific measures are used to determine how fit an individual is, as well as measures such as parsimony that are not problem-specific. First, the number of correct state transitions recorded and each correct transition are rewarded with a value of 100. Each incorrect transition is penalized with a value of 20. The reward and penalty values are summed as s. Secondly, the number of correct parameter values passed back to the SCSM is recorded. A correct parameter value is rewarded with a value of 100, and each incorrect value is penalized with a value of 20. The reward and penalty values are summed as p. These values are then used to compute a normalized fitness value as follows. Raw fitness r is given by $r=s+p$. Normalized fitness n is given by $n=k-r$ where k is a value that is dependent on the number of state transitions and message parameters in the problem being considered, such that for a 100 % fit individual $n=0$. It is this normalized fitness value that is used to select individuals during tournament selection.

Lastly, the size of program generated by GP has a direct impact on both the run-time performance of any real-world application, and the performance of the GP system. Therefore, a degree of pressure is placed on the evolving programs to reduce the size of the program by rewarding parsimonious programs that have a high fitness value.

Example of a service complex number translation

To illustrate how Genetic Programming can be used to evolve service logic, an example service is presented next. Number translation was chosen for this example since, although it is one of the simpler services, it is also the most common service implemented using the IN (Eberhagen 1998) and forms the basis of many more complex services.

The example uses GP to evolve service logic for a service where two database look-ups are required. An example of this scenario occurs in the real world where a service needs to route to one number during working hours and another number during out of work hours. The first database request in this represents the query that determines a time-based key for the subsequent request, which obtains the final number to route to. A message sequence chart was defined for this service. This was used to derive an external state diagram that forms the basis of the fitness function, and the state diagram is then encoded into a state machine which drives the fitness function.

For this example, the population size was chosen to be 500 and the number of generations was 200. When this problem was run 50 times using different random number seeds, GP created a 100 % correct program 49 times, and the probability of finding a correct solution at generation 200 was 72 %.

Some interesting results were observed. First, all 49 correct programs were different. The differences ranged from the selection of different variables to some counter-intuitive program trees when compared to programs that a human programmer might have devised. One of the less intuitive programs is shown in Figure 15.9. Note that the first function called is the route function. This operation relies purely on the ordering of evaluation of function arguments, rather than a procedural sequencing more commonly found in human-generated programs. This demonstrates rather nicely that one of the features of GP is that it explores and exploits hidden relationships and features which can lead to highly optimal solutions. The main point

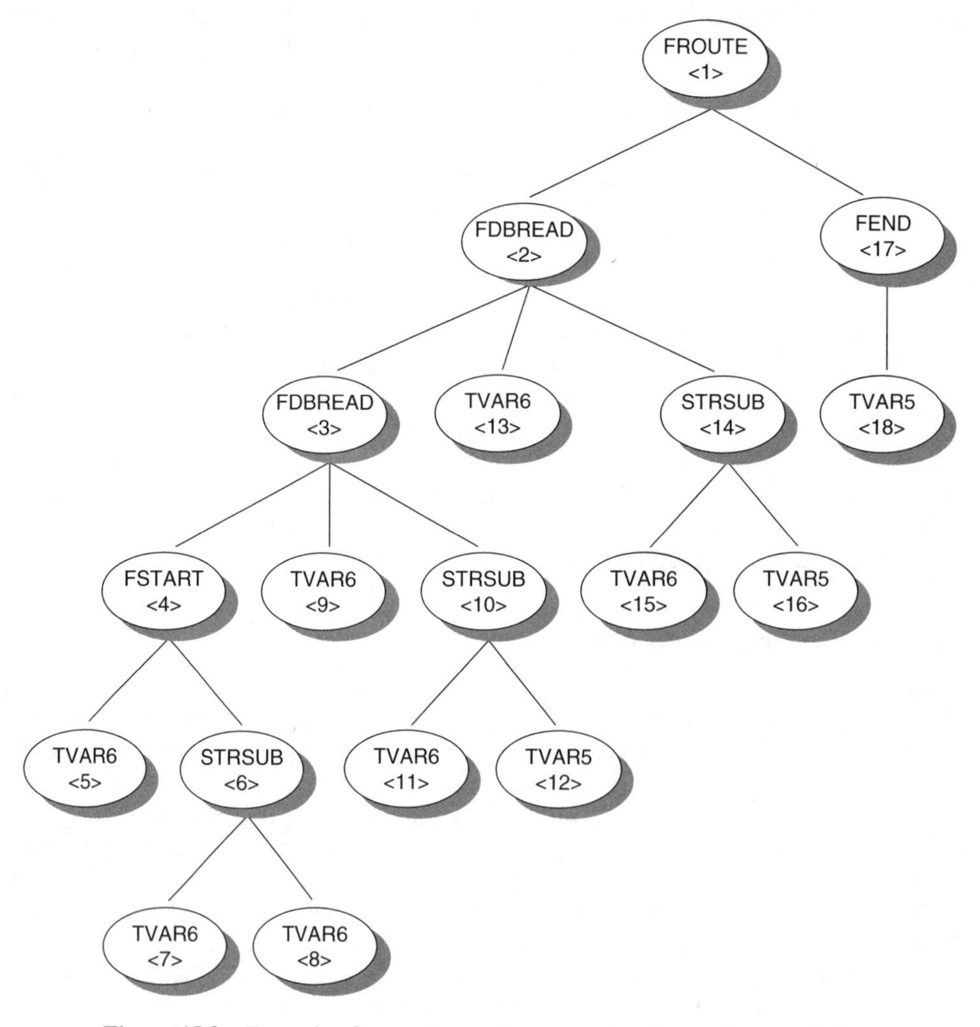

Figure 15.9 Example of a novel correct program tree for number translation.

is that using GP does not remove the need to capture requirements for a service, nor to design the basic operation. These two activities are crucial to formulating the fitness functions. What GP does have to offer is the automatic generation of the service logic.

In conclusion, using GP moves the focus of attention to the 'what' rather than the 'how' of a service, with the potential of delivering services to budget and on time.

15.5.3 Discussion

In terms of raw performance, GP service creation compares well to a human performing the same task, with GP taking minutes to find a 100 % correct individual, and a human taking around one hour to hand code a similar program and test it. However, a comparison made purely on time to complete a task does not tell the whole story. In the case of GP, one of the important tasks of the fitness function is to rank how well an individual comes to being able to solve a problem. The fitness function could also be the test case for the solution so a correct program could well be classified as 100 % tested, with the usual caveats concerning testing metrics.

For many years, automatic program generation has been a goal of software engineering. GP in its current state of maturity goes some way to achieving this goal. This could be viewed as a threat to traditional software engineering, but such a narrow view misses the broader benefits of GP. GP offers an alternative to the traditional coding phase of software development, facilitating the creation of quality tested software. What remain, of course, are the essential activities of requirements capture and system specification. Another question often raised when evolutionary techniques are discussed is how we can be sure that the evolved structures do not have any hidden surprises, for example that a program gives erroneous results under a set of conditions that were not expected. The principal argument against this objection is that every individual is tested against the fitness function. Since the fitness function in effect embodies the specification for the program, the fitness function can contain the test cases needed to ensure that the evolved program performs as required.

Finally, it is frequently remarked that understanding the evolved programs is often hard, since the programs do not always adhere to what a human programmer might consider good style. However, it is rare that an analytical understanding of the program is required. In any case, software engineering is currently happy to trust tools to generate code in other areas. A useful analogy here is to consider how CASE tools and high-level language optimizing compilers have replaced flow charts and hand-crafted assembler code in the vast majority of applications.

15.6 Conclusions

15.6.1 Summary of evolving technologies

This chapter has explored four advanced technologies that have shown promising results in the area of service engineering. There is a trend for new technologies to be based on processes and behaviors observed in nature, examples being Software Agents, ANNs and Genetic Programming. There is also a lot of interest in other techniques that aim to directly apply natural processes, including DNA Computing (DNA 2003, Adleman 1994), Molecular Computing (http:/ /www.molecularelectronics.com/) and Quantum Computing (http:/ /www.qbit.org/).

15.6.2 Barriers to the adoption of evolving technologies

The technologies discussed in this chapter are still mainly in the research labs, and transferring technologies from a research environment to a commercial environment is notoriously difficult. Amongst the barriers often encountered are a general mistrust of new technologies, a lack of understanding about the technology being considered, and uncertainty about the results that new approaches can give, as discussed in the Genetic Programming section. Attempts to transfer technology also fail because they are often oversold, and the hype rarely lives up to user experience.

Issues such as scalability and performance also affect the acceptance of a new technology. Many techniques have often been used to solve what many practitioners consider to be toy problems, and have not been proven on large-scale commercial problems. The result is that the first time a new technology is tried with hard real-world problems, it fails to live up to the hype it was presented with, and the technology is not adopted.

Another barrier in the past has been the lack of robust development tools backed by commercial vendors. However, vendors are reluctant to invest in the development of new tools until there is a proven market for them. This gap is now being filled by tools developed as part of university research programs that make tools freely available. With the general acceptance of licensing models such as the GNU GPL (http:/ /www.gnu.org/) and Open Source (http:/ /www.opensource.org/), this trend is set to increase.

15.6.3 Outlook for evolving technologies

While this chapter has considered several technologies in isolation, it is clear that a single technology is not the answer for all problems. Furthermore, combining multiple technologies can draw on the strengths of each whilst minimizing their weaknesses. For example, it is possible to combine neural networks with genetic algorithms, or agents with neural networks. An interesting example of this is the use of a Genetic Programming system to help with determining a good architecture and weights for a neural network (Cho and Shimohara 1998). Another promising area is developing Constraint Satisfaction strategies using evolutionary techniques (Terashima-Marin *et al.* 1999). Other technologies that could be usefully considered in a solution include Bayes nets, fuzzy logic, and data mining techniques. Web service and Grid technologies are likely to provide highly flexible and powerful computing environments for service development and operation.

All the technologies discussed in this chapter are beginning to make an impact in commercial applications, both in general software engineering and in service engineering in particular. As communication services become more complex, the expectation is that these and other technologies will continue to be developed to provide solutions to problems that are currently hard to solve using traditional techniques.

15.7 Further Reading

This chapter has been able to give only a brief overview of some of the evolving technologies that are being used for service engineering. For those who would like to explore these technologies in more depth, a number of books give more comprehensive details. These books have been chosen with a bias to the practical aspects of the technologies concerned.

A general introduction to agent technolgies can be found in Wooldridge (2002). The role of agents in telecommunication systems is discussed by Gaiti *et al.* (2002). A comprehensive introduction to Constraint Satisfaction is provided by Tsang (1993), while the use of Constraint Satisfaction together with agent technologies can be found in Yokoo (2000). Artificial Neural Networks are covered using a non-mathematical treatment by Bosque (2002). Finally, for those who want to explore Genetic Programming in more depth, a good broad-based coverage of the subject is given by Banzhaf *et al.* (1998).

16

Prospects

David J. Marples[1], Kenneth J. Turner[2] and Evan H. Magill[2]

[1]Telcordia Technologies Inc, USA [2]University of Stirling, UK

16.1 Introduction

The purpose of this chapter is to peek over the horizon at the kinds of capabilities that we should expect to see in a three to ten year timeframe. Here we consider both the services themselves, and their support and delivery environments, including the devices which deliver the services to the user. It is intended to build upon the material of the previous two chapters and hence highlight the overall direction, rather than explore the individual means that the previous chapters describe. Driven by the human need for interaction, there can be few areas of technological research that are moving faster than communications. This chapter reflects the individual views of the authors, and certainly shouldn't be interpreted as necessarily the views of their employers. In any case, the next few years promise to be the most exciting so far!

16.2 Technical Changes

The changes that we foresee in the services market will be, in significant part, driven by developments in the devices that present those services to the user. The range of devices that link us to the network are changing more rapidly than ever before. For most of the last century there was really only one type of terminal device, the familiar telephone. Although the latter decades of the century saw the handset extended with facilities such as voice recording and fax machines, it remained a network endpoint with little network support.

Today, although the telephone handset remains ubiquitous, the situation is much changed. In addition to the continuing trend of adding functionality to devices, there is also a move to increase the network support for terminals. The former trend follows in the tradition of the fax machine which applies the old rule of thumb, 'Dumb Network, Intelligent Endpoint'. This of course has strong market drivers, as it is always easier to upgrade an endpoint than an entire network infrastructure. However, the improving sophistication of devices also motivates the trend of improving network support. This is now gaining momentum and we now see more network support for intelligent endpoints than has previously been the case. Capabilities

Service Provision – Technologies for Next Generation Communications. Edited by Kenneth J. Turner, Evan H. Magill and David J. Marples
 ISBN: 0-470-85066-3

as simple as the voicemail icon on your mobile phone, or calling line ID on your fixed line, all require significant changes to the core network. This is recognition by the core network actors that the endpoints are now capable of processing more sophisticated information. In other words, to realize the full potential of more powerful devices, we require networks that offer these devices better support.

This begs the question, which networks can best provide the support? Clearly the commercial implications are profound. For voice, the traditional wired network is at a huge disadvantage. It still has to be capable of ringing 50 year old phones and so is hamstrung by backwards compatibility issues. An attempt to start again was made with the Integrated Services Digital Network (ISDN) but uptake of this in many territories has been limited by various regulatory, technical, political, and commercial considerations. In short, why would users use feature-limited legacy networks when the capabilities of a wireless terminal are so much more attractive? Wireless allows phones to be portable, personal, and much more available than a wired phone ever can be. However, the wired network has a few advantages:

- it is ubiquitous and largely standardized;
- it supports relatively high speeds for modem connections in comparison to what mobile terminals have been able to cost effectively offer so far;
- its pricing is much more competitive;
- Digital Subscriber Line (DSL) connections allow high speeds (8 Mb/s downstream, 640 KB/s upstream is a typical maximum for Asymmetric DSL)(Alcatel 2003).

In most markets we can expect many of the pricing and ubiquity arguments in favor of wired lines to fall away, and the only arguments for them are based around data. In short, the future is a largely wireless one, so let us start there.

16.2.1 Wireless service innovation

As we move forwards, what can we expect in the new wireless world? Certainly wholesale changes to the network to support intelligent terminals are out of the question in the short term, but initiatives such as Parlay do offer a reasonable approximation to this ideal. Whole new network signaling standards, required for 3rd Generation mobility and standardized by the 3GPP, will allow us to increase the minimum capability that is expected of a terminal and will be able to give it a more complex baseline functionality.

We will see the emergence of terminals that take advantage of this enhanced signaling available from the network. These might offer conditional diversion, depending on the particular task you are performing, or who is calling. The terminals might support video calls when negotiation identifies that both endpoints are video capable. However, this network signaling has to be available to all players in the market, so it is unlikely that terminals will differ much in the functionality that they can offer.

In addition, we will see new functionality become available in the endpoints themselves. This might incidentally make use of network capabilities, but this would not be its primary purpose. 3G mobile terminals are starting to emerge now, and their base functionality is expected to include video calls, email, and other such capabilities. However, it is appropriate to look further out at other typical terminals that may arrive and the capabilities that they will offer.

This assumes a continuing trend within national wide area wireless networks, but other wireless networks are being developed. A large perturbation can be expected from Bluetooth and other wireless Personal Area Network (PAN) technologies. Here, communication capabilities are distributed around a user's body and personal area. The move towards wireless headsets is already in progress, but we should also expect to see video projected into the user's eye, with cameras capturing exactly what the user sees and audio being delivered directly to an earpiece. Indeed, it is not difficult to imagine a user wearing glasses purely as a communication aid. Other possible developments include dialpads and phonebooks integrated into watches, sensors around the body for heart rate, temperature, radiation exposure, or a hundred other similar factors. Of course, once we reach this point the mobile phone with its keypad, earpiece, and microphone becomes pretty irrelevant, probably only used for the gateway between the local and wide area networks. It is only a matter of time before the mobile phone is simply a small polished block similar in look to old cigarette cases with no external connections at all, charged by induction. All of the input and output performed by devices that are associated with it.

So what services should we be expecting? Terminals will become more sophisticated to support this myriad of local devices and will, as a side effect, become programmable. Who is allowed to access this programmability is open to debate, and will probably vary according to the purpose of the offered service. For non-call-affecting capabilities (e.g. games, address books, calculators, new ancillary terminal functionality) there is unlikely to be any real restriction. Indeed, initiatives such as the Mobile Information Device Profile (MIDP)(Sun MIDP 2003), Mophun, Open Services Gateway Initiative (OSGi) and the Binary Runtime Environment for Wireless (BREW)(Qualcomm BREW 2003) all go some way to addressing this market.

Call affecting functionality, that can in some way manipulate the progress of a phone call, is liable to be somewhat more constrained. Neither terminal manufacturers nor network operators wish to be considered responsible for any loss of call handling performance when this might stem from some arbitrary third party application.

Most of the execution environments available today for mobile terminals are limited to sandbox operation. This prevents them from being able to manipulate resources outside of this, except through narrow interfaces. If no interface for call control/manipulation is provided then it will be impossible for truly sophisticated services affecting calls to be created. What will probably happen is that access to call-affecting interfaces, and thus the creation of call-affecting services, will initially be restricted to a select number of developers who need to get their applications certified before they can be deployed in the market.

16.2.2 Fixed line service innovation

Of course, the fact that wireless voice will continue to expand does not mean that other communications services will become unnecessary, far from it. Data communications, or direct machine-to-machine communications, are arguably already more important than voice communication (especially when considering that voice communication can be approximated by Instant Messaging). This importance is only likely to increase, eventually subsuming wired voice communication. Again, the change in the capabilities of the endpoint will drive this change, leading to innovation such as video calling, online gaming, and other sophisticated services. We already see still image transfer being adopted as an early step along this path, although it is the mobile phones, increasingly incorporating digital camera capabilities, that are taking the lead.

Video calling will be driven by the data industry rather than the voice communications industry. Consumer video calling is already happening with extensions to Instant Messaging capabilities from the likes of AOL and Yahoo! Online gaming is a prime example of a sophisticated endpoint driving user functionality. Gaming users are able to communicate directly with each other in the context of the game (with typed or voice commands), and already online gaming communities supporting such activities are starting to emerge.

16.2.3 Service delivery

Historically, users wanting access to a new voice communication service would need to call their service operator who would, after a period of time, eventually instantiate the service on the user's phone line. In the days where all of the (limited) functionality was concentrated in the network, this was perfectly acceptable.

These traditional ways do not work in the highly devolved and heterogeneous environments that are emerging today. Services will originate from a range of sources, and be deployed across a variety of platforms. The sources will range from communication companies, through content providers, to individual users. They might execute within networks, or on the periphery, or indeed both. Those on the periphery may operate on servers, or powerful end devices. The environment in which services are written, deployed, and executed is changing.

16.3 Service Environments

Until this point, this chapter has largely concentrated on the developments in the endpoint devices which connect to the network, and the enhancements in the services that they will enable. There is good reason for this viewpoint – the average user perceives the terminal device as the totality of the functionality available to them, and cannot differentiate between services implemented on the device and services implemented in the network behind the device.

However, at some point the services need to be considered explicitly as the first class citizens that they undoubtedly are. In this section, we consider the changes in service technology and implementation which will lead to new user functionality.

16.3.1 Location

Historically, communications services were embodied in the core switching software of the Common Office. Changes to this switching software, which potentially could impact some 100 000 users or so, were slow, time consuming, and expensive. As a result, a limited number of services were available. Innovations, such as the Intelligent Network, only had limited impact on this issue. Development was further limited by the capabilities of the signaling available via the endpoint devices. As a result, more intelligent endpoint devices were often used to introduce new services rather than the network. Voicemail is a good example of this trend, frequently implemented as an endpoint service, rather than a core network service.

As end devices become more powerful, this is becoming more common. Indeed, a point is being reached where the devices themselves are becoming *programmable* from the network, and can thus host services. This results in an increased flexibility in the services that can be delivered to the endpoint devices. This is largely happening in the mobile space simply because there is more functionality in a typical mobile terminal than in the normal landline terminal. The signaling capabilities of a mobile terminal are also more sophisticated. Initially, this is being exploited by games with Over the Air (OTA) provisioning, but in future it is not unreasonable to expect much of the functionality of a terminal to be downloaded from the network. Thus, the personality of an endpoint device would be set by a network operator rather than by an equipment provider. This is obviously in the network operator's interest and contrary to the equipment provider's interest, which leads to an interesting dynamic. Network-specific badging and operator-specific software is starting to emerge, and this will accelerate as operators exploit their position over their equipment providers (RCR News 2003).

Typical services that will become available when services can be dynamically downloaded into the terminal device include;

- *Operator personalization*: Installing the 'personality' of a specific operator onto a device, with operator-specific menus and capabilities in a manner akin to the branding of Web portals of the early 2000s such as those offered by AOL and Yahoo!
- *Vertical market*: Adding specific capabilities to the device for particular environments; a time-limited access token for the car that you've just hired, or perhaps a product pricing calculator for the salesperson on the road, which determines price and availability based on the actual stock at the factory at that point in time.
- *Security*: Since the mobile terminal is becoming as ubiquitous as a set of keys, then it makes sense to use it as a more capable keyring, offering personal tokens for access, identification, and payment.
- *Local interaction*: Museum guides, user interface to display/keyboardless devices and pervasive computing elements such as in the car, bathroom, kitchen, or on the train.
- *Medical*: The ability to process locally and to deliver information back to a remote location on a schedule or exception conditions is very useful for applications such as blood glucose monitoring.

Increasingly, the service, rather than the infrastructure over which it is provided, will become the prime concern of the user, and the infrastructure will fade into the background. It is the service that provides the user benefit, and is the chargeable element after all. With this in mind, it should be expected that these individual services will increasingly be charged for and that these services will be available across multiple infrastructures – perhaps HTTP, WAP, touch screens, gestures, voice response, and direct human/human interfaces. In short, a single service will be provided via any means that is secure, reliable, and capable enough to support it. The user interfaces to modern banks are the first example of this change of paradigm. The kind of service that they represent is much broader than the definition used elsewhere within the book. Unless the network operators can find a way into this widened definition of a service market they, like the equipment manufacturers before them, may find themselves pushed into a marginal position in the service value chain.

16.3.2 Structure

All of these environmental changes have already imposed many significant changes in the way in which communications applications are constructed. Historically, the code that ran in the core switching fabric of the network, although potentially modular in its internal construction, was a monolithic entity. It was replaced in its whole in order to upgrade the functionality that it provided to users. Over time this has changed so that individual modules of code can be replaced.

As terminal devices become more capable there is even more demand for such modularization and the ability to be able to place code at different points in the call processing chain. With this fragmentation, manageable update becomes essential. Both the OSGi and .NET platforms contain capabilities to do exactly this. It will increasingly be a requirement that micro version control capability be available on any platform where it is desired to run services.

We should also expect smaller and smaller devices to be able to be programmable. In times past this was difficult with Read Only Memory (ROM) imposing restrictions on the changes that could be made once a device had been manufactured. Today, with erasable memory cheaper than read only memory, there is little technical reason for not making devices field upgradeable.

When it is the core application that is being upgraded there is little implied effect on the structure of the software. Obviously the facility to allow for upgrade needs to be incorporated, but that is the limit of the changes that are required. It is when the software can be *extended* to incorporate new functionality or additional services, that this has implications for the structure of the software system. In this case, the capabilities of the base system must be made accessible to programmers wishing to extend it. One of the major reasons for the early success of PalmOS was down to the fact that it was user-accessible and extensible. After all, there were devices on the market long before the Palm which offered similar end-user functionality in their factory default configuration, but were not extensible in the way that PalmOS devices are.

Incorporating new capabilities into these devices has huge implications for feature interaction, which has been discussed in Chapter 13. If we believe that extensible feature sets and new capabilities deployed into user devices will become a way of life for the early part of the 21st century, then we also need to accept that feature interaction will become a much more significant problem.

16.3.3 Migration

The question needs to be asked why services will move away from their traditional locations in the core network. In general, many will move out towards endpoints, but this is not a universal rule. It is far easier to upgrade an endpoint than an entire core network, but there are other considerations to bear in mind when selecting the most appropriate location for services.

Upgrades to the core network imply a one-size-fits-all approach, or at most an approach that offers limited flexibility. In contrast, users are free to modify endpoint devices as they see fit, within the constraints of network signaling, regulatory frameworks, and the technical limitations of devices. This encourages movement towards configurable devices as the endpoints.

Services do not always require permanent network connectivity, and indeed it may be mandatory for them to continue to operate even when such connectivity is not available. This might be the case for a heart monitoring service, for example, which could use an emergency channel to dial out for an exception condition even if regular channels were not available. Again, this encourages movement of services towards the local devices, i.e. the endpoints.

Services may need to process significant amounts of data locally and only upload minimal amounts of information to the network. In any environment where access is charged according to the amount of information transferred, then there will always be a pressure to minimize the amount of data transfer.

In contrast, mobile devices in particular always suffer from performance/price/storage/ power consumption trade-offs. Performing processing locally will, in general, require more performance and power consumption than performing it remotely. A typical service that might take advantage of a network-based processing capability could be a speech-to-text capability, where the user speaks into a handset, the speech is forwarded over the network, processed remotely, and the resultant text is returned to the handset. For this kind of service, implementation within the network (or at least at a remote endpoint which does not suffer the same resource constraints) is obviously preferable, provided that the connecting network is capable of carrying the required data in a timely fashion.

Services will also be provided at new points along the communications path at 'semi-trusted' locations where services can do more than an end-user is capable of, yet cannot fully control all aspects of a communication session. Technologies for achieving this are starting to be adopted now, with Parlay, Computer–Telephony Integration (CTI) and Web services being examples of capabilities which support this trend. Each of these is important enough to be considered in turn.

Parlay

Parlay offers a set of Application Programming Interfaces (APIs) to the network for semi-trusted applications. These applications can access information that was previously held privately within the network and can, within certain constraints, influence the progress of communication sessions. Parlay was initially a way of allowing controlled access to public switches but has evolved a long way since then to be a more general set of communications APIs.

Computer–Telephony Integration

In many ways analogous to Parlay, CTI has been widely adopted in the Private Branch Exchange (PBX) market for connecting voice and data communications services together – popping the caller's name and address details on your PC screen as your phone rings being the most frequently, and trivially, quoted example. CTI brings a lot of enhancements to the progress of communications sessions and its true value is only just starting to be realized. Useful applications like ensuring a certain caller is always directed to the same agent in a sales group, or that lawyers log all outgoing billable calls correctly are examples of the way in which CTI is starting to be used. Voice over IP (VoIP) is often confused with CTI because VoIP systems generally have always shipped with advanced CTI capabilities, but the reality is that any system can have very powerful CTI facilities integrated into it, no matter what the transmission technology that is used.

Web services

Web services are intended for direct program-to-program communication, without the requirement for human beings to be involved in the transaction – the 'Web for programs' is one way of describing it. The adoption of Web services promises to offer a world in which companies and individuals can export services to make them available to other services across the Internet – service composition will become hugely more straightforward and will require considerably fewer bipartite agreements.

16.4 Market Changes

The break away from the conventional wired network environment allows services to be deployed in a huge range of new environments which have previously not been amenable to network connectivity. There are three main environments which are worthy of special consideration.

16.4.1 Vehicle

A modern vehicle already contains many communication services. Gone are the days of wires for every indicator lamp on the dashboard, replaced instead with Local Area Networks (LANs)(CanCIA 2003) which provide a cheap, effective signaling environment inside the vehicle. Extending this inherently software-based system to incorporate remote communications capabilities is really only a minor enhancement to what is already done today. Although there is considerable current interest in automobiles, this same principle applies to vehicles as diverse as aeroplanes and ships.

So, if this is the case, what new capabilities can we expect tomorrow's vehicles to contain? There are EU research projects underway to investigate exactly this issue, and it is reasonable to expect to see capabilities such as:

- Direct vehicle-to-vehicle communication for communicating issues that are relevant in the local area: black ice warning, a distributed cruise control amongst several vehicles on the motorway, or a warning of severe speed decrease (e.g. blockage up ahead).
- Accident Black Box recording to recover data relevant to an accident (e.g. direction, speed, indicator status) to help in a subsequent investigation.
- Vehicle flow control to provide information to control centers which can be used, in aggregate, to determine the flow of traffic in a vicinity and to re-route traffic if necessary.
- Vehicle performance control to control dynamically the performance of a vehicle according to certain parameters (price paid for insurance, local weather conditions, age/experience of driver, etc.).

While we may not think of the above as traditional communication services, they all require some level of device-to-device linkage in order to be able to deliver an end-user service, so it is not unreasonable to think of them in this context. Of course, the above does not even start to include the more traditional applications that are amenable to in-vehicle deployment, such as entertainment delivery and voice communications.

16.4.2 Home

Residential exposure to traditional communications services has historically always been via the home phone line. 91 % of UK households have a fixed line (Oftel 2003), a number that is little changed over the past few years, but home service providers are facing increasing challenges from the mobile operators. The dominance of the traditional home phone line should reasonably be expected to decrease over the next few years, challenged by the mobile phone.

The saving grace for the home phone line operators is the fact that this same extensive delivery infrastructure can be used to deliver high-speed data networking capabilities via Digital Subscriber Line (DSL). The main challenge to DSL comes from the cable operators with some other potential challengers such as power line delivery on the horizon, but the DSL environment does potentially offer some significant benefits to users which are not yet widely exploited. DSL offers link-level multiplexing which allows multiple Quality of Service (QoS) aware streams to be delivered over the same access link. Most of the current cable infrastructure deployed does not offer this same capability today, so DSL is arguably better for use in multiple media environments.

In any case, the true service opportunity for the home is not the means of delivery to it. It is the wide range of relatively powerful computational devices that exist within a modern home, which will only increase over the next few years. The amount of processing power in a satellite decoder, or a cable set-top box, exceeds that used to put a man on the moon, yet we accept this computational resource as just a normal part of our lives.

We will see services delivered over the network running in the home environment. These services may be Wide Area Collaborative (i.e. working in conjunction with resources outside of the home), Local Area Collaborative (i.e. coordinating resources within the home), or Local.

Wide Area Collaborative services might typically include:

- coordinating power consumption across multiple individual homes;
- providing content-based services to the home (e.g. multiple media content delivery);
- security and access services to/from the home;
- delivering services to the home for rendering to local devices (e.g. emergency alert services that use whatever devices are available to alert the user).

Local Area Collaborative services imply coordination between multiple devices to deliver an enhanced service to the user: automatically closing the curtains when the movie starts, sending the TV sound out of the hi-fi speakers, or setting the oven temperature and cooking profile from the Web browser displaying the recipe under construction.

Local services simply execute on a single device, but that service may itself have been downloaded from a network connection. This might be the case for a game downloaded from a local service provider for example.

There are two major technical issues with access into the home for the purpose of realizing services. First, due to addressing, security, privacy, and configuration considerations, it is unlikely that every device inside a given home will be uniquely addressable from the outside world. Secondly, it is not reasonable to expect services outside of the home to adjust their operation according to the specific devices that are available inside a single home. Services will need to be able to render themselves on the devices that are available in any one specific location.

For both of these reasons, it is likely that there will be a gateway device at the edge of the home, mediating access to it. This might itself run services, which one would expect would be dynamically loaded and unloaded on request using a capability such as that provided by OSGi. It would certainly provide authentication and security features. It is also reasonable to expect it to perform a mapping between the generic description of the requirements a particular service has and the resources in a particular home. Thus, the privacy of the home will be maintained while the service can still be delivered to it.

This is currently the subject of research (Telcordia 2003c). There is time for these issues to be resolved since it is unlikely there will be any useful service delivery into home networks for a number of years yet, given that home networks themselves are still in their infancy.

16.4.3 Personal Area Network

The Personal Area Network (PAN) is potentially the most pervasive, and most exciting, new environment for service deployment. Enabled by wireless, PANs allow services to be delivered directly to the human consumer, rather than to some arbitrary electronic artifact which might just happen to be close. The mobile phone can be considered as the first success of the PAN. Instead of a wired telephone ringing at the other end of the house and having to run to answer it, the device in your pocket rings.

Increasingly, we will see the trend, piloted by Steve Mann (Wearcam 2003) and others, towards wearing/integrating more and more network interface components onto/into our bodies. Already we do not think twice about this. We wear hearing aids, glasses, pacemakers, and watches to augment capabilities that (some of our) bodies are not so capable of delivering. Significant restrictions in terms of power consumption, processing capability and I/O capability still exist, but these are being addressed in innovative ways on a daily basis because it is realized that there is significant commercial opportunity. For example, new battery technology and fuel cells (USFCC 2003) promise to free us from many of the power restrictions we have faced until now, not to mention true innovation in exploiting the body's own power producing apparatus (SMH 2003). This is extremely exciting for application in deeply embedded devices such as pacemakers and internal monitoring equipment.

It is, however, the sociological implications of the PAN that are probably the most significant, and the most poorly understood, since predicting the actions of people is always extremely difficult. We have already started to see the first hints of social change with the extensive use of the GSM Short Messaging Service (SMS) by today's mobile generation.

PANs will make us always connected, always available, and potentially always leaking information about ourselves to a wider audience. How will that change our behavior, and what social/political/legal barriers will we put up to contain and limit this to give us some of our privacy back?

No longer do we need physical co-presence to have logical co-presence (how many times have you seen someone apparently talking to themselves only to find, when you get closer, that they have a mobile phone earpiece and microphone?) What does that mean for the geography and dynamics of the social groups that we form? For example, Dave's close work-social group consists of people from all over the world, co-located in an Instant Messaging availability panel. His associates group (which is slightly more distant in a logical sense) communicates by email from anywhere on the planet.

The services enabled by PAN devices also may change the way in which we perform introductions, with potentially dangerous ramifications. For example, there is a dating device in the form of a small keyfob (FindtheOne 2003) which will check your compatibility with the people around you, the intention being to identify suitable mates. Although undoubtedly useful in certain circumstances, pre-checking compatibility closes off exposure to new experiences and prevents our horizons being broadened by people that we would never have expected to have hit it off with on the basis of some checklist selection.

We already hear reports of our children's writing and math skills becoming poorer because they are propped up with tool support. Will the same happen with their social skills?

16.5 Commercial Changes

16.5.1 The Open Source movement

There is a huge change coming in the way in which software products are bought and sold, and that change is the Open Source movement. Born out of the belief that it is only by standing on the shoulders of others that we can become ever greater, the Open Source Software (OSS) movement believes that the source to software applications should be made available to users on request, as an addition to the executable form of the application itself. Users should then be free to modify this application to meet their own requirements.

Due to the structure of many OSS licenses, it becomes very difficult to charge for the software product itself. Instead the charges must be for the support activities relating to it. It is also very difficult for a company to survive when marketing an inferior product, or a product which does not adequately meet user needs, because someone could take the source code for the product and modify it to meet these needs more accurately, pushing the poorer product out of the market.

By only having to pay for ongoing support and maintenance of the product, rather than for access to the license itself, companies reduce the capital expense involved in moving to a new program. Of course, the training and other hidden costs of software ownership remain, but they should be largely on parity between the Open and Closed source alternatives provided that the support organizations are equally competent.

This change in the nature of software supply is probably the most fundamental shift in the last twenty years of software development, but it should be remembered that many very successful companies exist purely on the basis of the intellectual property embedded in their software, which it would be foolhardy for them to jeopardize. For this reason we also see a move towards open interfaces which promote collaborative development and teaming without endangering the huge investment in established software bases.

So, what effect does this have on services? If the code implementing services is released in Open Source form, then it is reasonable to assume that, due to the large number of potential coders that have access to it, it will improve and become more functional very rapidly. Thus, in an Open Source environment, highly functional services should quickly become available. However, for real users, it is not the services themselves that have utility, but the benefit that they glean from having these services available. This requires more than the code that implements the service. It also requires training, support when something goes wrong, an operational environment, feature interaction compatibility checking, and any number of other issues to be addressed to ensure that the service consumer gets the user benefit that they expect.

Since the user, in most cases, wishes to get ongoing benefit from the application of the service, then it is reasonable that they will pay some fee for these support activities. One of the things that increasing network connectivity and commercialization has made possible is that establishing a binding between a supplier and a customer is becoming cheaper and more pervasive; it is more difficult to fake or to break. This means that it is increasingly possible for suppliers to charge a customer a small, recurring amount of money for the ongoing use of a service, even if the code for that service is freely available.

In short, we should expect to pay rentals for services and their support, but expect the quality of that support to increase, since there is little lock-in to prevent a consumer from going to another service provider if that support fails to come up to scratch.

16.5.2 The role of copyright

There is controversy over copyright and the way it applies in the information age. This discussion is of fundamental importance for consideration of future services, since a fair proportion of these services will be required to deliver content of one form or another, and content is nearly always subject to some form of copyright restriction.

Much of the current controversy comes from the role of file sharing and distribution over high-capacity networks such as the Internet. Many half-baked defenses have been put forward as justifications for this activity. The fact is that the majority of perpetrators do so because they are obtaining a product with value (in soft form) without needing to pay for it, when that is against the wishes of the copyright holder. Whichever way you look at it, that is not right. Many do not agree with the way in which the music industry is developed or managed, but condoning unlawful (note not necessarily illegal) activity cannot be an acceptable solution to the problem.

The technology industry has responded by developing solutions to lock down content in the form of Digital Rights Management (DRM). Unfortunately, DRM is always liable to be cracked, and it only needs a small number of cracks before the dam breaks. This was seen with the Digital Versatile Disk (DVD) Content Scrambling System (CSS) crack, known as DeCSS, which has effectively made CSS transparent to any user who wishes to circumvent it. It is not unreasonable to suppose that equivalent cracks will eventually emerge for other mandatory DRM schemes that are constructed.

What is actually required is a fundamental change in the industry structure to not just allow new forms of content distribution, but to actively embrace them. Unfortunately, as with any revolution this inevitably means that some of the current players will find themselves without a role, and this causes a huge amount of understandable resistance to change. However, when an irresistible force meets an immovable object, something has to give. The irresistible force of public disrespect for the implementation of existing copyright law means that we can expect to see significant changes here over the next few years.

In terms of the effect on services, these changes should make content more accessible to users, on a wider variety of devices. We should be able to access services that allow content that we legitimately have the rights to use, available on any device capable of rendering it – no matter if that device is in our hotel room, home, or workplace. We might wish to forward recordings of TV programs to friends and colleagues, and to obtain archived content from established content providers (BBCNews 2003), not to mention being able to download books freely to read on demand.

All of this implies that established practices, procedures, payment processes, and money flows must change beyond all recognition. That will not happen overnight, and certainly will not happen in the ways in which we forecast it to. But it will happen, and the implications for the way in which we are able to exploit content in the services we deliver to end-users will be profound.

None of these changes automatically imply that industries will be destroyed or revenues reduced. Video tapes did not destroy the movie industry, nor should network delivery destroy cinemas, libraries, or record stores. Their role in the whole supply and user demand fulfillment chain will change.

16.5.3 Safety and security

Of course, all of these changes (new services, network delivered capabilities and the like) mean that the number of potential openings for intrusion, unwanted code execution, 'spy-ware' and any other number of a variety of 'malware' will increase significantly.

In future we can expect to see systems become tighter and more controlled, if only because of the user backlash at the amount of virus susceptibility of our current platforms. This will, of course, mean more inconvenience for legitimate users and it will be systems that balance this tightening of security with the requirement to make things acceptable to end-users that will gain acceptance. Couple this with the increasing requirement to be able to unambiguously identify an individual user (for DRM, anti-spam, payment, and any number of other reasons). It is then possible to see a future world in which, despite the reservations that have been expressed to date, some form of universal identification system for service users will become the norm.

This has huge implications for personal privacy, not to mention the issues when identity theft and 'misconfiguration' are taken into consideration. Imagine for one second the problem today when your credit card is refused due to a processing problem. Now multiply that a hundredfold and you cannot even use the phone to correct the mixup!

There is also an open issue of personal privacy which is most concerning. Scott McNealy of Sun is famously quoted as having said 'You have zero privacy anyway' (WiredPolitic 2003). There is often the question raised as to why legitimate users should be concerned about their movements being tracked. We must return to the origins of our species. Once out of sight of our brethren we are used to having a personal space which is our own and which is not overlooked. It would be of concern, and the mental and physical consequences uncertain, if the norms that thousands upon thousands of years of evolution have ingrained into us were upturned simply because we are able to do it.

16.6 And Finally...

It is not difficult to see a future world in which services collaborate actively, delivering the results of their collaboration directly to us for us to act upon. During a typical working day, for example, our agents might negotiate a set of meetings based on our preferences, expectations, and work schedule, projecting the details of each engagement directly to a head up display integrated into our glasses. At the end of the day they could make our dinner engagement, order the taxi to get us to it, and organize the payment afterwards. All of this is almost possible today, and certainly will be within a few years. The question becomes: at what point do we stop being served by these applications and instead start to serve them?

Appendix 1 Abbreviations

As with most of computer science, communications is replete with acronyms. Although these are explained as they appear in each chapter, a list is included here for reference.

1G	First Generation (cellular system)
2G	Second Generation (cellular system)
2.5G	Second-to-Third Generation (cellular system)
3G	Third Generation (cellular system)
3GPP	Third Generation Partnership Project
4G	Fourth Generation (cellular system)
AAL1	ATM Adaptation Layer 1
ACAP	Adaptive Communicating Applications Platform
ACL	Access Control List
ADL	Architecture Description Language
ADPCM	Adaptive Dynamic Pulse Code Modulation
ADSL	Asymmetric Digital Subscriber Line
ADT	Abstract Data Type
AF	Assured Forwarding
AH	Authentication Header
AIN	Advanced Intelligent Network
AMIC	Automotive Multimedia Interface Consortium
AMPS	Advanced Mobile Phone Service
ANN	Artificial Neural Network
ANSI	American National Standards Institute
APEX	Application Exchange
API	Application Programming Interface
ARPU	Average Revenue Per User
ASN	Abstract Syntax Notation
ASP	Active Server Pages
ASP	Application Service Provider
AT	Access Tandem
ATM	Asynchronous Transfer Mode
AuC	Authentication Center

Service Provision – Technologies for Next Generation Communications. Edited by Kenneth J. Turner, Evan H. Magill and David J. Marples
 ISBN: 0-470-85066-3

AWC	Area Wide Centrex
BCM	Basic Call Model
BCSM	Basic Call State Model
BER	Bit Error Rate
BICC	Bearer Independent Call Control
BS	Base Station
BSC	Base Station Controller
BTS	Base Transceiver Station
BWA	Broadband Wireless Access
CAMEL	Customized Applications for Mobile (station) Enhanced Logic
CASE	Computer Aided Software Engineering
CBR	Constant Bit Rate
CC	Country Code
CCAF	Call Control Agent Function
CCD	Charge-Coupled Device
CCF	Call Control Function
CCIR	International Consultative Committee for Radio
CCITT	International Consultative Committee for Telephony and Telegraphy
CCM	CORBA Component Model
CCS	Common Channel Signaling
CD	Compact Disk
CDMA	Code Division Multiple Access
CDPD	Cellular Digital Packet Data
Centrex	Central Exchange (services)
CIDL	Component Implementation Definition Language
CIM	Common Information Model
CNP	Contract Net Protocol
Codec	Coder–Decoder
CORBA	Common Object Request Broker Architecture
COTS	Commercial Off The Shelf
CPH	Call Party Handling
CPL	Call Processing Language
CPR	Call Processing Record
CRM	Customer Relationship Management
CS	Capability Set
CTI	Computer–Telephony Integration
CWM	Common Warehouse Model
D-AMPS	Digital AMPS
DAB	Digital Audio Broadcasting
DAML	DARPA Agent Markup Language
DARPA	Defense Advanced Research Projects Agency
DCE	Distributed Computing Environment
DCT	Discrete Cosine Transform
DECT	Digital Enhanced Cordless Telecommunications
DEN	Directory-Enabled Networks
DFP	Distributed Functional Plane

DiffServ	Differentiated Services
DMTF	Distributed Management Task Force
DP	Detection Point
DPCM	Differential Pulse Code Modulation
DPE	Distributed Processing Environment
DS	Differentiated Service
DSCP	Differentiated Service Code Point
DSCQS	Double Stimulus Continuous-Quality Scale
DSL	Digital Subscriber Line
DSP	Digital Signal Processor/Processing
DSS1	Digital Signaling System Number 1
DTMF	Dual-Tone Multiple Frequency
DVB	Digital Video Broadcasting (-C Cable, -T Terrestrial, -S Satellite)
DVD	Digital Versatile Disk
E-OTD	Enhanced Observed Time Difference
E911	Enhanced 911 (emergency service)
EDACS	Enhanced Digital Access Communication System
EDGE	Enhanced Data rates for GSM (or Global) Evolution
EDP	Event Detection Point
EF	Expedited Forwarding
EFSM	Extended Finite State Machine
EIA	Electronics Industry Association
EIR	Equipment Identity Register
EJB	Enterprise Java Beans
EMA	Electronic Messaging Association
EO	End Office
EPSRC	Engineering and Physical Sciences Research Council (UK)
ESP	Encapsulated Security Payload
ESTELLE	Extended Finite State Machine Language
ETI	Embed The Internet
ETSI	European Telecommunications Standards Institute
FCC	Federal Communications Commission
FDD	Frequency Division Duplex
FDMA	Frequency Division Multiple Access
FDT	Formal Description Technique
FIM	Feature Interaction Manager
FOMA	Freedom of Mobile Multimedia Access (Japanese 3G system)
FORCES	Forum for Creation and Engineering of Telecommunications Services
FSA	Finite State Automaton
FSM	Finite State Machine
FTP	File Transfer Protocol
GA	Genetic Algorithm
GFP	Global Functional Plane
GGSN	Gateway GPRS Support Node
GIF	Graphics Interchange Format
GIOP	General Inter-ORB Protocol

GMSC	Gateway MSC
GP	Genetic Programming
GPL	General Public License (GNU)
GPRS	General Packet Radio Service
GPS	Global Positioning System
GR	Generic Requirements
GSM	Global System for Mobile Communications
GUI	Graphical User Interface
HAPS	High Altitude Platform System
HAVi	Home Audio–Video Interoperability
HDTV	High-Definition Television
HLR	Home Location Register
HomePlug	Home Powerline Alliance
HSCSD	High-Speed Circuit-Switched Data
HTML	HyperText Markup Language
HTTP	HyperText Transfer/Transport Protocol
ICMP	Internet Control Message Protocol
IDL	Interface Definition Language
IEC	International Electrotechnical Commission
IETF	Internet Engineering Task Force
IM	Instant Messaging
IMEI	International Mobile Equipment Identity
IMPP	Instant Messaging and Presence Protocol
IMPS	Instant Messaging and Presence Service
IMSI	International Mobile Subscriber Identity
IMXP	Instant Messaging Exchange Protocol
IN	Intelligent Network
INAP	Intelligent Network Application Part
IntServ	Integrated Services
IP	Intelligent Peripheral
IP	Internet Protocol
IPSec	Internet Protocol Security
IR	Infrared
IrDA	Infrared Data Association
IS	Interim Standard
ISC	International Softswitch Consortium
ISDB	Integrated Services Digital Broadcasting
ISDN	Integrated Services Digital Network
ISO	International Organization for Standardization
ISP	Internet Service Provider
ISUP	ISDN User Part
ITU	International Telecommunications Union
ITU-R	International Telecommunications Union – Radio
ITU-T	International Telecommunications Union – Telecommunications
IVR	Interactive Voice Response
JAIN	Java API for Integrated Networks

JCC	Java Call Control
JFIF	JPEG File Interchange Format
JIAC	Java Intelligent Agent Componentware
JPEG	Joint Photographic Experts Group
JSP	Java Server Pages
LA	Location Area
LAN	Local Area Network
LBS	Location-Based Service
LCD	Liquid Crystal Display
LDAP	Lightweight Directory Access Protocol
LEO	Low Earth Orbit
LIDB	Line Information Database
LMDS	Local Multipoint Distribution Service
LNP	Local Number Portability
LOTOS	Language Of Temporal Ordering Specification
LSSGR	Local Switching Systems General Requirements
MAC	Medium Access Control
MAC	Message Authentication Code
MBone	Virtual Multicast Backbone on the Internet
MDA	Model Driven Architecture
MEO	Medium Earth Orbit
MGC	Media Gateway Controller
MGCP	Media Gateway Control Protocol
MHA	Message Handling Agent
MHEG	Multimedia and Hypermedia Information Coding Expert Group
MIDI	Musical Instrument/Integration Digital Interface
MIME	Multipurpose Internet Mail Extensions
MMS	Multimedia Messaging Service
MMUSIC	Multi-party MUltimedia SessIon Control
MOF	Meta-Object Facility
MP3	MPEG-1 layer 3
MPC	Multimedia Personal Computer
MPEG	Moving Picture Experts Group
MPLS	Multi-Protocol Label Switching
MPT	Ministry of Post and Telecommunications
MS	Mobile Station
MSC	Message Sequence Chart
MSC	Mobile Switching Center
MSF	Multi-Service Switching Forum
MSISDN	Mobile Station ISDN Number
MSRN	Mobile Station Roaming Number
MTI	Missed Trigger Interaction
MTP	Message Transfer Part
MUMC	Multiple User Multiple Component
MUSC	Multiple User Single Component
MVNO	Mobile Virtual Network Operator

NDC	National Destination Code
NGN	Next Generation Network
NIST	National Institute of Standards and Technology
NNI	Network–Network Interface
NPA	Network Provider Agent
NSF	National Science Foundation
OA&M	Operations, Administration, and Management
OBCM	Originating Basic Call Model
OCS	Originating/Outgoing Call Screening
ODP	Open Distributed Processing
OMA	Object Management Architecture
OMA	Open Mobile Alliance
OMC	Operations and Management Center
OMG	Object Management Group
ORB	Object Request Broker
OSA	Open Services Access
OSGi	Open Services Gateway Initiative
OSI	Open Systems Interconnection
OSS	Operation Support System
PAM	Presence and Availability Management
PAMR	Public Access Mobile Radio
PAN	Personal Area Network
PBX	Private Branch Exchange
PC	Personal Computer
PCA	Personal Communications Agent
PCF	Packet Control Function
PCM	Pulse Code Modulation
PCS	Personal Communication System
PDA	Personal Digital Assistant
PDB	Per-Domain Behavior
PDC	Personal Digital Cellular
PDSN	Packet Data Serving Node
PE	Physical Entity
PHB	Per-Hop Behavior
PHS	Personal Handyphone System
PIC	Point In Call
PIM	Personal Information Manager
PIM	Platform Independent Model
PIM	Presence and Instant Messaging
PLMN	Public Land Mobile Network
PMR	Private Mobile Radio
PNA	Phoneline Networking Alliance
POA	Portable Object Adapter
POS	Packet Over SONET
POTS	Plain Old Telephone Service
PP	Physical Plane

PRIM	Presence and Instant Messaging
PSM	Platform Specific Model
PSTN	Public Switched Telephone Network
QoS	Quality of Service
RAN	Radio Access Network
RF	Radio Frequency
RFC	Request For Comments
RM-ODP	Reference Model for Open Distributed Processing
RMI	Remote Method Invocation
RNC	Radio Network Controller
RPC	Remote Procedure Call
RSVP	Resource Reservation Protocol
RTCP	Real-Time Control Protocol
RTP	Real-time Transport Protocol
RTSP	Real-Time Streaming Protocol
RTT	Radio Transmission Technology
SAE	Society for Automotive Engineers
SAI	Sequential Action Interaction
SCE	Service Creation Environment
SCF	Service Control Function
SCP	Service Control Point
SD	Speed Dialing
SDF	Service Data Function
SDL	Specification and Description Language
SDP	Service Data Point
SGSN	Serving GPRS Support Node
SIB	Service Independent Building Block
SIG	Special Interest Group
SIM	Subscriber Identity Module
SIMPLE	SIP for Instant Messaging and Presence Leveraging Extensions
SIP	Session Initiation Protocol
SIP-T	SIP for Telephony
SLA	Service Level Agreement
SLEE	Service Logic Execution Environment
SLP	Service Location Protocol
SLP	Service Logic Program
SMF	Service Management Function
SMR	Specialist Mobile Radio
SMS	Service Management System
SMS	Short Message Service
SMSC	Serving MSC
SMTP	Simple Mail Transfer Protocol
SN	Service Node
SN	Subscriber Number
SNMP	Simple Network Management Protocol
SOAP	Simple Object Access Protocol

SOHO	Small Office-Home Office
SONET	Synchronous Optical Network
SP	Service Plane
SPA	Service Provider Access
SPA	Service Provider Agent
SPC	Stored Program Control
SRF	Specialized Resource Function
SRP	Specialized Resource Point
SS7	Signaling System Number 7
SSCP	Service Switching and Control Point
SSF	Service Switching Function
SSL	Secure Sockets Layer
SSM	Switching State Model
SSP	Service Switching Point
STB	Set-Top Box
STI	Shared Trigger Interaction
STP	Signal Transfer Point
SUMC	Single User Multiple Component
SUSC	Single User Single Component
SVoD	Subscription Video-on-Demand
TACS	Total Access Communication System
TBCM	Terminating Basic Call Model
TCAP	Transaction Capabilities Application Part
TCP	Transmission Control Protocol
TD-CDMA	Time Division CDMA
TD-SCDMA	Time Division Synchronous CDMA
TDD	Time Division Duplex
TDM	Time Division Multiplexing
TDMA	Time Division (or Demand) Multiple Access
TDOA	Time Difference of Arrival
TDP	Trigger Detection Point
Telco	Telephone Company
TETRA	TErrestrial Trunked RAdio
TIA	Telecommunications Industry Association
TINA	Telecommunication Information Networking Architecture
TINA-C	TINA Consortium
TIPHON	Telephony and Internet Protocol Harmonization Over Networks
TLS	Transport Layer Security
TMF	TeleManagement Forum
TMN	Telecommunication Management Network
ToS	Type of Service
TSC	Telematics Suppliers Consortium
TUP	Telephony User Part
UCM	Use Case Map
UDDI	Universal Description, Discovery, and Integration
UDP	User Datagram Protocol

UM	Unified Messaging
UML	Unified Modeling Language
UMTS	Universal Mobile Telecommunications System
UNI	User–Network Interface
UPnP	Universal Plug and Play
UPT	Universal Personal Telecommunications
Usenet	USErs' NETwork
UWB	Ultra-Wide Band
VLR	Visitor Location Register
VoATM	Voice over ATM
VoD	Video-on-Demand
VoIP	Voice over Internet Protocol
VPIM	Voice Profile for Internet Mail
VPN	Virtual Private Network
W3C	World Wide Web Consortium
WAN	Wide Area Network
WAP	Wireless Access/Application Protocol
WCDMA	Wideband CDMA
WLAN	Wireless Local Area Network
WLL	Wireless Local Loop
WML	Wireless Markup Language
WSDL	Web Services Description Language
WVMIMP	Wireless Village Mobile Instant Messaging and Presence
WWW	World Wide Web
xDSL	Digital Subscriber Line (of any type)
XML	Extensible Markup Language

Appendix 2 Glossary

Key terms are explained as they appear in each chapter, but a selected list is included here for reference.

Abstract Data Type: a data type that is defined, usually formally, independent of an implementation.
Access Control List: matrix of permissions in a system.
adversary: anyone outside the group of those with authorized access.
American National Standards Institute: the national body responsible for defining telecommunications standards within the US, typically in close collaboration with the International Telecommunications Union.
anonymity: communication without necessary identification of the participants.
architectural semantics: a bridge between the concepts of an architecture and a specification language.
Architecture Description Language: a language used to describe the high-level organization of a (software) system.
asynchronous key cipher: *see* public key cipher.
Asynchronous Transfer Mode: a connection-oriented, cell-based switching and transmission technology; now the world's most widely deployed backbone network.
authentication: the act of proving one's identity.
Authentication Header: IP Security method for digitally signing, but not encrypting, encapsulated data.
bank: from a security perspective, an authority responsible for issuing coins.
biometrics: authentication mechanisms keyed to an individual's physical characteristics.
blind signature: secure digital signatures created without direct access to the data.
block cipher: a cryptographic algorithm operating over sets of bits (usually 64) at a time, contrasted with stream cipher.
Centrex: a virtual Private Branch Exchange service provided by a public network local exchange, targeted at lower end business users where the expense of a Private Branch Exchange cannot be justified (e.g. due to size of business premises).
Class *N* Office: US terminology for exchanges, categorized as one of Class 1 to 5; a Class 1 office is an international gateway exchange, Classes 2, 3, and 4 are different types of trunk exchange, and Class 5 is a local exchange.

Service Provision – Technologies for Next Generation Communications. Edited by Kenneth J. Turner, Evan H. Magill and David J. Marples
 ISBN: 0-470-85066-3

coin: from a security perspective, a unit of electronic currency.

Common Object Request Broker Architecture: a standard interface definition between Object Management Group compliant objects.

communications service: a service concerned mainly with data traffic.

compromised: the state of a message or stream that can be read, modified, inserted into, or deleted by an adversary.

credential: a verifiable statement of relationship with another object.

cryptographic hash: a one-way encryption algorithm usually taking an input M of arbitrary size and resulting in a known, usually smaller, size output N.

dictionary attack: a brute force attempt to guess a password by trying words from a dictionary.

divisible: from a security perspective, able to be divided one or more times without compromising the integrity of the sum or the parts of the object.

Encapsulated Security Payload: IP Security method for digitally signing and encrypting the encapsulating payload.

European Telecommunications Standards Institute: the body responsible for defining telecommunications standards within Europe, typically in close collaboration with the International Telecommunications Union.

extended finite state machine: a finite state machine that has state variables as well as a major state.

Extended Finite State Machine Language: a standardized formal specification language based on extended finite state machines (ISO/IEC 9074).

feature interaction: the activities of one feature or service interfering with the behavior of another, often within the same call or session.

fingerprint: a unique identifier for a message, often created using a cryptographic hash.

Finite State Automaton/Machine: an abstract machine with a set of states and transitions between these.

firewall: device and/or software limiting access to the network based on rules for allowing in, and keeping out, network data for security reasons.

formal: referring to form and systematic treatment.

Formal Description Technique: a standardized formal method.

formal language: a language with a precise syntax, and possibly also precise semantics.

formal method: a systematic technique making use of a formal language and formally-defined rules for development with the language.

hash: *see* cryptographic hash.

implementation: a low-level and detailed description of a design.

insecure channel: *see* unsecure channel.

Integrated Services Digital Network: a circuit-switched digital network capable of providing both voice and data transmission capabilities; voice is carried at a rate of 64 kbit/s, and data may be carried at a rate of 1×64 kbit/s to 30×64 kbit/s.

Intelligent Network: a technology/architecture to promote fast feature-roll out via a centralized intelligent platform that hosts the feature-related application software (ITU Q.1200 series).

International Telecommunications Union: the primary international body responsible for defining telecommunications and radio standards for public networks.

Internet Control Message Protocol: a management protocol used at the Internet network layer.

Internet Engineering Task Force: the body responsible for defining the standards and protocols used on the Internet.

Internet Protocol: the network layer communications protocol used on the Internet.

IP Security: a collection of algorithms for layering cryptography onto the Internet Protocol stack.

local exchange: a telephone exchange that directly hosts users/subscribers, handling signaling from subscriber lines.

Local Switching Systems General Requirements: the North American standards for landline telephony services.

LOTOS (Language Of Temporal Ordering Specification): a standardized formal specification language based on abstract data types and process algebras (ISO/IEC 8807).

man-in-the-middle attack: from a security perspective, an attack wherein an adversary gets in the middle of communications between one or more recipients and masquerades as the proper objects.

message: from a security perspective, a set of digital information of arbitrary, but finite size, transmitted between objects.

Message Sequence Chart: a diagrammatic notation that shows the exchange of a message among system components (ITU Z.120).

message authentication code: a cryptographically computed message fingerprint.

Missed Trigger Interaction: a type of feature interaction where one feature causes another to miss a triggering event.

Mobile Switching Center: a special type of telephone exchange that handles the telephone calls for mobile phone users; controls a number of base stations that provide the radio network to connect to handsets.

modal logic: a specification language with various modes of expression such as talking about necessity, belief, or permission.

model-based language: a specification language that uses simple mathematical objects like sets and functions as the building blocks.

Multiple User Multiple Component: a classification of telephone calls used in feature interaction, signifying that two or more features are interacting on different platforms for two or more separate users.

Multiple User Single Component: a classification of telephone calls used in feature interaction, signifying that two or more features are interacting on the same platform for two or more separate users.

National Science Foundation: United States governmental organization and major funder for the development and growth of the Internet.

net: a form of finite state machine with tokens in places to mark its state.

nonce: a random collection of bits often attached to a message to make it unique in order to prevent a replay attack.

non-repudiation: removing the possibility of plausible denial from a system.

nym: any of a number of (pseudo-)anonymous identities untraceably tied back to a single user-object and that cannot be connected with one another.

nyms: key pairs that can be associated with each other without exposing the identity of the owner.

one-time password: a passphrase created to be used for a single instance, greatly reducing the risk of a replay attack.

Open Distributed Processing: a standardized architecture and set of standards governing how distributed systems can interwork (ISO/IEC 10746).

Open Systems Interconnection: a standardized architecture and set of standards governing how networked systems can communicate (ISO/IEC 7498).

passphrase: *see* password.

password: string associated with a user-object used for authentication.

Plain Old Telephone Service: basic landline telephone service with no custom calling features.

port: numeric addressing scheme for identifying applications and protocols running on top of Transmission Control Protocol and User Datagram Protocol.

Private Branch Exchange: a private exchange on customer premises, typically extremely feature-rich and connected to the public telephone network.

process algebra: a specification language that models a system as a collection of communicating processes.

public key cipher: a cryptography algorithm using separate keys for encryption and decryption, in which the encryption key is often well known; also known as an asynchronous key cipher.

Pulse Code Modulation: a technique by which an analogue signal is sampled, and its magnitude (with respect to a fixed reference) is quantized and digitized for transmission over a common transmission medium.

Quality of Service: the set of quantitative and qualitative characteristics that are necessary to achieve a level of functionality and end-user satisfaction with a service.

replay attack: attempted compromise of a message or stream through inserting data previously seen to work for a system, such as a password.

Resource Reservation Protocol: an Internet protocol to reserve network resources.

rigorous method: a formal method that does not use strict reasoning.

salt: a nonce used in password schemes.

secure: practical, if not theoretical, unbreakability of a design or implementation.

semantics: the rules giving meaning to a language specification.

Sequential Action Interaction: a type of feature interaction where features activated in sequence interfere with each other.

service: an abstraction of a component or layer interface.

Service Control Point: the network entity that supports service execution.

Service Level Agreement: a contract with a service provider specifying the nature of the service and associated costs.

Service Switching Point: the exchange that triggers a Service Control Point when a call meets pre-determined criteria.

Session Initiation Protocol: an Internet protocol to establish and maintain sessions.

Shared Trigger Interaction: a type of feature interaction where features interfere because they may be triggered by the same event.

signature: hash of a message cryptographically associated with a single key.

Simple Network Management Protocol: a standard for Internet operations, maintenance, and administration.

Single User Multiple Component: a classification of telephone calls used in feature interaction, signifying that two or more features are interacting on different platforms for a single user.

Single User Single Component: a classification of telephone calls used in feature interaction, signifying that two or more features are interacting on a single platform for a single user.

Specification and Description Language: a standardized formal specification language based on extended finite state machines (ITU Z.100).

spoofing: impersonation of a network device through creation of false credentials such as Internet Protocol headers or the like.

Stored Program Control: computer-based technology used to provide telephony control in digital telephone exchanges.

stream cipher: a cryptographic algorithm often used to protect data with an unknown size at run-time, contrasted with a block cipher.

streaming block cipher: a cryptographic algorithm operating over blocks of data and changing over time; a hybrid of block cipher and stream cipher.

Strowger switch: an electromechanical technology used as the basis for telephony switching for many years; named after its inventor, Almon B. Strowger, a Kansas City undertaker.

symmetric key cipher: a cryptography algorithm using the same key for encryption and decryption.

syntax: the grammatical rules of a language.

Telecommunication Management Network: a standard architecture from the International Telecommunications Union for operating, maintaining, and administering telecommunications networks.

telecommunications service: a service concerned mainly with voice traffic.

telnet: remote login to a computer or server.

temporal logic: a specification language that uses a modal logic with time-related operators.

token: from a security perspective, a small set of data.

transferable: able to be passed off to another user-object or system without compromising the integrity of the object.

Transmission Control Protocol: a session-based protocol used in conjunction with the Internet Protocol, often jointly labeled TCP/IP.

Transport Layer Security: the Internet standard based on, and successor to, the Secure Sockets Layer.

trunk exchange: a telephone exchange that does not directly host users/subscribers; the trunk exchange supports only inter-exchange trunk circuits, and provides the transit switching capability in a modern telephony network.

unsecure channel: any means of communication over which an adversary is able to read, change, re-order, insert, or delete the information transmitted.

Use Case Map: a graphical notation for representing sequences of actions and the causality among them.

Usenet: short for Users' Network, a distributed bulletin board system.

User Datagram Protocol: a packet-based protocol used in conjunction with the Internet Protocol.

user-object: from a security perspective, a data object noted by an identifier and associated other account information, such as name, address, and most importantly access permissions.

Voice over Internet Protocol: the transmission of voice using Internet technology; public voice networks are currently evolving from digital circuit-switched networks to Internet-based packet networks.

wallet: from a security perspective, a device used to hold coins.

Appendix 3 Websites

Major websites mentioned in chapters are collected below.Further websites, including specific articles, appear in the Bibliography, and in the Further reading sections at the ends of some chapters.

3G GSM Newsroom	http://www.3gnewsroom.com/
3G Partnership Programme	http://www.3gpp.org/
3G Partnership Programme – cmda2000	http://www.3gpp2.org/
ANSI (American National Standards Institute)	http://www.ansi.org/
AR Greenhouse (Telcordia Networked Home)	http://www.argreenhouse.com/
Biometric Digest	http://www.biodigest.com/
Center for Quantum Computation	http://www.qbit.org/
Digital Signatures and the Law	http://www.epic.org/crypto/dss/
DMTF (Distributed Management Task Force)	http://www.dmtf.org
DMTF Directory-Enabled Network Initiative	http://www.dmtf.org/standards/standard_den.php
ETSI (European Telecommunications Standards Institute)	http://www.etsi.org/
European Multimedia Forum	http://www.e-multimedia.org/
FIPA (Foundation for Intelligent Physical Agents)	http://www.fipa.org/
Formal Methods Activity List	http://archive.comlab.ox.ac.uk/formal-methods.html
Genetic Programming	http://www.genetic-programming.org/
GNU (GNU's Not Unix)	http://www.gnu.org/
GSM Association	http://www.gsmworld.com/
IETF (Internet Engineering Task Force)	http://www.ietf.org/
IETF Differentiated Services Working Group	http://www.ietf.org/html.charters/diffserv-charter.html
IETF Multi-protocol Label Switching Working Group	http://www.ietf.org/html.charters/mpls-charter.html
IETF Session Initiation Protocol Working Group	http://www.ietf.org/html.charters/sip-charter.html

Service Provision – Technologies for Next Generation Communications. Edited by Kenneth J. Turner, Evan H. Magill and David J. Marples
 ISBN: 0-470-85066-3

IFIP WG 6.1 (International Federation for Information Processing, Working Group 6.1)	http://www-run.montefiore.ulg.ac.be/IFIP-WG/IFIP-description.html
INNS (International Neural Network Society)	http://www.inns.org/
International Biometric Society Journal	http://stat.tamu.edu/Biometrics/
ITU (International Telecommunications Union)	http://www.itu.int/
ITU Telecommunications Sector	http://www.itu.int/ITU-T/
JPEG (Joint Photographic Experts Group)	http://www.jpeg.org/
Kerberos (MIT)	http://web.mit.edu/kerberos/www/
Molecular Computing	http://www.molecularelectronics.com/
MPEG (Moving Pictures Experts Group)	http://www.mpeg.org/
MP3 (MPEG Audio Layer 3)	http://www.mp3.com/
MSF (Multi-Service Switching Forum)	http://www.msforum.org/
Multimedia PC Working Group	http://www.spa.org/mpc
OMA (Open Mobile Alliance)	http://www.openmobilealliance.org/
OpenSource	http://www.opensource.org/
OSGi (Open Service Gateway Initiative)	http://www.osgi.org
PAM Forum	http://www.pamforum.org/
Parlay Group	http://www.parlay.org/
PHS (Personal Handyphone System) Forum	http://www.phsmou.or.jp/
Public Key Cryptography Standards	http://www.rsasecurity.com/rsalabs/pkcs/
Quantum Computing	http://www.qbit.org/
SDL Forum	http://www.sdl-forum.org/
SIP (Session Initiation Protocol)	http://www.cs.columbia.edu/sip/
TD-SCDMA (Time Division-Synchronous Code Division Multiple Access) Forum	http://www.tdscdma-forum.org/
Telecommunications Management Network Standards	http://www.itu.int/TMN/
WELL (Worldwide Environment for Learning LOTOS)	http://www.cs.stir.ac.uk/well/

Bibliography

The citations from all chapters are collected below, in addition, several further works of interest have been included. References to major websites appear in Appendix 3.

3GPP (2001) *Third Generation Partnership Project*, http://www.3gpp.org/.

3GPP (2002a) *Open Service Access (OSA); Application Programming Interface (API) Part 1: Overview (Release 5), V5.1.0.* 29.198-1, 3rd Generation Partnership Project; Technical Specification Group Core Network.

3GPP (2002b) *Open Service Access (OSA); Application Programming Interface (API) Part 2: Common data (Release 5), V5.1.1.* 29.198-2, 3rd Generation Partnership Project; Technical Specification Group Core Network.

3GPP (2002c) *Open Service Access (OSA); Application Programming Interface (API) Framework (Release 5), V5.1.0.* 29.198-3, 3rd Generation Partnership Project; Technical Specification Group Core Network.

3GPP (2002d) *Open Service Access (OSA); Application Programming Interface (API) Part 4: Call Control; Sub-part 1: Call Control Common Definitions (Release 5), V5.1.0.* 29.198-4-1, 3rd Generation Partnership Project; Technical Specification Group Core Network.

3GPP (2002e) *Open Service Access (OSA); Application Programming Interface (API) Part 4: Call Control; Sub-part 2: Generic Call Control SCF (Release 5), V5.1.0.* 29.198-4-2, 3rd Generation Partnership Project; Technical Specification Group Core Network.

3GPP (2002f) *Open Service Access (OSA); Application Programming Interface (API) Part 4: Call Control; Sub-part 3: Multiparty Call Control SCF (Release 5), V5.1.0.* 29.198-4-3, 3rd Generation Partnership Project; Technical Specification Group Core Network.

3GPP (2002g) *Open Service Access (OSA); Application Programming Interface (API) Part 4: Call Control; Sub-part 4: Multimedia Call Control SCF (Release 5), V5.1.0* 29.198-4-4, 3rd Generation Partnership Project; Technical Specification Group Core Network.

3GPP (2002h) *Open Service Access (OSA); Application Programming Interface (API) Part 5: Generic User Interaction (Release 5), V5.1.0.* 29.198-5, 3rd Generation Partnership Project; Technical Specification Group Core Network.

3GPP (2002i) *Open Service Access (OSA); Application Programming Interface (API) Part 6: Mobility (Release 5), V5.1.0.* 29.198-6, 3rd Generation Partnership Project; Technical Specification Group Core Network.

Service Provision – Technologies for Next Generation Communications. Edited by Kenneth J. Turner, Evan H. Magill and David J. Marples

ISBN: 0-470-85066-3

3GPP (2002j) *Open Service Access (OSA); Application Programming Interface (API) Part 7: Terminal Capabilities (Release 5), V5.2.0.* 29.198-7, 3rd Generation Partnership Project; Technical Specification Group Core Network.

3GPP (2002k) *Open Service Access (OSA); Application Programming Interface (API) Part 8: Data Session Control (Release 5), V5.1.0.* 29.198-8, 3rd Generation Partnership Project; Technical Specification Group Core Network.

3GPP (2002l) *Open Service Access (OSA); Application Programming Interface (API) Part 11: Account Management (Release 5), V5.1.0.* 29.198-11, 3rd Generation Partnership Project; Technical Specification Group Core Network.

3GPP (2002m) *Open Service Access (OSA); Application Programming Interface (API) Part 12: Charging (Release 5), V5.1.0.* 29.198-12, 3rd Generation Partnership Project; Technical Specification Group Core Network.

3GPP (2002n) *Open Service Access (OSA); Application Programming Interface (API) Part 13: Policy Management (Release 5), V5.1.0.* 29.198-13, 3rd Generation Partnership Project; Technical Specification Group Core Network.

3GPP (2002o) *Open Service Access (OSA); Application Programming Interface (API) Part 14: Presence and Availability Management (Release 5), V5.1.0.* 29.198-14, 3rd Generation Partnership Project; Technical Specification Group Core Network.

Abernethy, T. W. and Munday, C. A. (1995) Intelligent Networks, Standards and Services. *BT Technology Journal*, **2**, 9–20.

Abrial, J.-R. (1996) *The B-Book: Assigning Programs to Meanings*, Cambridge University Press, Cambridge, UK.

Accorsi, R., Areces, C., Bouma, L. G. and de Rijke, M. (2000) Features as constraints, in Calder, M. H. and Magill, E. H. (eds) *Proc. 6th Feature Interactions in Telecommunications and Software Systems*, pp. 210–225, IOS Press, Amsterdam, the Netherlands.

Adleman, L. (1994) Molecular computation of solutions to combinatorial problems, *Science*, **266**, 1021–1024.

AFX News (2001) Mobile Messaging Standards Coming, AFX News Ltd., 26th April 2001, http://www. mbizcentral.com/story/m-allNews/MBZ20010426S0004.

Ahmed, W., Hussain, A. and Shah, S. (2001) Location Estimation in Cellular Networks using Neural Networks, *Proc. International NAISO Congress on Information Science Innovations* (ISI'2001), Dubai.

Aho, A. V., Gallagher, S., Griffeth, N. D., Schell, C. R. and Swayne, D. F. (1998) SCF3/ Sculptor with Chisel: Requirements engineering for communications services, in Kimbler, K. and Bouma, W. (eds) *Proc. 5th Feature Interactions in Telecommunications and Software Systems*, pp. 45–63, IOS Press, Amsterdam, The Netherlands.

Ahonen, T. and Barrett. J. (eds) (2002) *Services for UMTS*. John Wiley & Sons, Chichester, UK.

Aiyarak, P., Saket, A. and Sinclair, M. (1997) Genetic programming approaches for minimum cost topology optimisation of optical telecommunication networks, in *2nd International Conference on Genetic Algorithms In Engineering Systems: Innovations And Applications*, University of Strathclyde, Glasgow, UK.

Alcatel (2003) http://www.cid.alcatel.com/doctypes/techpaper/dsl/dsl_types.jhtml.

Allen, S. and Wolkowitz, C. (1987) *Homeworking myths and realities*, Macmillan Education.

Amyot, D., Andrade, R., Logrippo, L. M. S., Sincennes, J. and Yi, Z. (1999a) Formal methods for mobility standards, in *IEEE 1999 Emerging Technology Symposium on Wireless Communications and Systems*, Richardson, Texas, USA.

Amyot, D., Buhr, R. J. A., Gray, T. and Logrippo, L. M. S. (1999b) Use case maps for the capture and validation of distributed systems requirements, in *Proc. 4th IEEE International Symposium on Requirements Engineering*, pp. 44–53, Institution of Electrical and Electronic Engineers Press, New York, USA.

Amyot, D., Charfi, L., Gorse, N., Gray, T., Logrippo, L. M. S., Sincennes, J., Stepien, B. and Ware, T. (2000) Feature description and feature interaction analysis with use case maps and LOTOS, in Calder, M. H. and Magill, E. H. (eds) *Proc. 6th Feature Interactions in Telecommunications and Software Systems*, pp. 274–289, IOS Press, Amsterdam, The Netherlands.

Amyot, D. and Logrippo, L. (eds) (2003) *Feature Interaction in Telecommunications and Software Systems VII*, IOS Press, Amsterdam, The Netherlands.

Andersson, K. and Hjerpe, T. (1998) Modeling constraint problems in CML, *Proc. PAPPACT98*, London, UK.

Ang, P. H. and Nadarajan, B. (1997) Issues in the regulation of Internet quality of service. *Proceedings of INET 97 Conference*.

APM Ltd. (1993) *ANSAware 4.1 Application Programming in ANSAware. Document RM.102.02. A.P.M.* Cambridge Limited, Castle Park, Cambridge, UK.

Appium (2003) *Appium Parlay simulator*, http://www.appium.com/products_and_services/application_platform. html.

Appleby, S. and Steward, S. (1994) Mobile software agents for control in telecommunications networks, *BT Technology Journal*, **12**(2), 104–113.

Armstrong, E., Bodoff, S., Carson, D., Fisher, M., Fordin, S., Green, D., Haase, K. and Jendrock, E. (2002) *The Java Web Services Tutorial*, http://www.java.sun.com/webservices/docs/1.0/tutorial/.

Arnold, K., O'Sullivan, B., Scheifler, R., Waldo, J. and Wollrath, A. (1999) *The Jini Specification*. Addison Wesley.

Ateniese, G. and Tsudik, G. (1999) *Group Signatures a la Carte*. ACM Symposium of Discrete Algorithms, January 1999.

Au, P. K. and Atlee, J. M. (1997) Evaluation of a state-based model of feature interactions, in Dini, P., Boutaba, R. and Logrippo, L. M. S. (eds,) *Proc. 4th International Workshop on Feature Interactions in Telecommunication Networks*, pp. 153–167, IOS Press, Amsterdam, The Netherlands.

Austin, S. and Parkin, G. I. (1993) Formal methods: A survey, Technical report, National Physical Laboratory, Teddington, Middlesex, UK.

Awduche, D. *etal.* (2002) *RSVP-TE: Extensions to RSVP for LSP Tunnels*, RFC 3209, Internet Society, New York, USA.

Baeten, J. C. M. and Weijland, W. P. (1990) *Process Algebra*, Cambridge University Press, Cambridge, UK.

Baker, F., Iturralde, C., Le Faucheur, F. and Davie, B. (2001) *Aggregation of RSVP for IPv4 and IPv6 Reservations*, RFC 3175, Internet Society, New York, USA.

Bakker, J.-L., McGoogan, J. R., Opdyke, W. F. and Panken, F. J. (2000) Rapid Development and Delivery of Converged Services Using APIs, *Bell Labs Technical Journal*, **5**(3), 12–29.

Bakker, J.-L., Tweedie, D. and Unmehopa, M. R. (2002) Evolving Service Creation; New Developments in Network Intelligence, *Telektronikk*, **98**(4), 58–68.

Banzhaf, W. (1993) Genetic programming for pedestrians, MERL Technical Report 93-03, Mitsubishi Electric Research Labs, Cambridge, Massachusetts, USA.

Banzhaf, W., Nordin, P., Keller, R. E. and Francone, F. D. (1998) *Genetic Programming – An Introduction*, Morgan Kaufmann, Heidelberg, Germany.

BBC News (2003) http://news.bbc.co.uk/1/hi/entertainment/tv_and_radio/3177479.stm.

Belina, F., Hogrefe, D. and Sarma, A. (eds) (1992) *SDL with Applications from Protocol Specification*, Prentice Hall.

Bellcore (1993a) *AIN Release 1*, SR-NWT-002247, Revision 1, Bellcore, Morristown, New Jersey, USA.

Bellcore (1993b) *AINGR: Switching Systems*, GR-1298-CORE, Issue 1, Bellcore, Morristown, New Jersey, USA.

Bellcore (1994) *AINGR: Switching Systems*, GR-1298-CORE, Issue 2, Bellcore, Morristown, New Jersey, USA.

Bellcore (2001a) *AINGR: Switch – Service Control Point (SCP)/Adjunct Interface*, GR-1299-CORE, Issue 7, Bellcore, Morristown, New Jersey, USA.

Bellcore (2001b) *AINGR: Switch – Intelligent Peripheral Interface (IPI)*, GR-1129-CORE, Issue 6, Bellcore, Morristown, New Jersey, USA.

Berndt, H., Hamada, T. and Graubmann, P. (2000), TINA: Its Achievements and its Future Directions, *IEEE Communications Surveys & Tutorials*, **3**(1).

Berners-Lee, T. (1994) *Universal Resource Identifiers in WWW: A Unifying Syntax for the Expression of Names and Addresses of Objects on the Network as used in the World-Wide Web*. RFC 1630, Internet Society, New York, USA.

Berners-Lee. T., Hendler, J. and Lassila, O. (2001) The Semantic Web, *Scientific American*, May 2001.

Biswas, J., Lazar, A. A., Huard, J.-F., Lim, K., Mahjoub, S., Pau, L.-F., Suzuki, M., Torstensson, S., Wang, W. and Weinstein, S. (1998) The IEEE P1520 Standards Initiative for Programmable Network Interfaces, *IEEE Communications Magazine*, **36**(10), 64–70.

Blair, G. S., Coulson, G., Anderson, A., Blair, L., Clarke, M., Costa, F., Duran-Limon, H., Fitzpatrick, T., Johnston, L., Moreira, R., Parlavantzas, N. and Saikoski, K. (2001) The Design and Implementation of Open ORB v2, Special Issue of *IEEE Distributed Systems Online on Reflective Middleware*, http://www.computer. org/dsonline/.

Blair, G., Blair, L., Bowman, H. and Chetwynd, A. (1998) *Formal Specification of Distributed Multimedia Systems*, UCL Press, London, UK.

Blair, G. S. and Stefani, J. B. (1998) *Open Distributed Processing and Multimedia*, Addison Wesley.

Blair, L., Blair, G., Bowman, H., and Chetwynd, A. (1995) Formal specification and verification of multimedia systems in Open Distributed Processing, *Computer Standards and Interfaces*, **17**(5–6), 413–436.

Blake, S. (1998) An Architecture for Differentiated Services, Internet Engineering Task Force, RFC 2475, Internet Society, New York, USA.

Blom, J., Bol, R. and Kempe, L. (1995) Automatic detection of feature interactions in temporal logic, in Cheng, K. E. and Ohta, T. (eds) *Proc. 3rd International Workshop on Feature Interactions in Telecommunications*, pp. 1–19, IOS Press, Amsterdam, The Netherlands.

Boehm, B. W. (1984) Verifying and validating software requirements and design specification, *IEEE Transactions on Software Engineering*, **1**(1), 75–88.

Bonabeau, E., Heanaux, F., Guerin, S., Snyers, D., Kuntz, P. and Theraulaz, G. (1998) Routing in telecommunications networks with smart ant-like agents, in *Intelligent Agents for Telecommunications Applications*, LNAI 1437, Springer Verlag, Berlin, Germany.

Bosque, M. (2002) *Understanding 99% of Artificial Neural Networks: Introduction & Tricks*, Writers Club Press.

Bouma, L. G. and Velthuijsen, H. (1994a), Introduction, in Bouma, L. G. and Velthuijsen, H. (eds) *Proc. 2nd International Workshop on Feature Interactions in Telecommunications Systems*, pp. vii–xiv, IOS Press, Amsterdam, The Netherlands.

Bouma, L. G. and Velthuijsen, H. (eds) (1994b) *Proc. 2nd International Workshop on Feature Interactions in Telecommunications Systems*, IOS Press, Amsterdam, The Netherlands.

Bowen, J. P. and Hinchey, M. G. (1995a) Seven more myths of formal methods, *IEEE Software*, **12**(4), 34–41.

Bowen, J. P. and Hinchey, M. G. (1995b) Ten commandments of formal methods, *IEEE Computer*, **28**(4), 56–63.

Bowman, H., Derrick, J., Linington, P. and Steen, M. W. A. (1996) Cross Viewpoint Consistency in Open Distributed Processing, *Software Engineering Journal*, **11**(1), 44–57.

Boyd, C. (2001) Why Strategy Must Change, *Management Issues*, MGT 487, SMU.

Braden, B., Clark, D. and Shenker, S. (1994) *Integrated Services in the Internet Architecture: An Overview*, RFC 1633, Internet Society, New York, USA.

Braden, B. *et al.* (1997) *Resource ReSerVation Protocol (RSVP) – Version 1 Functional Specification*, RFC 2205, Internet Society, New York, USA.

Bradner, S. (1996) The Internet Standards Process – Revision 3, RFC 2026, Internet Society, New York, USA.

Bradshaw, J. (1997) *Software Agents*, AAAI Press/The MIT Press.

Bredereke, J. (2000) Families of formal requirements in telephone switching, in Calder, M. H. and Magill, E. H. (eds) *Proc. 6th Feature Interactions in Telecommunications and Software Systems*, pp. 257–273, IOS Press, Amsterdam, The Netherlands.

Brunner, M. and Stadler, R. (2000) Service Management in Multi-Party Active Networks, IEEE *Communications Magazine*, Special Issue on Active and Programmable Networks, **38**(3).

Buhr, R. J. A. and Casselman, R. S. (1996) *Use Case Maps for Object-Oriented Systems*, Prentice Hall.

Buhr, R. J. A., Amyot, D., Elammari, M., Quesnel, D., Gray, T. and Mankovski, S. (1998) Feature-interaction visualization and resolution in an agent environment, in Kimbler, K. and Bouma, W. (eds) *Proc. 5th Feature Interactions in Telecommunications and Software Systems*, pp. 135–149, IOS Press, Amsterdam, The Netherlands.

Caffery, J. and Stuber, G. (1998) Overview of Wireless Location in CDMA Systems, *IEEE Communications Magazine*, **36**, 38–45.

Calder, M. H. (1998) What use are formal design and analysis methods to telecommunications services? In Kimbler, K. and Bouma, W. (eds) *Proc. 5th Feature Interactions in Telecommunications and Software Systems*, pp. 23–31, IOS Press, Amsterdam, The Netherlands.

Calder, M. H. and Magill, E. H. (eds) (2000) *Proc. 6th Feature Interactions in Telecommunications and Software Systems*, IOS Press, Amsterdam, The Netherlands.

Calder, M. H., Magill, E. H. and Marples, D. J. (1999) A hybrid approach to software interworking problems: Managing interactions between legacy and evolving telecommunications software, *IEE Software*, **146**(3), 167–180.

Calder, M. H., Kolberg, M., Magill, E. H. and Reiff-Marganiec, S. (2003) Feature interaction: A critical review and considered forecast, *Computer Networks*, **41**, 115–141.

Camarillo, G. (2002) *SIP Demystified*, McGraw Hill.

Cameron, J. and Lin, F. J. (1998) Feature interactions in the new world, in Kimbler, K. and Bouma, W. (eds) *Proc. 5th Feature Interactions in Telecommunications and Software Systems*, pp. 3–9, IOS Press, Amsterdam, The Netherlands.

Cameron, E. J., Griffeth, N. D., Lin, Y.-J., Nilson, M. E., Schnure, W. K. and Velthuijsen, H. (1993) A feature-interaction benchmark for IN and beyond, *IEEE Communications Magazine*, pp. 64–69.

Cameron, E. J., Griffeth, N., Lin, Y.-J., Nilson, M. E. and Schnure, W. K. (1994) A feature interaction benchmark for IN and beyond, in Bouma, L. G. and Velthuijsen, H. (eds) *Proc. 2nd International Workshop on Feature Interactions in Telecommunications Systems*, pp. 1–23, IOS Press, Amsterdam, The Netherlands.

Cameron, J., Cheng, K., Lin, F. J., Liu, H. and Pinheiro. B. (1997) A formal AIN service creation, feature interactions analysis and management environment: An industrial application, in Dini, P., Boutaba, R. and Logrippo, L. M. S. (eds) *Proc. 4th International Workshop on Feature Interactions in Telecommunication Networks*, pp. 342–346, IOS Press, Amsterdam, The Netherlands.

Cameron, J., Cheng, K., Gallagher, S., Lin, F. J., Russo, P. and Sobirk, D. (1998) Next generation service creation: Process, methodology and tool integration, in Kimbler, K. and Bouma, W. (eds) *Proc. 5th Feature Interactions in Telecommunications and Software Systems*, pp. 299–304, IOS Press, Amsterdam, The Netherlands.

Campbell, A. T., Kounavis, M. E., Villela, D. A., Vicente, J. B., de Meer, H. B., Miki K. and Kalaichelvan, K. S. (1999) Spawning networks, *IEEE Network Magazine*, **13**(4), 16–29.

CanCIA (2003) http://www.can-cia.org.

Capellmann, C., Combes, P., Pettersson, J., Renard, B. and Ruiz, J. L. (1997) Consistent interaction detection – A comprehensive approach integrated with service creation, in Dini, P., Boutaba, R. and Logrippo, L. M. S. (eds) *Proc. 4th International Workshop on Feature Interactions in Telecommunication Networks*, pp. 183–197, IOS Press, Amsterdam, The Netherlands.

Carpenter, G. and Grossberg, S. (1988) The ART of adaptive pattern recognition by a self-organizing neural network, *IEEE Computer*, **21**(3), 77–88.

Carro, M. and Hernenegildo, M. (1998) Some design issues in the visualisation of constraint program execution, in *Proc. Joint Conference on Declarative Programming (AGP'98).*

Cassez, F., Ryan, M. D. and Schobbens, P.-Y. (2001) Proving feature noninteraction with alternating-time temporal logic, in Gilmore, S. T. and Ryan, M. D. (eds) *Language Constructs for Describing Features (FIREworks Workshop)*, pp. 85–103, Springer-Verlag, Berlin, Germany.

Champion, M., Ferris, C., Newcomer, E. and Orchard, D. (2002) *Web Services Architecture.* W3C Working Draft, World Wide Web Consortium, Geneva, Switzerland.

Chaum, D. (1983) Blind Signatures for Untraceable Payments, *Proc. CRYPTO'8*, Plenum Press, New York, USA.

Chaum, D. (1985) Security without Identification: Transaction Systems to Make Big Brother Obsolete, *Communications of the ACM*, **28**(10).

Chaum, D. and Pedersen, T. (1992) Transferred Cash Grows in Size, *Proc. EUROCRYPT'92*, pp. 390–407, LNCS 658, Springer-Verlag, Berlin, Germany.

Chen, W. and Mulgrew, B. (1992) Application of the functional-link technique for channel equalization, *Signal Processing*, **28**, 91–107.

Cheng, K. E. and Ohta, T. (eds) (1995) *Proc. 3rd International Workshop on Feature Interactions in Telecommunications*, IOS Press, Amsterdam, The Netherlands.

Chess, D. (1998) Security Issues in Mobile Code Systems, in Lecture Notes in Computer Science 1419, Springer-Verlag, Berlin, Germany.

Chinnici, R., Gudgin, M., Moreau, J. and Weerawarana, S. (2003) *Web Services Description Language (WSDL) Version 1.2*. W3C Working Draft, World Wide Web Consortium, Geneva, Switzerland.

Cho, S. B. and Shimohara, K. (1998) Evolutionary learning of modular neural networks with genetic programming, *Applied Intelligence*, **9**(3), 191–200.

Christensen, C. M. (1997) *The Innovator's Dilemma: When New Technologies Cause Great Firms to Fail*, Harvard Business School Press.

Clarke, E. and Wing, J. (1996) Formal methods: State of the art and future directions, Technical Report CMU-CS-96-178, Carnegie-Mellon University, Pittsburgh, USA.

Clarke, E. M., Emerson, E. A. and Sistla, A. P. (1986) Automatic verification of finite state concurrent systems using temporal logic specification, *TOPLAS*, **8**(2), 244–263.

Clocksin, W. F. and Mellish, C. S. (1994) *Programming in Prolog*, Springer-Verlag, Berlin, Germany.

Cochinwala, M. (2002) Using Objects for Next Generation Communication Services, *Proc. 14th European Conference on Object-Oriented Programming*, Sophia Antipolis and Cannes, France, June 2000.

Cochinwala, M., Shim, H. S., *et al.* (2003) Adaptive Resource Management of a Virtual Call Center Using a Peer-to-Peer Approach, *Proc. 8th International Symposium on Integrated Network Management*, pp. 425–437, March 2003.

Conn, P. (1995) Time Affordances, *Proc. CHI-95*, ACM Press, http://www.acm.org/sigchi/chi95/proceedings/papers/apc_bdy.htm.

Coulson, G. (2000) What is Reflective Middleware? Introduction to the reflective middleware subarea, *Distributed Systems Online Journal*, IEEE Computer Society, http://boole.computer.org/dsonline/middleware/RMArtlicle1. htm.

Courtiat, J.-P., Dembinski, P., Holzmann, G. J., Logrippo, L. M. S., Rudin, H. and Zave, P. (1995) Formal methods after 15 years: Status and trends, *Computer Networks and ISDN Systems*, **28**, 1845–1855.

Courtney, D. (2001) Instant messages: They're not just from humans anymore, *ZDNet*, 1st May 2001.

Craigen, D., Gerhart, S. and Ralston, T. (1993a) An international survey of industrial applications of formal methods: Volume 1 – Purpose, approach analysis and conclusions, Technical Report NISTGCR 93/626, National Institute of Standards and Technology, Gaithersburg, USA.

Craigen, D., Gerhart, S. and Ralston, T. (1993b) An international survey of industrial applications of formal methods: Volume 2 – Case studies, Technical Report NISTGCR 93/626, National Institute of Standards and Technology, Gaithersburg, USA.

Crowcroft, A., Handley, A. and Wakeman I. (1998) *Internetworking multimedia*, UCL Press, London, UK.

Dacker, B. (1995) The development and use of Erlang concurrent functional programming in industry, in Lovrek, I. and Sinkovic, V. (eds) *ConTel 95*, pp. 1–15, Croatian Telecommunications Society, Zagreb, Croatia.

Dahl, O. C. and Najm, E. (1994) Specification and detection of IN service interference using LOTOS, in Tenney, R. L., Amer, P. D. and Uyar, M. Ü. (eds) *Proc. Formal Description Techniques VI*, pp. 53–70. North-Holland.

Darwin, C. (1859) On the Origin of Species by Means of Natural Selection, or the Presentation of Favoured Races in the Struggle for Life, Murray (London), First Edition.

Davie, B. *et al.* (2002) An Expedited Forwarding PHB (Per-Hop Behavior), RFC 3246, Internet Society, New York, USA.

Davies, I. and McBain, A. (1988) ISPBXs and terminals, *Computer Communications*, **11**(4), 203–207.

Davis, R. and Smith, R. G. (1983) Negotiation as a metaphor for distributed problem solving, *Artificial Intelligence*, **20**(1), 63–109.

DECT Forum (1997) *DECT The Standard Explained*, http://www.dect.ch/publicdocs/Technical Document.pdf.

Derrick, J. and Bowman, H. (eds) (2001) *Formal Methods for Distributed Processing: An Object Oriented Approach*, Cambridge University Press, Cambridge, UK.

Dierks, T. and Allen, C. (1999) *The TLS Protocol*, RFC 2246, Internet Society, New York, USA.

Diffie, W. and Hellman, M. (1976) New Directions in Cryptography, *IEEE Trans. Information Theory*, **22**, 644–654.

Dini, P., Boutaba, R. and Logrippo, L. M. S. (eds) (1997) *Proc. 4th International Workshop on Feature Interactions in Telecommunication Networks*, IOS Press, Amsterdam, The Netherlands.

Disabatino, J. (2001) IM group works toward interoperability, *Computer World*, 8th February 2001.

DMTF (1999) *Common Information Model Specification (CIM)*, Version 2.2, DSP0004, Distributed Management Task Force.

DNA (2003) Introduction to DNA computing. http://www.liacs.nl/home/pier/webPages DNA/.

Dong, J. S. and Duke, R. (1993) An object-orientated approach to the formal specification of ODP trader, in de Meer, J., Mahr, B. and Spaniol, O. (eds) *Proc. International Conference on Open Distributed Processing*, pp. 312–322. Gesellschaft für Mathematik und Datenverarbeitung, Berlin, Germany.

Downes, L. and Mui, C. (1998) *Unleashing the Killer App – Digital Strategies for Market Dominance*, Harvard Business School Press.

Dssouli, R., Somé, S., Guillery, J.-W. and Rico, N. (1997) Detection of feature interactions with REST, in Dini, P., Boutaba, R. and Logrippo, L. M. S. (eds) *Proc. 4th International Workshop on Feature Interactions in Telecommunication Networks*, pp. 271–283, IOS Press, Amsterdam, The Netherlands.

du Bousquet, L., Ouabdesselam, F., Richier, J.-L. and Zuanon, N. (1998), Incremental feature validation: A synchronous point of view, in Kimbler, K. and Bouma, W. (eds) *Proc. 5th Feature Interactions in Telecommunications and Software Systems*, pp. 262–275, IOS Press, Amsterdam, The Netherlands.

du Bousquet, L., Ouabdesselam, F., Richier, J.-L. and Zuanon, N. (2000) Feature interaction detection using a synchronous approach and testing, *Computer Networks*, **32**(4), 419–431.

Dunlop, J., Girma, G. and Irvine, J. (1999) *Digital Mobile Communications and the TETRA System*, John Wiley & Sons, Chichester, UK.

Dupuy, F., Nilsson, G. and Inoue, Y. (1995) The TINA Consortium: Toward Networking Telecommunications Information Services, *IEEE Communications Magazine*, **33**(11), 78–83.

Eberhagen, S. (1998) Considerations for a Successful Introduction of Intelligent Networks from a Marketing Perspective, in *Proc. 5th International Conference on Intelligent Networks*, Bordeaux, France. ICIN, Adera, France.

Elfe, C., Freuder, E. and Lesaint, D. (1998) Dynamic Constraint Satisfaction for Feature Interaction, *BT Technology Journal*, **16**(3), 38–45.

Emmerich, W. (2000) *Engineering Distributed Objects*, John Wiley & Sons, Chichester, UK.

Ericsson (2002) Ericsson Parlay simulator, see http://www.ericsson.com/mobilityworld/sub/open/technologies/parlay.

Ernst & Young (1998) *E-Mail beats Telephone As Top Means Of Workplace Communication according To Ernst & Young/Hrfocus Survey*, http://markbushstudio.com/hk/ernieweb2/About/ey10.html.

ETSI (1994) *Core INAP CS 1*, ETS 300–374, European Telecommunications Standards Institute, Sophia Antipolis, France.

ETSI (1998) *Extensions to IN CS 1 INAP for CAMEL – Protocol Specification*, EN 301 152-1, European Telecommunications Standards Institute, Sophia Antipolis, France.

ETSI (2003a) *Application Programming Interface (API) Part 4: Call Control; Subpart 5: Conferencing Call Control SCF*, V1.1. 1202 915-4-5, European Telecommunications Standards Institute, Sophia Antipolis, France.

ETSI (2003b) *Application Programming Interface (API) Part 9: Generic Messaging SCF*, V1.1. 202 915-9, European Telecommunications Standards Institute, Sophia Antipolis, France.

ETSI (2003c) *Application Programming Interface (API) Part 10: Connectivity Manager SCF*, V1.1.1. 202 915-10, European Telecommunications Standards Institute, Sophia Antipolis, France.

ETSI (2003d) *Voice over IP (TIPHON) Overview*, European Telecommunications Standards Institute, Sophia Antipolis, France, http://www.etsi.org/frameset/home.htm?/technical activ/VOIP/voip.htm.

EURESCOM (2001a) Agent Based Computing: A Booklet for Executives, http://www.eures com.de/pub-deliverables/p800-series/P815/booklet/!AGENT_B.pdf.

EURESCOM (2001b) P712 Project Public Web Page, http://www.eurescom.de/public/projects/P700-series/P712.htm.

EURESCOM (2001c) P815 Project Public Web Page, http://www.eurescom.de/public/projects/P800-series/P815/default.asp.

Faci, M., Logrippo, L. M. S. and Stepien, B. (1997) Structural models for specifying telephone systems, *Computer Networks*, **29**(4), 501–528.

Fagan, M. E. (1979) Design and code inspections to reduce errors in programs, *IBM Systems Journal*, **15**(3), 7/1–7/26.

Farley, T. (2001) TelecomWriting.com's Telephone History Series, http://www.privateline.com/TelephoneHistory/History1.htm.

Faynberg, I., Gabuzda, L. R., Jacobson, T. and Lu, H. L. (1997) The development of the Wireless Intelligent Network (WIN) and its relation to the international Intelligent Network Standard, *Bell Labs Technical Journal*, **3**(2), 57–80.

Feller, W. (1968) *An introduction to probability theory and its applications*, Vol.1, Third edition, John Wiley & Sons.

Fielding, R., Gettys, J., Mogul, J., Frystyk, H., Masinter, L., Leach, P. and Berners-Lee, T. (1999) *Hypertext Transfer Protocol – HTTP/1.1*, RFC 2616, Internet Society, New York, USA.

FindtheOne (2003) http://www.findtheone.com/FTO Mobile Dating.html.

Ford, N. and Ford, J. (1993), *Introducing Formal Methods*, McGraw-Hill, New York, USA.

Frappier, M., Mili, A. and Desharnais, J. (1997) Detecting feature interactions on relational specifications, in Dini, P., Boutaba, R. and Logrippo, L. M. S. (eds) *Proc. 4th International Workshop on Feature Interactions in Telecommunication Networks*, pp. 123–137, IOS Press, Amsterdam, The Netherlands.

Fraser, M. D. and Vaishnavi, V. K. (1997) A formal specifications maturity model, *Communications of the ACM*, **40**(12), 95–103.

Frei, C. and Faltings, B. (2000) Abstraction and constraint satisfaction techniques for planning bandwidth allocation, in *INFOCOM* (1), pp. 235–244.

Frühwirth, T. and Brisset, P. (1997) Optimal planning of digital cordless telecommunication systems, in *Proc. 3rd International Conference on The Practical Application Of Constraint Technology*, pp. 165–176, Blackpool, The Practical Application Ltd.

Fu, Q., Harnois, P., Logrippo, L. M. S. and Sincennes, J. (2000) Feature interaction detection: A LOTOS-based approach, *Computer Networks*, **32**(4), 433–448.

Gaiti, D. and Martikainen, O. (eds) (2002) *Intelligent Agents for Telecommunication Environments*, Kogan Page Science.

Gamma, E., Helm, R., Johnson, R. and Vlidisses, J. (1994) *Design Patterns: Elements of Reusable Object-Oriented Software*, Addison Wesley.

Gan, W., Soraghan, J. and Durranu, T. (1992) Application of the functional-link technique for channel equalization, *Electronics Letters*, **28**(17), 1164–1643.

Garlan, D., Monroe, R. T. and Wile, D. (1997) ACME: An architecture description interchange language, in *Proc. CASCON'97*, pp. 169–183, Toronto, Canada.

Genesereth, M. and Ketchpel, S. (1994) Software Agents, *Communications of the ACM*, **37**(7), 49–53.

Ghosale, N. M., Green, J. A., Hernandez-Herrero, J., Huang, G. G. and Parikh, P. S. (2003) On implementing a High-Performance Open API with Java, *Bell Labs Technical Journal*, Special issue on Wireless Networks, **7**(2).

Gibson, J. P., Hamilton, G. and Méry, D. (2000) A taxonomy for triggered interactions using fair object semantics, in Calder, M. H. and Magill, E. H. (eds) *Proc. 6th Feature Interactions in Telecommunications and Software Systems*, pp. 193–209, IOS Press, Amsterdam, The Netherlands.

Goguen, J. A., Kirchner, C., Kirchner, H., Mégrelis, A. and Meseguer, J. (1988) An introduction to OBJ 3, in Jouannaud, J. P. and Kaplan, S. (eds) *Proc. Conditional Term Rewriting*, in Lecture Notes in Computer Science 308, pp. 258–263, Springer-Verlag, Berlin, Germany.

Graham, G. and Denning, P. (1972) Protection – principles and practice, in *Proc. Spring Joint Computer Conference*, AFIPS Press, Montvale, NJ, USA.

Greenhouse, A. R. (2000a) SIP Extensions for Communicating with Networked Appliances, http://www. argreenhouse.com/iapp/draft-tsang-sip-appliances-do-00.txt.

Greenhouse, A. R. (2000b) Framework Draft for Networked Appliances Using the Session Initiation Protocol, http: //www.argreenhouse.com/iapp/draft-moyer-sip-appliances-frame work-01.pdf.

Greenhouse, A. R. (2001) *AR Greenhouse*, http://www.argreenhouse.com/iapp/.

Griffeth, N. D. and Velthuijsen, H. (1994) The negotiating agents approach to runtime feature interaction resolution, in Bouma, L. G. and Velthuijsen, H. (eds) *Proc. 2nd International Workshop on Feature Interactions in Telecommunications Systems*, pp. 217–235, IOS Press, Amsterdam, The Netherlands.

Gudgin, M., Madley, M., Mendelsohn, N., Moreau, J. and Nielsen, H. F. (2002) *SOAP Version 1.2 Part 1: Messaging Framework*, W3C Candidate Recommendation, World Wide Web Consortium, Geneva, Switzerland.

Hall, A. (1990) Seven myths of formal methods, *IEEE Software*, **7**(5), 11–19.

Hall, R. J. (1998) Foreground/background models, in Kimbler, K. and Bouma, W. (eds) *Proc. 5th Feature Interactions in Telecommunications and Software Systems*, pp. 232–246, IOS Press, Amsterdam, The Netherlands.

Hall, R. J. (2000a) Feature combination and interaction detection via foreground/background models, *Computer Networks*, **32**(4), 449–469.

Hall, R. J. (2000b) Feature interactions in electronic mail, in Calder, M. H. and Magill, E. H. (eds) *Proc. 6th Feature Interactions in Telecommunications and Software Systems*, pp. 67–82, IOS Press, Amsterdam, The Netherlands.

Handley, M. and Jacobson, V. (1998) SDP: Session Description Protocol, RFC 2327, Internet Society, New York, USA.

Harel, D. and Gery, E. (1996) Executable object modeling with Statecharts, in *Proc. 18th International Conference on Software Engineering*, pp. 246–257, Institution of Electrical and Electronic Engineers Press, New York, USA.

Harte, L., Levine, R. and Kikta, R. (2001) *3G Wireless Demystified*, McGraw Hill.

Heinanen, J., Baker, F., Weiss, W. and Wroclawski, J. (1999) *Assured Forwarding PHB Group*, RFC 2597, Internet Society, New York, USA.

Hertz, J., Krogh, A. and Palmer, R. G. (1990) *Introduction to The Theory of Neural Computation*, Perserus Publishing.

Hicks, M. (1999) Messaging Systems of the World, unit, *PC Week*, 4th October.

Hill, I. D. (1982) Wouldn't it be nice if we could write computer programs in ordinary English – or would it? *Computer Bulletin*, pp. 306–312.

Hoare, C. A. R. (1985) *Communicating Sequential Processes*, Prentice Hall.

Holland, J. (1975) *Adaptation in natural artificial systems*, University of Michigan Press, Ann Arbor, USA.

Holma, H. and Toskala, A. (2002) *WCDMA for UMTS*, Second edition, John Wiley & Sons, Chichester, UK.

Holzman, G. J. and Pehrson, B. (1995) *The Early History of Data Networks*, IEEE Press, Los Alamitos, California, USA.

Howes, T. and Smith, M. (1997) *LDAP: Programming Directory-Enabled Applications with Lightweight Directory Access Protocol*, Macmillan Technical Publishing.

Hussain, A., Soraghan, J. and Durrani, T. (1997) A new Adaptive Functional-Link Neural Network Based DFE for Overcoming Co-channel Interference, *IEEE Transactions on Communications*, **45**(11), 1358–1362.

IEEE (2003) 802.16 Working Group on Broadband Wireless Access Standards, http://grouper.ieee.org/groups/802/16/.

IETF (1996) *RTP: A transport protocol for real-time application*, RFC 1889, Internet Society, New York, USA.

IETF (1998) *Real Time Streaming Protocol (RTSP)*, RFC 2326, Internet Society, New York, USA.

IETF (1999a) *Session Initiation Protocol*, RFC 2543, Internet Society, New York, USA (obsoleted by RFC 3261 in 2002).

IETF (1999b) *Media Gateway Control Protocol*, RFC 2705, Internet Society, New York, USA (obsoleted by RFC 3435 in 2003).

IETF (2000a) *Media Gateway Control Protocol Architecture and Requirements*, RFC 2805, Internet Society, New York, USA.

IETF (2000b) *Megaco Protocol version 1.0*, RFC 3015, Internet Society, New York, USA.

IETF (2000c) *Stream Control Transmission Protocol*, RFC 2960, Internet Society, New York, USA.

IETF (2000d) Applying Contract Net Protocol to Mobile Handover State Transfer, http://www.ietf.org/internet-drafts/draft-neumiller-seamoby-cnpmobility-00.txt.

IETF (2001a) *ISDN Q.921-User Adaptation Layer*, RFC 3057, Internet Society, New York, USA.

IETF (2001b) *MIME Media Types for ISUP & QSIG Objects*, RFC 3204, Internet Society, New York, USA.

IETF (2001c) *Control of Service Context using SIP*, RFC 3087, Internet Society, New York, USA.

IETF (2001d) Pico SIP, http://www.ietf.org/internet-drafts/draft-odoherty-pico-sip-00.txt.

IETF (2002a) *Version 2 of the Protocol Operations for the Simple Network Management Protocol*, RFC 3416, Internet Society, New York, USA.

IETF (2002b) *SS7 MTP3-User Adaptation Layer*, RFC 3332, Internet Society, New York, USA.

IETF (2002c) *SIP-T Context & Architectures*, RFC 3372, Internet Society, New York, USA.

IETF (2002d) *SIP: Session Initiation Protocol*, RFC 3261, Internet Society, New York, USA.

IETF (2002e) The SIP Negotiate Method, http://www.ietf.org/internet-drafts/draft-spbs-sip-negotiate-01.txt.

Indulska, J., Bearman, M. and Raymond, K. (1993) A type management system for an ODP trader, in de Meer, J., Mahr, B. and Spaniol, O. (eds) *Proc. International Conference on Open Distributed Processing*, pp. 141–151. Gesellschaft für Mathematik und Datenverarbeitung, Berlin, Germany.

International Messaging Consortium (2001) Unified Messaging Tutorial, http://www.iec.org/tutorials/unified_mess/.

International Softswitch Consortium (2002a) Softswitch Frequently Asked Questions, www.softswitch.org/educational/tac_faq.asp-tac1.

International Softswitch Consortium (2002b) Softswitch Reference Architecture, www.softswitch.org/attachments/Reference_Architecture_5-02.pdf.

ISO/IEC (1986a) *Information Processing Systems – Open Systems Interconnection – Connection-Oriented Transport Protocol Specification*, ISO/IEC 8073, International Organization for Standardization, Geneva, Switzerland.

ISO/IEC (1986b) *Information Processing Systems – Open Systems Interconnection – Transport Service Definition*, ISO/IEC 8072, International Organization for Standardization, Geneva, Switzerland.

ISO/IEC (1989) *Information Processing Systems – Open Systems Interconnection – LOTOS – A Formal Description Technique based on the Temporal Ordering of Observational Behaviour*, ISO/IEC 8807, International Organization for Standardization, Geneva, Switzerland.

ISO/IEC (1990a) *Information Processing Systems – Open Systems Interconnection – Formal Description in LOTOS of the Connection-Oriented Session Service*, ISO/IEC TR 9571, International Organization for Standardization, Geneva, Switzerland.

ISO/IEC (1990b) *Information Processing Systems – Open Systems Interconnection – Formal Description in LOTOS of the Connection-Oriented Transport Service*, ISO/IEC TR 10023, International Organization for Standardization, Geneva, Switzerland.

ISO/IEC (1990c) *Information Processing Systems – Open Systems Interconnection – Guidelines for the Application of ESTELLE, LOTOS and SDL*, ISO/IEC TR 10167, International Organization for Standardization, Geneva, Switzerland.

ISO/IEC (1992) *Information Technology – coding of moving pictures and associated audio for digital storage media at up to about 1.5 Mbit/s*, ISO/IEC 11172, International Organization for Standardization, Geneva, Switzerland.

ISO/IEC (1994a) *Information Technology – digital compression and coding of continuous-tone still images*, ISO/IEC 10918, International Organization for Standardization, Geneva, Switzerland.

ISO/IEC (1994b) *Information Technology – generic coding of moving pictures and associated audio information*, ISO/IEC 13818, International Organization for Standardization, Geneva, Switzerland.

ISO/IEC (1994c) *Information Processing Systems – Open Systems Interconnection – Basic Reference Model*, ISO/IEC 7498, International Organization for Standardization, Geneva, Switzerland.

ISO/IEC (1995) *Information Processing Systems – Open Distributed Processing –Basic Reference Model*, ISO/IEC 10746, International Organization for Standardization, Geneva, Switzerland.

ISO/IEC (1996a) *Information Processing Systems – Open Distributed Processing – ODP Trading Function*, ISO/IEC 13235, International Organization for Standardization, Geneva, Switzerland.

ISO/IEC (1996b) *Open Distributed Processing – Basic Reference Model – Part 4: Architectural Semantics*, ISO/IEC 10746–4, International Organization for Standardization, Geneva, Switzerland.

ISO/IEC (1997a) *Information Technology – coding of multimedia and hypermedia information*, ISO/IEC 13522, International Organization for Standardization, Geneva, Switzerland.

ISO/IEC (1997b) *Information Technology – Open Systems Interconnection – ESTELLE: A Formal Description Technique based on an Extended State Transition Model*, ISO/IEC 9074, International Organization for Standardization, Geneva, Switzerland.

ISO/IEC (1999) *Information Technology – Open Distributed Processing – Interface Definition Language*, ISO/IEC 14750, International Organization for Standardization, Geneva, Switzerland.

ISO/IEC (2000a) *Information Technology – JPEG 2000 image coding system – Part 1: Core coding system*, ISO/IEC 15444-1, International Organization for Standardization, Geneva, Switzerland.

ISO/IEC (2000b) *High-Level Petri Net Standard*, ISO/IEC 15909, International Organization for Standardization, Geneva, Switzerland.

ISO/IEC (2001a) *Information Technology – coding of audio-visual objects*, ISO/IEC 14496, International Organization for Standardization, Geneva, Switzerland.

ISO/IEC (2001b) *Information Processing Systems – Open Systems Interconnection – Enhanced LOTOS – A Formal Description Technique based on the Temporal Ordering of Observational Behaviour*, ISO/IEC 15437, International Organization for Standardization, Geneva, Switzerland.

ISO/ITU-T (1995a) *Open Distributed Processing – Reference Model – Part 1: Overview*, ISO/IEC 10746-1/ITU X.901, International Organization for Standardization, Geneva, Switzerland.

ISO/ITU-T (1995b) *Open Distributed Processing – Reference Model – Part 2: Foundations*, ISO/IEC 10746-2/ITU X.902, International Organization for Standardization, Geneva, Switzerland.

ISO/ITU-T (1995c) *Open Distributed Processing – Reference Model – Part 3: Architecture*, ISO/IEC 10746-3/ITU X.903, International Organization for Standardization, Geneva, Switzerland.

ISO/ITU-T (1995d) *Open Distributed Processing – Reference Model – Part 4: Architectural Semantics*, ISO/IEC 10746-4/ITU X.904, International Organization for Standardization, Geneva, Switzerland.

ISO/ITU-T (1997) *Open Distributed Processing – Trading Function: Specification*, ITU-T X.950, International Telecommunications Union, Geneva, Switzerland.

ITU (1980) *Signalling System Number 7 – Telephone User Part*, ITU-T Q.721–Q.724, International Telecommunications Union, Geneva, Switzerland.

ITU (1984a) *ISDN User-Network Interface Layer 3 Specification for basic call control*, ITU-T Q.931, International Telecommunications Union, Geneva, Switzerland, (subsequently revised in 1993, 1998).

ITU (1984b) *The International Public Telecommunication Numbering Plan*, ITU-T E.164, International Telecommunications Union, Geneva, Switzerland (subsequently revised in 1988, 1991, 1997).

ITU (1988) *Signalling System Number 7 – ISDN User Part*, ITU-T Q.761–Q.765, International Telecommunications Union, Geneva, Switzerland (subsequently revised in 1992, 1996, 2000).

ITU (1990) *40-, 32-, 24-, and 16 kbit/s Adaptive Differential Pulse Code Modulation (ADPCM)*, ITU-T G.726, International Telecommunications Union, Geneva, Switzerland.

ITU (1991) *Principles for a Telecommunications Management Network Working Party IV, Report 28*, ITU-T M.3010, International Telecommunications Union, Geneva, Switzerland.

ITU (1992a) *Digital compression and coding of continuous-tone still images – Requirements and guidelines*, ITU-T T.81, International Telecommunications Union, Geneva, Switzerland.

ITU (1992b) *Coding of speech at 16 kbit/s using Low-delay Code Excited Linear Prediction (LD-CELP)*, ITU-T G.728, International Telecommunications Union, Geneva, Switzerland.

ITU (1992c) *Guidelines for the Application of ESTELLE, LOTOS and SDL*, International Telecommunications Union, Geneva, Switzerland.

ITU (1993a) *Video Codec for Audiovisual Services at 64 kbits*, ITU-T H.261, International Telecommunications Union, Geneva, Switzerland.

ITU (1993b) *Introduction to CCITT Signalling System Number 7*, ITU-T Q.700, International Telecommunications Union, Geneva, Switzerland.

ITU (1993c) *Formats And Codes Of The ISDB User Part Of Signalling System Number 7*, ITU-T Q.763, International Telecommunications Union, Geneva, Switzerland.

ITU (1993d) *Digital Subscriber Signalling System No. 1 (DSS 1) – ISDN User-Network Interface Layer 3 Specification For Basic Call Control*, ITU-T Q. 931, International Telecommunications Union, Geneva, Switzerland.

ITU (1993e) *Introduction to Intelligent Network Capability Set 1*, ITU-T Q.1211, International Telecommunications Union, Geneva, Switzerland.

ITU (1993f) *The Directory: Overview of Concepts, Models and Service*, ITU-T X.500, International Telecommunications Union, Geneva, Switzerland.

ITU (1993g) *Intelligent Network – Global Functional Plane Architecture*, ITU-T Q.1203, International Telecommunications Union, Geneva, Switzerland.

ITU (1993h) *Intelligent Network – Q.120x Series Intelligent Network Recommendation Structure*, ITU-T Q.1200 Series, International Telecommunications Union, Geneva, Switzerland.

ITU (1993i) *Intelligent Networks Capability Set 1*, ITU-T Q.1210–Q.1218, International Telecommunications Union, Geneva, Switzerland (subsequently revised in 1995).

ITU (1993j) *General Recommendations on Telephone Switching and Signaling: Intelligent Network Physical Plane Architecture*, ITU-T Q.1205, International Telecommunications Union, Geneva, Switzerland.

ITU (1995a) ITU-T Q.1210 Series, International Telecommunications Union, Geneva, Switzerland.

ITU (1995b) *Distributed Functional Plane for Intelligent Network CS 1*, ITU-T Q.1214, International Telecommunications Union, Geneva, Switzerland.

ITU (1995c) *Global Functional Plane for Intelligent Network CS 1*, ITU-T Q.1213, International Telecommunications Union, Geneva, Switzerland.

ITU (1995d) *Physical Plane for Intelligent Network CS 1*, ITU-T Q.1219, International Telecommunications Union, Geneva, Switzerland.

ITU (1995e) *Interface Recommendation for Intelligent Network CS 1*, ITU-T Q.1218, International Telecommunications Union, Geneva, Switzerland.

ITU (1996a) *Coding of speech at 8 kbit/s using Conjugate Structure Algebraic-Code-Excited Linear-Prediction (CS-ACELP)*, ITU-T G.729, International Telecommunications Union, Geneva, Switzerland.

ITU (1996b) *Packet-based Multimedia Communication System*, ITU-T H.323, International Telecommunications Union, Geneva, Switzerland.

ITU (1996c) *Open Distributed Processing – Basic Reference Model*, ITU X.900 Series, International Telecommunications Union, Geneva, Switzerland.

ITU (1996d) *Open Distributed Processing – Basic Reference Model – Part 4: Architectural Semantics*, ITU X.904, International Telecommunications Union, Geneva, Switzerland.

ITU (1996e) *Call Signaling Protocols and Media Stream Packetization for Packet Based Multimedia Communications Systems*, ITU-T H.225, International Telecommunications Union, Geneva, Switzerland, (subsequently revised in 1998, 1999, 2000).

ITU (1996f) *Control Protocol for Multimedia Communication*, ITU-T H.245, International Telecommunications Union, Geneva, Switzerland, (subsequently revised in 1998, 1999, 2000).

ITU (1997a) *Vocabulary of terms for broadband aspects of ISDN*, ITU-T I.113, International Telecommunications Union, Geneva, Switzerland.

ITU (1997b) ITU-T Q.1220 Series, International Telecommunications Union, Geneva, Switzerland.

ITU (1997c) *Intelligent Networks Capability Set 2*, ITU-T Q.1220–Q.1228, International Telecommunications Union, Geneva, Switzerland.

ITU (1998a) *Generic functional protocol for the support of supplementary services in H.323*, ITU-T H.450.1, International Telecommunications Union, Geneva, Switzerland.

ITU (1998b) *Video coding for low bit rate communication*, ITU-H.263, International Telecommunications Union, Geneva, Switzerland.

ITU (1999) *Infrastructure of Audiovisual Services – Systems and Terminal Equipment for Audiovisual Services*, ITU-T H.320, International Telecommunications Union, Geneva, Switzerland.

ITU (2000a) *Framework Recommendation for Multimedia Services*, ITU-T F.700, International Telecommunications Union, Geneva, Switzerland.

ITU (2000b) *Infrastructure of Audiovisual Services – Systems and Terminal Equipment for Audiovisual Services*, ITU-T H.323, International Telecommunications Union, Geneva, Switzerland.

ITU (2000c) *One-way transmission delay*, ITU-T G.114, International Telecommunications Union, Geneva, Switzerland.

ITU (2000d) *Guideline Recommendation for Identifying Multimedia Service Requirements*, ITU-T F.701, International Telecommunications Union, Geneva, Switzerland.

ITU (2000e) *Gateway Control Protocol*, ITU-T H.248, International Telecommunications Union, Geneva, Switzerland.

ITU (2000f) *Message Sequence Chart (MSC)*, ITU-T Z.120, International Telecommunications Union, Geneva, Switzerland.

ITU (2000g) *SDL Combined with UML*, ITU-T Z.109, International Telecommunications Union, Geneva, Switzerland.

ITU (2000h) *Specification and Description Language*, ITU-T Z.100, International Telecommunications Union, Geneva, Switzerland.

ITU (2000i) *Bearer Independent Call Control – Capability Set 1*, ITU-T Q.1902.1–Q1902.6, International Telecommunications Union, Geneva, Switzerland.

ITU (2000j) *Bearer Independent Call Control – Capability Set 2*, ITU-T Q.1902.1–Q1902.6, International Telecommunications Union, Geneva, Switzerland.

ITU (2001) *Transmission Systems and Media, Digital Systems and Networks, Quality of Service and Performance, End-user Multimedia QoS*, ITU-T G.1010, International Telecommunications Union, Geneva, Switzerland.

ITU (2002) *Methodology for the subjective assessment of the quality of television pictures*, ITU-R BT.500, International Telecommunications Union, Geneva, Switzerland.

IWAN (2000) *Proc. 2nd International Working Conference on Active Networks*, Tokyo, Japan.

Jackson, M. and Zave, P. (1998) Distributed feature composition: A virtual architecture for telecommunications services, IEEE *Transactions on Software Engineering*, **24**(10), 831–847.

JAIN (2001) *JAIN TCAP*, http://www.jcp.org/en/jsr/detail?id=11.

JAIN (2002a) *JAIN JCC*, http://www.jcp.org/en/jsr/detail?id=21.

JAIN (2002b) *JAIN JCC RI/TCK*, http://www.argreenhouse.com/ngnapi/.

JAIN (2002c) *JAIN SIP*, http://www.jcp.org/en/jsr/detail?id=32.

JAIN (2003a) *JAIN JCAT*, http://www.jcp.org/en/jsr/detail?id=122.

JAIN (2003b) *Certified Products, 2003*, http://java.sun.com/products/jain/certified_products.html.

JAIN (2003c) *JAIN SCE*, http://www.jcp.org/en/jsr/detail?id=100.

JAIN (2003d) *JAIN SLEE*, http://www.jcp.org/en/jsr/detail?id=22.

Jain, R., Anjum, F. and Bakker, J.-L. (2003) *Call Control: Programming Interfaces for Next Generation Networks*, John Wiley & Sons, Hoboken, USA.

Jainschigg, J. (2001) Agents of Change, *Computer Telephony*, April 2001.

Jakobssen, M. and Yung, M. (1996) Revocable and Versatile Electronic Money (Extended Extract), http://www.bell-labs.com/~markusj/revoke.ps.

Jampel, M. (1996) Over-Constrained Systems, LNCS 1106. Springer-Verlag Berlin, Germany.

Jones, C. B. (1990) *Systematic Software Development using VDM*, second edition, Prentice Hall.

Jonsson, B., Margaria, T., Naeser, G., Nyström, J. and Steffen, B. (2000) Incremental requirements specification for evolving systems, in Calder, M. H. and Magill, E. H. (eds) *Proc. 6th Feature Interactions in Telecommunications and Software Systems*, pp. 145–162, IOS Press, Amsterdam, The Netherlands.

JTAPI (2003) *Generic JTAPI and JCC, 2003*, http://gjtapi.sourceforge.net/.

Kaaranen, H., Ahtiainen, A., Laitinen, L., Naghian, S. and Niemi, V. (2001) *UMTS Networks: Architecture, Mobility and Services*, John Wiley & Sons, Chichester, UK.

Kamoun, J. and Logrippo, L. M. S. (1998) Goal-oriented feature interaction detection in the Intelligent Network model, in Kimbler, K. and Bouma, W. (eds) *Proc. 5th Feature Interactions in Telecommunications and Software Systems*, pp. 172–186, IOS Press, Amsterdam, The Netherlands.

Kantschik, W., Dittrich, P., Brameier, M. and Banzhaf, W. (1999) Meta-evolution in graph GP, in Poli, R., Nordin, P., Langdon, W. and Fogarty, T. (eds) *Proc. Euro Genetic Programming*, LNCS 1598, pp. 15–28, Göteborg, Sweden. Springer-Verlag, Berlin, Germany.

Karlsson, K. and Vetterli, M. (1989) Packet video and its integration into network architecture. *IEEE JSAC*, **7**(5), 739–751.

Keck, D. O. (1998) A tool for the identification of interaction-prone call scenarios, in Kimbler, K. and Bouma, W. (eds) *Proc.* 5th *Feature Interactions in Telecommunications and Software Systems*, pp. 276–290, IOS Press, Amsterdam, The Netherlands.

Keck, D. O. and Kühn, P. J. (1998) The feature and service interaction problem in telecommunications systems: A survey, *IEEE Transactions on Software Engineering*, pp. 779–796.

Kerckhoffs, A. (1883a) La cryptographie militaire, in *Journal des Sciences Militaries*, Series IX, pp. 5–38, Jan.

Kerckhoffs, A. (1883b) La cryptographie militaire, in *Journal des Sciences Militaries*, Series IX, pp. 161–191, Feb.

Kilkki, K. (1999) *Differentiated Services for the Internet*, Macmillan Technical Publishing.

Kimbler, K. (1997) Addressing the interaction problem at the enterprise level, in Dini, P., Boutaba, R. and Logrippo, L. M. S. (eds) *Proc. 4th International Workshop on Feature Interactions in Telecommunication Networks*, pp. 13–22, IOS Press, Amsterdam, The Netherlands.

Kimbler, K. and Bouma, W. (eds) (1998) *Proc. 5th Feature Interactions in Telecommunications and Software Systems*, IOS Press, Amsterdam, The Netherlands.

Kitahara, Y. (1982) *Information Network System*, The Telecommunications Association, Japan.

Kitawaki, N. and Itoh, K. (1991) Pure delay effects on speech quality telecommunications, *IEEE JSAC*, **9**(4), 586–593.

Klein, C., Prehofer, C. and Rumpe, B. (1997) Feature specification and refinement with state transition diagrams, in Dini, P., Boutaba, R. and Logrippo, L. M. S. (eds) *Proc. 4th International Workshop on Feature Interactions in Telecommunication Networks*, pp. 284–297, IOS Press, Amsterdam, The Netherlands.

Klensin, J. *et al.* (2001) *Simple Mail Transfer Protocol*, RFC 2821, Internet Society, New York, USA.

Kolberg, M., Sinnott, R. O. and Magill, E. H. (1999a) Engineering of interworking TINA-based telecommunications services, *Proc. IEEE Telecommunications Information Networking Architecture Conference*, IEEE Press.

Kolberg, M., Sinnott, R. O. and Magill, E. H. (1999b) Experiences modelling and using formal object-oriented telecommunication service frameworks, *International Journal of Computer and Telecommunications Networking*, **31**(23), 2577–2592.

Kon, F., Román, M., Liu, P., Mao, J., Yamane, T., Magalhães, L. C. and Campbell, R. H. (2000) Monitoring, Security and Dynamic Configuration with the dynamicTAO Reflective ORB, *Proc. IFIP International Conference on Distributed Systems Platforms and Open Distributed Processing*, IBM Palisades, New York, USA.

Koza, J. (1992) *Genetic programming: On the programming of computers by means of natural selection*, MIT Press, Cambridge, Massachusetts, USA.

Kristiansen, L. (1997) *Service Architecture*, TINA Consortium, Version 5.0.

Kwok, T. C. (1995) A vision for residential broadband services: ATM-to-the-home, *IEEE Network*, pp. 14–28, September.

Kwok, T. C. (1997) Residential broadband Internet services and application requirements, *IEEE Communications Magazine*, **35**(6), 76–83.

Lagerberg, K., Plas, D.-J. and Wegdam, M. (2002) Web Services in 3G Service Platforms, *Bell Labs Technical Journal*, Special issue on Wireless Networks, **7**(2), 167–183.

Lamport, L. (1993) The temporal logic of actions, Technical Report 79, Systems Research Center, Digital Equipment Corporation.

Lampson, B. (1971) Protection, *Proc. 5th Princeton Symposium on Information Sciences and Systems*, reprinted in *ACM Operating Systems Review*, **8**(1).

Le Charnier, B. and Flener, P. (1998) Specifications are necessarily informal, or: Some more myths of formal methods. *Systems Software*, **40**, 275–296.

Le Faucheur, F. I. (2002) Multi-Protocol Label Switching (MPLS) Support of Differentiated Services, RFC 3270, Internet Society, New York, USA.

Lin, F. J. and Lin, Y.-J. (1994) A building block approach to detecting and resolving feature interactions, in Bouma, L. G. and Velthuijsen, H. (eds) *Proc. 2nd International Workshop on Feature Interactions in Telecommunications Systems*, pp. 86–119. IOS Press, Amsterdam, The Netherlands.

Logrippo, L. M. S. (2000) Immaturity and potential of formal methods: A personal view, in Calder, M. H. and Magill, E. H. (eds) *Proc. 6th Feature Interactions in Telecommunications and Software Systems*, pp. 9–13, IOS Press, Amsterdam, The Netherlands.

Lucent (2003) Lucent Parlay simulator, http://www.lucent.com/developer/milife/.

Lucidi, F., Tosti, A. and Trigila, S. (1996) Object oriented modelling of advanced IN services with SDL-92, in Brezocnik, Z. and Kapus, T. (eds) *Applied Formal Methods in System Design*, pp. 17–26, Maribor, Slovenia.

Luck, I., Vogel, S. and Krumm, H. (2002) Model-Based Configuration of VPNs, *Proc. 8th IEEE/IFIP Network Operations and Management Symposium*, pp. 589–602, Florence, Italy, April.

Luckham, D. C., Augstin, L. M., Kenney, J. J., Veera, J., Bryan, D. and Mann, W. (1995) Specification and analysis of system architectures using Rapide, *IEEE Transactions on Software Engineering*, **21**(4).

Lukschandl, E., Borgvall, H., Nohle, L., Nordahl, M. and Nordin, P. (1999) Evolving routing algorithms with the JBGP-System, in Poli, R., Voigt, H.-M., Cagnoni, S., Corne, D., Smith, G. and Fogarty, T. (eds) *Proc. Evolutionary image analysis, signal processing and telecommunications*, Göteborg, Sweden. LNCS 1596, pp. 193–202, Springer-Verlag, Berlin, Germany.

Lupu, E. and Sloman, M. (1997) A policy based role model, in Milosevic, Z. (ed) *Proc. 1st International Enterprise Distributed Object Computing Workshop*, pp. 36–47, Institution of Electrical and Electronic Engineers Press, New York, USA.

Lysyanskaya, A. and Ramzan, Z. (1998) Group Blind Digital Signatures: A Scalable Solution to Electronic Cash, MIT, Cambridge, Massachusetts, USA.

Marples, D. (2002) *PhD Thesis*, University of Strathclyde, UK.

Marples, D. and Kriens, P. (2001) *The Open Services Gateway Initiative: An Introductory Overview*, IEEE *Communication Magazine*, **39**(12).

Marples, D. and Magill, E. H. (1998) The use of rollback to prevent incorrect operation of features in Intelligent Network based systems, in Kimbler, K. and Bouma, W. (eds) *Proc. 5th Feature Interactions in Telecommunications and Software Systems*, pp. 115–134, IOS Press, Amsterdam, The Netherlands.

Martin, P. (2000) Genetic programming for service creation in intelligent networks, in Poli, R., Banzhaf, W., Langdon, W., Miller, J., Nordin, P. and Fogarty, T. (eds) *Proc. Euro Genetic Programming*, Edinburgh, LNCS 1802, pp. 106–120, Springer-Verlag, Berlin, Germany.

McBain, A. (2001) Our Vision of an IP Future, ITU Telecom Africa 2001.

McCulloch, W. and Pitts, W. (1943) A logical calculus of ideas immanent in nervous activity, *Bulletin of Mathematical Biophysics*, **5**, 115–133.

McLean, J. (1994) Security Models, in Marciniak, J. (ed.) *Encyclopedia of Software Engineering*, John Wiley & Sons, New York, USA.

Medvidovic, N. and Taylor, R. N. (1997) A framework for classifying and comparing architecture description languages, in *Proc. 6th European Software Engineering Conference/Proc. 5th Symposium on the Foundations of Software Engineering*, pp. 60–76, Zurich, Switzerland.

Menezes, A. J., van Oorschot, P. C. and Vanstone, S. A. (1997) *Handbook of Applied Cryptography*, CRC Press, New York, USA.

Microsoft (2000) TAPI (Telephony Application Programming Interface), http://www.microsoft.com/communications/tapilearn30.htm.

Milner, A. J. R. G. (1989) *Communication and Concurrency*, Addison Wesley, Reading, Massachusetts, USA.

Minton, S. (1996) Automatically configuring constraint satisfaction programs: A case study, *Constraints*, **1**(1/2), 7–43.

Mitchell, R. (ed.) (1987) *Formal Methods: Present and Future in Industrial Software Technology*, Peter Peregrinus.

Mittal, S. and Falkenhainer, B. (1990) Dynamic constraint satisfaction problems, in *Proc. AAAI90*, pp. 25–32, Boston, Massachusetts, USA.

Moerdijk, A.-J. and Klostermann, L. (2003) Opening the networks with Parlay/OSA: Standards and aspects behind the APIs, *IEEE Network Magazine*, **17**(3).

Moreira, A. M. D. and Clark, R. G. (1996) Adding rigour to object-oriented analysis, *Software Engineering Journal,* **11**(5), 270–280.

Muller, C., Magill, E. H. and Smith, D. G. (1992) The application of constraint satisfaction to feature interactions, in Velthuijsen, H., Griffith, N. and Lin, Y.-J. (eds) *Proc. 1st International Workshop on Feature Interactions in Telecommunications Software Systems*, pp. 164–165, Florida, USA.

Nakamura, M., Kikuno, T., Hassine, J. and Logrippo, L. M. S. (2000) Feature interaction filtering with Use Case Maps at requirements stage, in Calder, M. H. and Magill, E. H. (eds) *Proc. 6th Feature Interactions in Telecommunications and Software Systems*, pp. 163–178, IOS Press, Amsterdam, The Netherlands.

Naur, P. (1982) Formalization in program development, Technical Report BIT 22 (1982) 437 453, Datalogisk Institut, Denmark.

Nichols, K., Blake, S., Baker, F. and Black, D. (1998) *Definition of the Differentiated Services Field (DS Field) in the IPv4 and IPv6 Headers*, RFC 2474, Internet Society, New York, USA.

Nielson, J. (1994) Usability Engineering, Chapter 5 of *Usability Heuristics*, Morgan Kaufmann.

Nielson, J. (1997) The need for speed, http://www.useit.com/alertbox/9703a.html.

NIST (2003) National Institute of Standards and Technology, Mobile Agent Security Project, http://www.csrc.nist.gov/mobileagents.

Nordin, P. (1997) *Evolutionary program induction of binary machine code and its applications*, PhD thesis, Universität Dortmund am Fachbereich Informatik.

Oftel (2003) *Oftel report, Consumers' use of fixed and mobile telephony*, 13 May 2003–31 July 2003.

Ohrtman, F. and Roeder, K. (2003) *WiFi Handbook: Building 802.11b Wireless Networks*, McGraw Hill.

OMG (1997) *Control and Management of A/V Streams Specification*, ftp://ftp.omg.org/pub/docs/formal/00-01-03.pdf, Object Management Group.

OMG (1998a) *Notification Service Specification*, ftp://ftp.omg.org/pub/docs/dtc/00-12-02.pdf, Object Management Group.

OMG (1998b) *CORBA/TCAP Interworking Specification*, ftp://ftp.omg.org/pub/docs/telecom/98-10-03.pdf, Object Management Group.

OMG (1998c) *CORBA/TMN Interworking Specification*, ftp://ftp.omg.org/pub/docs/telecom/98-10-10.pdf, Object Management Group.

OMG (1999) *CORBA Telecom Log Service Specification*, ftp://ftp.omg.org/pub/docs/formal/00-01-04.pdf, Object Management Group.

OMG (2000a) *Management of Event Domain Specification*, ftp://ftp.omg.org/pub/docs/formal/01-06-03.pdf, Object Management Group.

OMG (2000b) *CORBA/FTAM-FTP Interworking Specification*, ftp://ftp.omg.org/pub/docs/telecom/00-02-04.pdf, Object Management Group.

OMG (2000c) *Wireless Access and Terminal Mobility Specification*, ftp://ftp.omg.org/pub/docs/dtc/01-05-01.pdf, Object Management Group.

OMG (2000d) *Telecom Service Access and Subscription Specification*, ftp://ftp.omg.org/pub/docs/dtc/00-10-03.pdf, Object Management Group.

OMG (2001a) *Telecommunications Domain Task Force*, http://www.omg.org/telecom/telecom_info.htm.

OMG (2001b) The OMG's Model Driven Architecture, http://www.omg.org/mda/.

OMG (2002) Common Object Request Broker Architecture: Core Specification, CORBA/IIO Specification.

Önder, E. (2001) Beyond Service Creation Stalemate, *Communications Solutions*, August.

OpenAPI (2003) Open API Solution, http://www.openapisolutions.com/home.htm.

OPENARCH (2001) *Proc. IEEE Conference on Open Architectures and Network Programming*, Anchorage, Alaska, USA.

Open Mobile Alliance (2001) *WAP-205 Multimedia Messaging Service architecture overview*, The Open Mobile Alliance, Piscataway, NJ, USA.

OPENSIG (2001) *OPENSIG 2001 Workshop*, http://www.dse.doc.ic.ac.uk/Events/opensig-2001.

Orfali, R. and Harkey, D. (1998) *Client/Server Programming with Java and CORBA*, Second edition, John Wiley & Sons, New York, USA.

OSGi (2002) *OSGi Service Platform*, (Release 2), IOS Press, Amsterdam, The Netherlands.

PacketCable (1999) PacketCable Network Based Call Signalling Protocol, PKT-SP-EC-MGCP-I01-990312, www.packetcable.com, (subsequently updated in 1999, 2001, 2003).

PAM Forum (2001) The PAM Forum, http://www.pamforum.org/, also http://www.parlay.org/about/pam/index.asp.

Parkin, G. I. and Austin, S. (1994) Overview: Survey of formal methods in industry, in Tenney, R. L., Amer, P. D. and Uyar, M. Ü. (eds) *Proc. Formal Description Techniques VI*, pp. 189–204. North Holland.

Parlay Group (2001) Parlay Overview. http://www.parlay.org/.

Parlay Group (2002) Parlay X White Paper, http://www.parlay.org/specs/library.

Parlay Group (2003) Parlay X Specification, http://www.parlay.org/specs.

PayCircle (2003) http://www.paycircle.org.

Perkins, C. (1996) *IP Mobility Support*, RFC 2002, Internet Society, New York, USA.

Petri, C. A. (1962) *Kommunikation mit Automaten*, PhD thesis, Institut für Instrumentelle Mathematik, Bonn, Germany.

Plath, M. C. and Ryan, M. D. (1998) Plug-and-play features, in Kimbler, K. and Bouma, W. (eds) *Proc. 5th Feature Interactions in Telecommunications and Software Systems*, pp. 150–164, IOS Press, Amsterdam, The Netherlands.

Postel, J. (1980) *User Datagram Protocol*, RFC 768, Internet Society, New York, USA.

PROFITS (2003) http://www.tmforum.org/browse.asp?catID = 1375&sNode = 1375&Exp = Y&linkI D = 27249.

Qualcomm BREW (2003) http://www.qualcomm.com/brew.

Ramzan, Z. (1999) *Group Blind Digital Signatures: Theory and Applications*, Masters Thesis, MIT, Boston, Mass- achusetts, USA.

RCR News 2003 http://rcrnews.com/cgi-bin/news.pl?newsId = 14052.

Riese, M. (1993) Diagnosis of extended finite automata as a dynamic constraint satisfaction problem, *Proc. International Workshop on Principles of Diagnosis* (DX'93), pp. 60–73. Aberystwyth, Wales, UK.

Rivest, R., Shamir, A. and Adleman, L. (1978) A Method for Obtaining Digital Signatures and Public-Key Cryptosystems, *Communications of the ACM*, **21**(2), 120–126.

RM (2000) Archive of the Workshop on Reflective Middleware, http://www.comp.lancs.ac.uk/computing/RM2000/.

RMI (2002) Java Remote Method Invocation (RMI), http://java.sun.com/products/jdk/rmi/.

Robrock II, R.B. (1991) The Intelligent Network – Changing the Face of Telecommunications, *Proc. of the IEEE*, **79**(1), 7–20.

Rojas, P. (1996) *Neural Networks – A Systematic Introduction*, Springer-Verlag.

Rosen, E., Viswanathan, A. and Callon, R. (2001) *Multiprotocol Label Switching Architecture*, RFC 3031, Internet Society, New York, USA.

Rosen, K. H. (1995) *Discrete Mathematics and Its Applications*, McGraw Hill, New York, USA.

Rosenberg, J., Schulzrinne, H., Camarillo, G., Johnson, A., Peterson, J., Sparks, R., Handley, M. and Schooler, E. (eds) (2002) *SIP: Session Initiation Protocol*, RFC 3261, Internet Society, New York, USA.

Rosenblatt, F. (1957) *The perceptron: A perceiving and recognizing automaton (project PARA)*, Technical Report 85-460-1, Cornell Aeronautical Laboratory, USA.

Rossi, F. and Sperduti, A. (1998) Learning solution preferences in constraint problems, *J. Experimental and Theoretical Artificial Intelligence*, **10**(1), 103–116.

RSA Laboratories (1993) *PKCS #1: RSA Encryption Standard.*

Ruggles, C. (ed) (1990) *Formal Methods in Standards*, Springer-Verlag.

Rumelhart, D., Hinton, G. and Williams, R. (1986) Learning internal representation by error propagation, Chapter 8 of Rumelhart, D. and McClelland, J. (eds) *Parallel Distributed Processing*, MIT Press, Cambridge, MA.

Sabin, D., Sabin, M., Russell, R. and Freuder, E. (1995) A Constraint-Based Approach to Diagnosing Software Problems in Computer Networks, in Montanari, U. and Rossi, F. (eds) *Proc. 1st International Conference on Constraint Programming*, Springer-Verlag.

SalCentral (2001) http://www.salcentral.com/.

SAML (2003) http://www.oasis-open.org/committees/tc_home.php?wg_abbrev=security.

Santos, E. (1993) On modeling time and uncertainty for diagnosis through linear constraint satisfaction, in *Proc. International Congress on Computer Systems and Applied Mathematics Workshop on Constraint Processing*, pp. 93–106.

Scheiderman, B. (1984) Response time and display rate in human performance with computers, *Computing Surveys*, **16**, 265–285.

Schessel, L. (1992) Administrable feature interaction concept, in Hugo Velthuijsen, Nancy Griffith and Yow-Jian Lin (eds) *Proc. 1st International Workshop on Feature Interactions in Telecommunications Software Systems*, Florida, USA.

Schneier, B. (1994) The Blowfish Encryption Algorithm, *Dr Dobb's Journal*, **19**(4).

Schoonderwoerd, R., Holland, O., Bruten, J. and Rothkrantz, L. (1997) Ant-base Load Balancing in Telecommunications Networks, *Adaptive Behaviour*, **5**(2), 169–207.

Schulte, C. (1997) Oz Explorer: A visual constraint programming tool, in Naish, L. (ed) *Proc. 14th International Conference on Logic Programming*, pp. 286–300, Leuven, Belgium. MIT Press, Cambridge, Massachusetts, USA.

Schulzrinne, H. (1996) *RTP: A Transport Protocol for Real-Time Applications*, RFC 1889, Internet Society, New York, USA.

Seigel, J. (2001) *Quick CORBA 3*, John Wiley & Sons, New York, USA.

Shami, S., Kirkwood, I. and Sinclair, M. (1997) Evolving simple fault-tolerant routing rules using genetic programming, *Electronics Letters*, **33**(17), 1440–1441.

Shaw, M. and Garlan, D. (1996) *Software Architecture: Perspectives on an Emerging Discipline*, Prentice Hall.

Shen, F. and Clemm, A. (2002) Profile-Based Subscriber Service Provisioning, *Proc. 8th IEEE/IFIP Network Operations and Management Symposium*, pp. 561–574, Florence, Italy, April.

Shenker, S., Partridge, C. and Guerin, R. (1997), *Specification of Guaranteed Quality of Service*, RFC 2212, Internet Society, New York, USA.

SID (2002) *TMF Shared Information Data Model*, TMF GB922, www.tmforum.org.

Simonis, H. (1997) Visual CHIP – a visual language for defining constraint programs, in *Proc. CCL Workshop*.

Sinnott, R. O. (1999a) Specifying aspects of multimedia in LOTOS, in Verma, B. (ed) *Proc. Conference on Computational Intelligence and Multimedia Applications*, New Delhi, India.

Sinnott, R. O. (1999b) Specifying multimedia configurations in Z, in Verma, B. (ed) *Proc. Conference on Computational Intelligence and Multimedia Applications*, New Delhi, India.

Sinnott, R. O. and Kolberg, M. (1999a) Creating telecommunication services based on object-oriented frameworks and SDL, in Raynal, M., Kikuno, T. and Soley, R. (eds) *Proc. 2nd IEEE International Symposium on Object-Oriented Real-Time Distributed Computing*, St Malo, France.

Sinnott, R. O. and Kolberg, M. (1999b) Engineering telecommunication services with SDL, in Ciancarini, P., Fantechi, A. and Gorrieri, R. (eds) *Proc. Formal Methods for Open Object-Based Distributed Systems*, Florence, Italy.

Sinnott, R. O. and Turner, K. J. (1995) Applying formal methods to standard development: The Open Distributed Processing experience, *Computer Standards and Interfaces*, **17**, 615–630.

Sinnott, R. O. and Turner, K. J. (1997) Applying the architectural semantics of ODP to develop a trader specification, *Computer Networks*, **29**(4), 457–471.

SMH (2003) http://www.smh.com.au/articles/2003/08/03/1059849278131.html.

Smith, B. (2001) Yes, Mom, I'll Take Your Call, *Wireless Week*, April.

Smith, D., Frank, J. and Jonsson, A. (2000) Bridging the gap between planning and scheduling, *Knowledge Engineering Review*, **15**(1).

Smith, G. L. and Jones, E. W. (1995) Design of multimedia services, *BT Technology Journal*, **13**(4), 21–31.

Smith, J. T. (1995) AIN System Development: The Customer Centered Service Context Profile, *Proc. Telecommunications Information Networking Architecture 95*.

Smith, R. (1980) The contract net protocol: High-level communication and control in a distributed problem solver, *IEEE Trans. on Computers*, **29**(12), 1104–1113.

Spivey, J. M. (1992) *The Z Notation: A Reference Manual*, Second edition, Prentice Hall.

Stallings, W. (1995) *Network and Internetwork Security: Principles and Practice*, Prentice Hall.

Stallings, W. (2002) *Wireless Communications and Networks*, Prentice Hall.

Stepien, B. and Logrippo, L. (1995) Representing and verifying intentions in telephony features using abstract data types, in Cheng, K. E. and Ohta, T. (eds) *Proc. 3rd International Workshop on Feature Interactions in Telecommunications*, pp. 141–155, IOS Press, Amsterdam, The Netherlands.

Stevens, P. and Pooley, R. (2000) *Using UML*, Addison Wesley, Reading, Massachusetts, USA.

Stinson, D. R. (1995) *Cryptography – Theory and Practice*, CRC Press, New York, USA.

Strasser, J. and Baker, F. (1999) *Directory Enabled Networking*, Macmillan Technical Publishing.

Sun Microsystems (2003) JAIN (Java APIs for The Integrated Network), http://java.sun.com/products/jain/.

Sun MIDP (2003) http://java.sun.com/products/midp.

Svensson, M. and Andersson, M. (1998) Analysis of feature interactions in mobile terminals, in Kimbler, K. and Bouma, W. (eds) *Proc. 5th Feature Interactions in Telecommunications and Software Systems*, pp. 318–324, IOS Press, Amsterdam, The Netherlands.

Szyperski, C. (1998) *Component Software: Beyond Object-Oriented Programming*, Addison Wesley.

Tapscott, D. (1996) *The Digital Economy: Promise and Peril in the Age of Networked Intelligence*, McGraw Hill.

Telcordia (1995) *Common Channel Signaling Network Interface Specification Supporting AIN*, GR-2863, Telcordia Technologies Inc., Piscataway, NJ, USA.

Telcordia (2001) GR-1298-CORE, Issue 7, Telcordia Technologies Inc., Piscataway, NJ, USA.

Telcordia (2003a) *ISCP SPACE System Service Creation Guide*, Telcordia Technologies, Piscataway, NJ, USA.

Telcordia (2003b) Intelligent Network, International Engineering Consortium Web ProForum Tutorials, http://www.iec.org/online/tutorials/in/index.html, 2003.

Telcordia (2003c) http://www.argreenhouse.com/demos/#iapp.

Telektronikk (2002) *XML Web Services*, Telenor, **98**(4).

Tennenhouse, D., Smith, J., Sincoskie, W.D., Weatherall, D. and Minden, G. (1997) Survey of Active Network Research, *IEEE Communications Magazine*, **35**(1).

Terashima-Marin, H., Ross, P. and Valenzuela-Rendon, M. (1999) Evolution of constraint satisfaction strategies in examination timetabling, in Banzhaf, W., Daida, J., Eiben, A., Garzon, M., Honavar, V., Jakiela, M. and Smith, R. (eds) *Proc. Genetic and Evolutionary Computation Conference*, Volume 1, pp. 635–642, Morgan Kaufmann.

Thayer, R., Doraswamy, N. and Glenn, R. (1998) *IP Security Document Roadmap*, RFC 2411, Internet Society, New York, USA.

Thomas, M. H. (1997) Modelling user views of telecommunications services for feature interaction detection and resolution, in Dini, P., Boutaba, R. and Logrippo, L. M. S. (eds) *Proc. 4th International Workshop on Feature Interactions in Telecommunication Networks*, pp. 168–182, IOS Press, Amsterdam, The Netherlands.

TINA Consortium (1997) *Telecommunications Information Networking Architecture*, TINA Business Model and Reference Points, Release 4.0.

Tisal, J. (2001) *The GSM Network*, Second edition, John Wiley & Sons, Chichester, UK.

Tsang, E. (ed) (1993) *The Constraint Satisfaction Problem*, Academic Press.

Tsang, S. and Magill, E. H. (1998a) Learning to Detect and Avoid Run-Time Feature Interactions in Intelligent Networks, *IEEE Transactions on Software Engineering*, **24**(10), 818–830, http://www.computer.org/tse/ts1998/extoc.htm.

Tsang, S. and Magill, E. H. (1998b) The Network Operator's Perspective: detecting and resolving Feature Interaction Problems, *Computer Networks and ISDN Systems*, **30**(15), 1421–1441.

Tsang, S., Magill, E. H. and Kelly, B. (1996) An investigation of the feature interaction problem in networked multimedia services, *Proc. 3rd Communication Network Symposium*, pp. 58–61.

Tsang, S., Magill, E. H. and Kelly, B. (1997) The feature interaction problem in networked multimedia services – present and future, *BT Technology Journal*, **15**(1), 235–246.

Turner, K. J. (ed) (1993) *Using Formal Description Techniques – An Introduction to ESTELLE, LOTOS and SDL*, John Wiley & Sons, Chichester, UK.

Turner, K. J. (1997a) An architectural foundation for relating features, in Dini, P., Boutaba, R. and Logrippo, L. M. S. (eds) *Proc. 4th International Workshop on Feature Interactions in Telecommunication Networks*, pp. 226–241, IOS Press, Amsterdam, The Netherlands.

Turner, K. J. (1997b) Relating architecture and specification, *Computer Networks*, **29**(4), 437–456.

Turner, K. J. (1998) An architectural description of Intelligent Network features and their interactions, *Computer Networks*, **30**(15), 1389–1419.

Turner, K. J. (2000a) Formalising the Chisel feature notation, in Calder, M. H. and Magill, E. H. (eds) *Proc. 6th Feature Interactions in Telecommunications and Software Systems*, pp. 241–256, IOS Press, Amsterdam, The Netherlands.

Turner, K. J. (2000b) Relating architecture and specification, in Kent, A., Williams, J. and Hall, C. M. (eds) *Encyclopaedia of Microcomputers*, volume 25, pp. 337–374, Marcel Dekker, New York, USA.

Turner, K. J. and van Sinderen, M. (1995) LOTOS specification style for OSI, in Bolognesi, T., van de Lagemaat, J. and Vissers, C. A. (eds) *The LOTOSPHERE Project*, pp. 137–159, Kluwer Academic Publishers, London, UK.

Tvede, L., Pircher, P. and Bodenkamp, J. (2001) *Data Broadcasting: Merging Digital Broadcasting with the Internet*, John Wiley & Sons, Chichester, UK.

UDDI (2002) *Universal Description, Discovery and Integration (UDDI)*, http://www.uddi.org.

USFCC (2003) http://www.usfcc.com/.

Utas, G. S. F. (1995) An Architecture for the Evolving Network; *Proc. ISS'95*, Berlin, Germany.

van der Linden, R. (1993) *An Overview of ANSA. Document APM.1000.01.* APM, Cambridge Limited, Castle Park, Cambridge, UK.

van der Linden, R. (1994) Using an architecture to help beat feature interaction, in Bouma, L. G. and Velthuijsen, H. (eds) *Proc. 2nd International Workshop on Feature Interactions in Telecommunications Systems*, pp. 24–35, IOS Press, Amsterdam, The Netherlands.

Vaughan, T. (1996) *Multimedia: Making it work*, Third edition, McGraw Hill.

Velthuijsen, H., Griffith, N. and Lin, Y.-J. (eds) (1992) *Proc. 1st International Workshop on Feature Interactions in Telecommunications Software Systems*, Florida, USA.

Vissers, C. A. (1993) What makes industries believe in formal methods, in Danthine, A. A. S., Leduc, G. and Wolper, P. (eds) *Proc. Protocol Specification, Testing and Verification XIII*, pp. 3–26. North Holland.

Vissers, C. A. and Logrippo, L. M. S. (1986) The importance of the concept of service in the design of data communications protocols, in Diaz, M. (ed) *Proc. Protocol Specification, Testing and Verification V*, pp. 3–17. North Holland.

Vissers, C. A., Scollo, G. and van Sinderen, M. (1991) Architecture and specification style in formal descriptions of distributed systems, *Theoretical Computer Science*, **89**, 179–206.

Vogel, A., Kerherve, B., Bochmann, G. and Gecsei, J. (1995) Distributed multimedia applications and quality of service: A survey. *IEEE Multimedia*, **2**(2), 10–18.

W3C (2001) *Web Services Description Language (WSDL) 1.1*, W3C Note 15, http://www.w3.org/TR/2001/NOTE-wsdl-20010315.

W3C (2003) *XML Signature Working Group*, W3C, http://www.w3.org/Signature/.

Wahl, M., Howes, T. and Kille, S. (1997) *Lightweight Directory Access Protocol* (v3), RFC 2251, Internet Society, New York, USA.

Wang, N., Schmidt, D.C. and O'Ryan, C. (2001) Overview of the CORBA Component Model. Component-Based Software Engineering: Putting the Pieces Together, in Heineman, G. and Councill, B. (eds) Addison Wesley.

Wearcam (2003) http://www.wearcam.org/.

Weinberg, G. M. and Freedman, D. P. (1984) Reviews, walkthroughs and inspections, *IEEE Transactions on Software Engineering*, **SE-10**(1), 68–72.

Weinstein, C. J. and Forgie, J. (1983) Experience with speech communication in packet networks, *IEEE JSAC*, **1**(6), 963–980.

Whyte, W. S. (1995) The many dimensions of multimedia communications, *BT Technology Journal*, **13**(4), 9–12.

Wikström, Å. (1987) *Functional Programming using Standard ML*, Prentice Hall.

WiredPolitic (2003) http://www.wired.com/news/politics/0,1283,17538,00.html.

Wooldridge, M. (2002) *Introduction to Multi Agent Systems*, John Wiley & Sons, Chichester, UK.

Wooldridge, M. and Jennings, N. (1995) Intelligent Agents: Theory and Practice, *Knowledge Engineering Review*, **10**(2), 115–152.

Wroclawski, J. (1997a) *The Use of RSVP with IETF Integrated Services*, RFC 2210, Internet Society, New York, USA.

Wroclawski, J. (1997b) *Specification of the Controlled-Load Network Element Service*, RFC 2211, Internet Society, New York, USA.

WSI (2003) *Web Services Interoperability Organization*. http://www.ws-i.org/.

XACML (2003) http://www.oasis-open.org/committees/tc_home.php?wg_abbrev=xacml.

xMethods (2003) http://www.xMethods.net.

XML-RPC (2003) *XML-RPC.com*, http://www.xmlrpc.com/directory/1568/services.

XML-security (2003) *XML Security page*, http://www.nue.et-inf.unisiegen.de/~geuer-pollmann/xml_security.html.

Yokoo, M. (2000) *Distributed Constraint Satisfaction: Foundations of Cooperation in Multi-Agent Systems*, Springer-Verlag.

Yoneda, T. and Ohta, T. (1998) A formal approach for definition and detection of feature interactions, in Kimbler, K. and Bouma, W. (eds) *Proc. 5th Feature Interactions in Telecommunications and Software Systems*, pp. 202–216, IOS Press, Amsterdam, The Netherlands.

Yoshida, J. and Hara, Y. (1998) iReady, Seiko develop Internet-ready LCDs, *EE Times*, 26th October.

Yourdon, E. (1989) *Modern Structured Analysis*, Prentice Hall.

Zave, P. (1998) Architectural solutions to feature-interaction problems in telecommunications, in Kimbler, K. and Bouma, W. (eds) *Proc. 5th Feature Interactions in Telecommunications and Software Systems*, pp. 10–22, IOS Press, Amsterdam, The Netherlands.

Zave, P. and Jackson, M. (2000) New feature interactions in mobile and multimedia telecommunications services, in Calder, M. H. and Magill, E. H. (eds) *Proc. 6th Feature Interactions in Telecommunications and Software Systems*, pp. 51–66, IOS Press, Amsterdam, The Netherlands.

Index

Service Provision – Technologies for Next Generation Communications. Edited by Kenneth J. Turner, Evan H. Magill and David J. Marples

ISBN: 0-470-85066-3